AF443202

Formal and Analytic Solutions of
Differential Equations

Formal and Analytic Solutions of
Differential Equations

Editors

Galina Filipuk
University of Warsaw, Poland

Alberto Lastra
University of Alcalá, Spain

Sławomir Michalik
Cardinal Stefan Wyszyński University in Warsaw, Poland

 **World Scientific**

NEW JERSEY · LONDON · SINGAPORE · BEIJING · SHANGHAI · HONG KONG · TAIPEI · CHENNAI · TOKYO

Published by

World Scientific Publishing Europe Ltd.

57 Shelton Street, Covent Garden, London WC2H 9HE

Head office: 5 Toh Tuck Link, Singapore 596224

USA office: 27 Warren Street, Suite 401-402, Hackensack, NJ 07601

Library of Congress Cataloging-in-Publication Data

Names: Filipuk, Galina, editor. | Lastra, Alberto, editor. | Michalik, Sławomir, editor.
Title: Formal and analytic solutions of differential equations / editors, Galina Filipuk,
    University of Warsaw, Poland, Alberto Lastra, University of Alcalá, Spain,
    Sławomir Michalik, Cardinal Stefan Wyszyński University in Warsaw, Poland.
Description: New Jersey : World Scientific, [2022] | Includes bibliographical references and index.
Identifiers: LCCN 2021038738 | ISBN 9781800611351 (hardcover) |
    ISBN 9781800611368 (ebook for institutions) | ISBN 9781800611375 (ebook for individuals)
Subjects: LCSH: Differential equations.
Classification: LCC QA371 .F68 2022 | DDC 515/.35--dc23
LC record available at https://lccn.loc.gov/2021038738

British Library Cataloguing-in-Publication Data
A catalogue record for this book is available from the British Library.

Copyright © 2022 by World Scientific Publishing Europe Ltd.

All rights reserved. This book, or parts thereof, may not be reproduced in any form or by any means, electronic or mechanical, including photocopying, recording or any information storage and retrieval system now known or to be invented, without written permission from the Publisher.

For photocopying of material in this volume, please pay a copying fee through the Copyright Clearance Center, Inc., 222 Rosewood Drive, Danvers, MA 01923, USA. In this case permission to photocopy is not required from the publisher.

For any available supplementary material, please visit
https://www.worldscientific.com/worldscibooks/10.1142/Q0335#t=suppl

Desk Editors: Soundararajan Raghuraman/Michael Beale/Shi Ying Koe

Typeset by Stallion Press
Email: enquiries@stallionpress.com

© 2022 World Scientific Publishing Europe Ltd.
https://doi.org/10.1142/9781800611368_fmatter

Preface

This volume aims to provide the reader with the landscape of the actual state of research in the field of study of functional equations and related topics, embracing the research of formal and analytic solutions of functional equations, such as differential, partial differential, difference, q-difference, q-difference-differential equations, etc. Some related studies are also considered, in order to link this with close areas which benefit from their techniques. It consists of selected contributions from the virtual conference "Formal and Analytic Solutions of Differential Equations on the Internet" (FASnet20), held from June 29th to July 3rd, 2020.

The difficult situation worldwide due to the global pandemic made the organization of a face-to-face conference unrealizable. In this context, a virtual conference was organized to overcome these difficulties.

The volume is divided into different chapters describing the advances at different points of the field of research.

The contribution by A. D. Bruno deals with the normal form of the Hamiltonian function related to an analytic Hamiltonian system near a stationary solution, near a periodic solution and near an invariant torus, studying families of periodic solutions and families of invariant tori adjoining the above-mentioned initial objects. The work by J. Sekiguchi is concerned with the relations appearing between the algebraic potentials having tri-Hamiltonian structure of the reflection group of different types and polynomial potentials of certain type.

In the contribution by A. Batkhin, the author considers an autonomous system of ordinary differential equations with non-degenerate linear part near its stationary point in the general case and in Hamiltonian case and proves results on the existence of invariant coordinate subspaces with explicit conditions. The non-standard discretizations of a cubic, non-autonomous Hamiltonian system related to the fourth Painlevé equation is considered in the contribution by G. Filipuk and T. Kecker. F. Zullo presents results on the distribution of the zeros of solutions of Airy's non-homogeneous equation, generalizing previous results on the homogeneous equation.

In their contribution, S. Carillo and C. Schiebold study third-order nonlinear evolution equations, termed KdV-type, with the so-called Bäcklund transformations. This tool is used in Abelian and non-Abelian nonlinear evolution equations.

Y. Takei discusses the generalization to the case of the hypergeometric differential equation of type $(1, 1, 3)$ of previous results relating Voros coefficients in the exact WKB analysis and the free energy in the topological recursion introduced by Eynard and Orantin in the case of the confluent family of the Gauss hypergeometric differential equations. In the work by T. Takahashi and M. Tanda, the authors consider the Gauss hypergeometric differential equation with a large parameter deformed to a differential equation with a simple pole at the origin. The Borel sums of the WKB solutions and Kummer's solutions in the neighborhood of the regular singular point 1 are also related.

K. S. Charak, R. Korhonen and G. Kumar present a unified way to consider uniqueness problems concerning a fixed meromorphic function through a general class of functions associated with the initial one, which includes operators of such a function and its derivatives.

The theory of summability of formal solutions is a central point of the study of analytic solutions to functional equations in the complex domain. There are several works related to this theory in the volume. M. Yoshino studies the solutions of partial differential equations which appear in the study of a movable singularity of a Hamiltonian system or a blowup phenomenon of equations in mathematical physics via global Borel summability. In the contribution by K. Ichinobe, the author considers Cauchy problems to some linear q-difference-differential equations with entire Cauchy data and describes novel results regarding summability of the formal solution.

N. Honda, S. Kamimoto and L. Prelli take a step further in the algebraic study of asymptotics regarding multi-specializations and minimal asymptotics. The study of special functions is quite related to the theory of formal and analytic solutions of functional equations. In this regard, the connection problem studied in the paper by S. Adachi is concerned with the generalized hypergeometric differential equation satisfied by the generalized hypergeometric series $_nF_{n-1}$. Also, the theory of orthogonal polynomials helps the development of the theory when applied to functional equations. Concerning the theory of orthogonal polynomials, C. Hermoso and A. Soria-Lorente present a new family of Sobolev-type orthogonal polynomials as solutions of certain second-order q-difference equations, involving an arbitrary number of first-order q-derivatives on a mass-point out of the orthogonality interval. In the contribution by M. D. N. Rebocho, a study of general divided-difference operators under certain fundamental properties, which is of usual consideration in the theory of orthogonal polynomials, is made.

The estimation of the coefficients of formal solutions of functional equations is a first step on summability of such solutions. S. A. Carrillo and C. A. Hurtado describe the Gevrey character, in an analytic function P, of formal power series solutions to certain first-order holomorphic partial differential equations, by means of techniques based on Nagumo norms and a division algorithm. G. Chen, A. Lastra and S. Malek give some clues on the parametric asymptotics related to certain families of singularly perturbed partial differential equations in two complex time variables by means of a truncated Laplace transform.

The proceedings volume provides methods, techniques, different mathematical tools and recent results in the study of summability of formal solutions to the mentioned functional equations, from the formal and analytic point of view.

The conference FASnet20 aimed to bring together experts in the field of formal and analytic solutions of functional equations, such as differential, partial differential, difference, q-difference, q-difference-differential equations. This meeting provided the opportunity for the exchange of recent results in the field and to exhibit different possible directions for future research. One of its main objectives was to promote both new and existing scientific collaborations of the researchers in these topics. The topics of the conference were:

- Ordinary differential equations in the complex domain. Formal and analytic solutions. Stokes multipliers.
- Formal and analytic solutions of partial differential equations.
- Formal and analytic solutions of difference equations (including q-difference and differential-difference equations).
- Special functions (hypergeometric functions and others), orthogonal polynomials, continuous and discrete Painlevé equations.
- Integrable systems.
- Holomorphic vector fields. Normal forms.
- Asymptotic expansions, Borel summability.

The scientific and organizing committees helped on the successful plan and organization of FASnet20.

Scientific Committee

- Alberto Lastra (University of Alcalá, Spain)
- Stephane Malek (University of Lille, France)
- Jorge Mozo (University of Valladolid, Spain)
- Yasunori Okada (Chiba University, Japan)
- Javier Sanz (University of Valladolid, Spain)
- Hidetoshi Tahara (Sophia University, Japan)

Organizing Committee

- Galina Filipuk (University of Warsaw, Poland)
- Javier Jiménez-Garrido (University of Cantabria, Spain)
- Alberto Lastra (University of Alcalá, Spain)

We refer to the conference web-page at http://www3.uah.es/fasnet for further information on the conference.

We would like to express our gratitude to the participants of the conference and the authors who have made the effort to write their contributions amidst these difficult circumstances worldwide. All contributions have been peer reviewed by anonymous referees chosen among the experts in the field. We also express our deep gratitude to them for their invaluable collaboration. We also acknowledge World Scientific Publishing for making this volume possible.

We express our sincere gratitude to all for making the conference FASnet20 a way to overcome the difficulty of a face-to-face meeting, and the possibility to continue spreading mathematical advances in the current situation of the world pandemic.

Galina Filipuk
Institute of Mathematics
University of Warsaw
Warsaw, Poland

Alberto Lastra
Departamento de Física y Matemáticas
University of Alcalá
Alcalá de Henares, Spain

Sławomir Michalik
College of Science
Cardinal Stefan Wyszyński University
Warsaw, Poland

© 2022 World Scientific Publishing Europe Ltd.
https://doi.org/10.1142/9781800611368_fmatter

About the Editors

 **Galina Filipuk** is an Assistant Professor at the University of Warsaw (Poland). She previously worked at the Polish Academy of Sciences, Loughborough University (UK), and the Belarusian State University. She held the Humboldt Research Fellowship for experienced researchers at TU Dresden and KU Eichstaett-Ingolstadt (Germany) and the Japan Society for the Promotion of Sciences Fellowship at Kumamoto University. In addition, she held several short-term visiting positions abroad. She is an editor for the journal *Advances in Difference Equations* and has also edited special issues in journals *Integral Transforms and Special Functions*, *Random Matrices: Theory and Applications*. She has co-authored a two-volume book titled *Analysis with Mathematica* published by de Gruyter and edited four conference proceedings volumes published by de Gruyter, Springer, and the Banach Center Publications. She was a co-organizer and a member of the Scientific Committee of several international conferences in Poland and abroad. Her research interests are in differential equations in the complex plane and special functions. She has more than 65 papers in peer-review journals and more than 15 book chapters and refereed conference proceedings.

Alberto Lastra currently works at the University of Alcalá (Spain). He is a researcher interested in the study of asymptotic analysis of functional equations in the complex domain and related topics. He has a constant and active ongoing research activity followed by publications of his results, visiting several foreign research centers, and organizing and being a member of the Scientific Committee of several international conferences, besides being the editor of two conference proceedings.

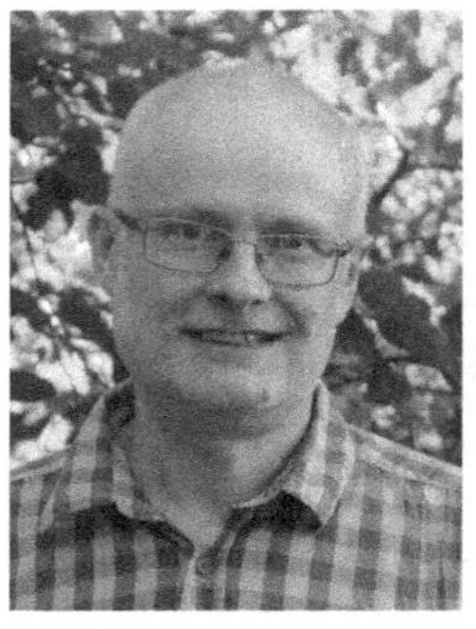

Sławomir Michalik works at the Cardinal Stefan Wyszyński University in Warsaw (Poland). His research interests concern the analytic theory of differential equations and the theory of summability. He has organized a number of international conferences devoted to complex differential equations, especially their formal and analytic solutions. And he has also been the editor of several conference proceedings.

© 2022 World Scientific Publishing Europe Ltd.
https://doi.org/10.1142/9781800611368_fmatter

Contents

© 2022 World Scientific Publishing Europe Ltd.
https://doi.org/10.1142/9781800611368_0001

Chapter 1

Families of Periodic Solutions and Invariant Tori of Hamiltonian Systems

Alexander Dmitrievich Bruno

Keldysh Institute of Applied Mathematics of RAS
Miusskaya sq. 4, Moscow 125047, Russia
abruno@keldysh.ru

Near a stationary solution, near a periodic solution and near an invariant torus of an analytic Hamiltonian system we consider the normal form of its Hamiltonian function. Usually, the normalizing transformation diverges in the whole neighborhood of each mentioned initial object, but there exist convergent transformations which normalize the Hamilton function only in some sets adjoining the initial object. The sets are analytic, include all formal families of periodic solutions and under a condition on small divisors, they include some formal families of invariant tori with similar bases of frequencies. So generically the real Hamiltonian system with n degrees of freedom has: (a) one-parameter families of periodic solutions, (b) one-parameter families of n-dimensional invariant irreducible tori and (c) $(l + 1)$-parameter families of $k(< n)$ dimensional irreducible invariant tori with exactly $2l$ eigenvalues having zero real parts, and for all of them imaginary parts are commensurable with frequencies.

1.   Introduction

We consider the real analytic autonomous Hamiltonian system with a finite number of degrees of freedom and without parameters near its

1

stationary solution (Section 2), near its periodic solution (Section 3) and near its invariant torus (Section 4).

Our aim is to study families of periodic solutions and families of invariant tori adjoining the mentioned above initial objects.

For that, we reduce the Hamiltonian system to its normal form. Using it, we select such formal sets, for which there exist convergent normalizing transformations and so the sets are analytic. They are the set $\tilde{\mathcal{A}}$, if small divisors are absent, and the set $\mathcal{B}$, if small divisors are present and satisfy a condition. The set $\tilde{\mathcal{A}}$ contains all families of periodic solutions.

It appeared that in general case a periodic solution lies at a one-parameter family of periodic solutions with near periods and invariant irreducible torus of the maximal possible dimension also lies at a one-parameter family of such tori with similar bases of frequencies, but formal families of invariant tori of non-maximal dimension are analytic only if their eigenvalues satisfy an additional condition: all eigenvalues with zero real part have imaginary parts commensurable with set of frequencies.

- Huygens in 17th century began to study periodic solutions.
- Kolmogorov [1] began to study analytic invariant tori.
- His successors Arnold, Moser and others developed a theory named KAM.

Here another approach from the Part II of Bruno [2] is used. In the present work, there are three new methods:

- a division of a formal set $\mathcal{A}^k$ into components $\mathcal{A}^k_I$;
- an introduction of the reduced normal form, not depending on time;
- writing the sets $\mathcal{A}^k_I$ and $\mathcal{B}^k$ for the reduced normal form.

2. A Neighborhood of a Stationary Solution

2.1. *Normal form*

Consider the Hamiltonian system

$$\dot{\xi}_j = \frac{\partial \gamma}{\partial \eta_j}, \quad \dot{\eta}_j = -\frac{\partial \gamma}{\partial \xi_j}, \quad j = 1, \ldots, n, \tag{1}$$

with n degrees of freedom in a neighborhood of the stationary point

$$\boldsymbol{\xi} = (\xi_1, \ldots, \xi_n) = 0, \quad \boldsymbol{\eta} = (\eta_1, \ldots, \eta_n) = 0. \tag{2}$$

If the Hamiltonian $\gamma(\boldsymbol{\xi}, \boldsymbol{\eta})$ is analytic at this point, then it can be expanded in the power series

$$\gamma(\boldsymbol{\xi}, \boldsymbol{\eta}) = \sum \gamma_{\mathbf{pq}} \boldsymbol{\xi}^{\mathbf{p}} \boldsymbol{\eta}^{\mathbf{q}}, \tag{3}$$

where $\mathbf{p} = (p_1, \ldots, p_n)$, $\mathbf{q} = (q_1, \ldots, q_n) \in \mathbb{Z}^n$, $\mathbf{p}, \mathbf{q} \geqslant 0$, $\boldsymbol{\xi}^{\mathbf{p}} = \xi_1^{p_1} \xi_2^{p_2} \cdots \xi_n^{p_n}$, $\gamma_{\mathbf{pq}}$ are constant coefficients.

Since point (2) is stationary, expansion (3) begins with quadratic terms. These terms are associated with the linear part of system (1). The eigenvalues of the matrix of the linear system are organized in pairs

$$\lambda_{j+n} = -\lambda_j, \quad j = 1, \ldots, n.$$

Let $\boldsymbol{\lambda} = (\lambda_1, \ldots, \lambda_n)$. The canonical changes of coordinates

$$\boldsymbol{\xi}, \boldsymbol{\eta} \longrightarrow \mathbf{x}, \mathbf{y} \tag{4}$$

preserve the Hamiltonian property of the system.

Theorem 1 ([3], §12). *There exists a canonical invertible formal transformation (4) that reduces initial system (1) to the normal form*

$$\dot{x}_j = \frac{\partial g}{\partial y_j}, \quad \dot{y}_j = -\frac{\partial g}{\partial x_j}, \quad j = 1, \ldots, n, \tag{5}$$

where the series

$$g(\mathbf{x}, \mathbf{y}) = \sum g_{\mathbf{pq}} \mathbf{x}^{\mathbf{p}} \mathbf{y}^{\mathbf{q}} \tag{6}$$

contains only resonant terms with

$$\langle \mathbf{p} - \mathbf{q}, \boldsymbol{\lambda} \rangle = 0. \tag{7}$$

Here $\langle \mathbf{p}, \boldsymbol{\lambda} \rangle = p_1 \lambda_1 + \cdots + p_n \lambda_n$ is the scalar product.

For the real original system (1), the coefficients $g_{\mathbf{pq}}$ of the complex normal form (6) satisfy special real type relations, and under the standard canonical linear change of coordinates $(\mathbf{x}, \mathbf{y}) \to (\mathbf{X}, \mathbf{Y})$, system (5) takes the real form ([4], Ch. I).

4   *A. D. Bruno*

Let $I = \{i_1, \ldots, i_m\}$ be a set of increasing natural indices $i \leqslant n$. Here $1 \leqslant m \leqslant n$. Consider the coordinate subspace

$$K_I = \{\mathbf{x}, \mathbf{y} : x_j = y_j = 0 \text{ for all } j \notin I\}. \tag{8}$$

If $I = \{1, \ldots, n\}$, then K_I is the space $\mathbb{C}^{2n}$ with coordinates $\mathbf{x}, \mathbf{y}$, which we denote as K_n.

Let us mention four numerical characteristics of a coordinate subspace K_I:

(1) Its *half-dimension $m_I = m$.*
(2) Its *multiplicity of resonances $\varkappa_I$*, as a number of linearly independent integer relations $\sum_{i \in I} p_i \lambda_i = 0$ with integer p_i.
(3) Its *degree of irrationality $\sigma_I = m_I - \varkappa_I$.*
(4) Its *subsets of eigenvalues $\boldsymbol{\lambda}_I = \{\lambda_i, i \in I\}$ and $\boldsymbol{\alpha}_I = \{\alpha_i, i \in I\}$.*

In what follows we will consider the case where all eigenvalues $\lambda_1, \ldots, \lambda_n$ are pure imaginary: $\lambda_j = i\alpha_j$, $\alpha_j \in \mathbb{R}$, and all $\lambda_j \neq 0$.

2.2.   *Convergence of the normalizing transformation*

Condition ω

Let $\omega_k = \min |\langle \mathbf{p}, \boldsymbol{\lambda} \rangle|$ over $\langle \mathbf{p}, \boldsymbol{\lambda} \rangle \neq 0$, $||\mathbf{p}|| < 2^k$, $\mathbf{p} \in \mathbb{Z}^n$, where $||\mathbf{p}|| = \sum |p_j|$. Then

$$-\sum_{k=1}^{\infty} \frac{\log \omega_k}{2^k} < \infty.$$

It is a very weak numerical restriction on the eigenvalues $\boldsymbol{\lambda}$. It satisfies for almost all sets $\boldsymbol{\lambda}$. In particular, it satisfies if all λ_j are pairwise commensurable, then the multiplicity of resonances $\varkappa_n$ for the whole space K_n is equal to $n - 1$. Condition ω is a restriction on "small divisors" arising in the normalizing transformation.

Condition A

There exists such a power series $a(\mathbf{x}, \mathbf{y})$ that in the normal form (5)

$$\frac{\partial g}{\partial y_j} = \lambda_j x_j a, \quad \frac{\partial g}{\partial x_j} = \lambda_j y_j a, \quad j = 1, \ldots, n.$$

It is a very hard restriction on the right parts of the normal form (5). It satisfies very seldom.

Theorem 2 ([5]). *If eigenvalues* $\boldsymbol{\lambda}$ *satisfy Condition* ω *and the normal form* (5) *satisfies Condition* **A**, *then the normalizing transformation converges in a neighborhood of the stationary point* $\mathbf{x} = \mathbf{y} = 0$.

According to [[3], §12] Condition **A** is equivalent to the condition that the Hamiltonian $g(\mathbf{x}, \mathbf{y})$ of the normal form (5) is a power series of one variable $\sum\limits_{j=1}^{n} \lambda_j x_j y_j$.

2.3. *Set* $\mathcal{A}$

Suppose that the functions $f_1(\mathbf{x}, \mathbf{y}), \ldots, f_r(\mathbf{x}, \mathbf{y})$ are analytic and vanish at the point $\mathbf{x} = \mathbf{y} = 0$. Then the system of equations

$$f_j(\mathbf{x}, \mathbf{y}) = 0, \quad j = 1, \ldots, r \tag{9}$$

defines an *analytic set* $\mathcal{N}$ containing the point $\mathbf{x} = \mathbf{y} = 0$. If $f_1, \ldots, f_r$ are formal power series, we say that (9) defines a *formal set* $\mathcal{N}$.

Problem 1. *Which invariant formal sets of the system* (1) *are analytic?*

The fact is that it is comparatively easy to calculate formal invariant sets using the normal form (5)–(7). We need to select among them only those that are analytic for the initial system (1).

From the normal form (5)–(7), we form the formal set

$$\mathcal{A} = \left\{ \mathbf{x}, \mathbf{y} : \frac{\partial g}{\partial y_j} = \lambda_j x_j a, \ \frac{\partial g}{\partial x_j} = \lambda_j y_j a, \ j = 1, \ldots, n \right\}, \tag{10}$$

where a is a free parameter. We can exclude it from the equations and obtain a representation of the set $\mathcal{A}$ in the form (9).

All solutions from the set $\Re\mathcal{A}$ are conditionally-periodic (including periodic and stationary solutions). In fact, the value of the parameter a is constant in each solution and we have

$$x_j = x_j^0 \exp \lambda_j a t, \quad y_j = y_j^0 \exp\left(-\lambda_j a t\right), \quad j = 1, \ldots, n.$$

Definition 1. Let K_I be a coordinate subspace in $\mathbf{x}, \mathbf{y}$. The *component* $\mathcal{A}_I$ *of the set* $\mathcal{A}$ in this subspace is defined by the system of equations

$$\frac{\partial g}{\partial y_j} = \lambda_j x_j a, \quad \frac{\partial g}{\partial x_j} = \lambda_j y_j a, \quad j \in I, \tag{11}$$

and all $x_j, y_j \neq 0$ for $j \in I$. In particular, there is the component $\mathcal{A}_n$ for the whole space K_n.

Theorem 3. *If in the normal form* (5) *Hamiltonian* $g(\mathbf{x}, \mathbf{y})$ *is analytic, then each component* $\Re\mathcal{A}_I$ *is a family of irreducible invariant tori of dimension* σ_I *with frequencies* $a\boldsymbol{\alpha}_I$. *Generically, the family is either one-parametric (along a) or it is empty. In degenerate cases it can have more than one parameter.*

A torus of dimension 1 is a periodic solution.

A coordinate subspace $K_{\tilde{I}}$ is called *rational* if the corresponding eigenvalues $\boldsymbol{\lambda}_{\tilde{I}}$ are pairwise commensurable. Let $\widetilde{\mathcal{K}}$ be the union of all rational subspaces, and $\tilde{\mathcal{A}} = \mathcal{A} \cap \tilde{\mathcal{K}}$.

Theorem 4 ([2], Part II, §3). *There exists an analytic canonical transformation* $\boldsymbol{\xi}, \boldsymbol{\eta} \to \mathbf{x}, \mathbf{y}$ *which transforms initial system* (1) *to normal form in the set* $\tilde{\mathcal{A}}$ *and the set is analytic.*

2.4.   Set $\mathcal{B}$

Let $\Lambda = \{\lambda_1, \ldots, \lambda_n\}$ be a diagonal matrix. In the set $\mathcal{A}$ we consider the $2n \times 2n$ matrix

$$B = \begin{pmatrix} \dfrac{\partial^2 g}{\partial \mathbf{y}\partial \mathbf{x}} - \Lambda a & \dfrac{\partial^2 g}{\partial \mathbf{y}\partial \mathbf{y}} \\ -\dfrac{\partial^2 g}{\partial \mathbf{x}\partial \mathbf{x}} & -\dfrac{\partial^2 g}{\partial \mathbf{x}\partial \mathbf{y}} + \Lambda a \end{pmatrix}, \tag{12}$$

where a is the same parameter as in equations (10). We define the formal set $\mathcal{B}$ as that subset of the set $\mathcal{A}$ in which the matrix B is nilpotent, that is,

$$\mathcal{B} = \{\mathbf{x}, \mathbf{y} : \quad \mathbf{x}, \mathbf{y} \in \mathcal{A}, \quad B^{2n} = 0\}. \tag{13}$$

Theorem 5 ([2], Part II). *Generically,* $\mathcal{B} = \mathcal{A}_n$.

Theorem 6 ([2], Part II). *If eigenvalues* $\boldsymbol{\lambda}$ *satisfy Condition* ω, *then there exists an analytic canonical transformation* $\boldsymbol{\xi}, \boldsymbol{\eta} \to \mathbf{x}, \mathbf{y}$ *which transforms initial system* (1) *to normal form in the set* $\mathcal{B}$ *and the set is analytic.*

2.5. *Example*

Suppose that $\lambda_1, \ldots, \lambda_n$ are linearly independent over integral numbers, that is, the equation $\langle \mathbf{p}, \boldsymbol{\lambda} \rangle = 0$ has only the trivial solution $\mathbf{p} = 0$ for the integers $\mathbf{p}$. Then in the normal form (5)–(7),

$$g = f\left(\rho_1, \ldots, \rho_n\right), \quad \text{where } \rho_j = x_j y_j, \quad j = 1, \ldots, n.$$

We therefore have

$$\mathcal{A} = \left\{ \mathbf{x}, \mathbf{y} : x_j \frac{\partial f}{\partial \rho_j} = \lambda_j x_j a, \quad y_j \frac{\partial f}{\partial \rho_j} = \lambda_j y_j a, \quad j = 1, \ldots, n \right\}.$$

We consider the set $\mathcal{A}$ in the Cartesian coordinates $\boldsymbol{\rho} = (\rho_1, \ldots, \rho_n)$. In the general case, each coordinate subspace (with respect to $\boldsymbol{\rho}$) contains one one dimensional (with respect to $\boldsymbol{\rho}$) component of the set $\mathcal{A}$ that does not lie in a smaller coordinate subspace. Consequently, the set $\mathcal{A}$ consists of $2^n - 1$ such components; for each $d \leqslant n$ there are exactly $n!/[d!(n-d)!]$ of these components situated in d-dimensional (with respect to $\boldsymbol{\rho}$) coordinate subspaces. In particular, there is one component,

$$\mathcal{A}_n = \left\{ \boldsymbol{\rho} : \frac{\partial f}{\partial \rho_j} = \lambda_j a, \quad j = 1, \ldots, n \right\} \tag{14}$$

situated outside the coordinate subspaces. $B^2 = 0$ in this component.

In fact,

$$B = \begin{pmatrix} R - \Lambda a & 0 \\ 0 & -R + \Lambda a \end{pmatrix} + \begin{pmatrix} U & 0 \\ 0 & V \end{pmatrix} \begin{pmatrix} S & S \\ -S & -S \end{pmatrix} \begin{pmatrix} V & 0 \\ 0 & U \end{pmatrix},$$

where the blocks are the following $n \times n$ matrices:

$$R = \left\{ \frac{\partial f}{\partial \rho_1}, \ldots, \frac{\partial f}{\partial \rho_n} \right\}, \quad U = \{x_1, \ldots, x_n\}, \quad V = \{y_1, \ldots, y_n\}$$

these are diagonal, and

$$S = \left(\frac{\partial^2 f}{\partial \rho_j \partial \rho_k} \right).$$

In the set $\mathcal{A}$, we have

$$B = \begin{pmatrix} U & 0 \\ 0 & V \end{pmatrix} \begin{pmatrix} S & S \\ -S & -S \end{pmatrix} \begin{pmatrix} V & 0 \\ 0 & U \end{pmatrix}$$

that is, $B^2 = 0$. In the general case, the matrix B is not nilpotent in coordinate subspaces, since there for some derivatives $\partial f/\partial \rho_j \neq \lambda_j a$. Thus, $\mathcal{B} = \mathcal{A}_n$.

8   *A. D. Bruno*

Moreover, the n components

$$\mathcal{A}'_j = \{\boldsymbol{\rho}: \quad \rho_i = 0, \quad i \neq j, \quad 1 \leqslant i \leqslant n\}, \quad j = 1, \ldots, n,$$

are coordinate axes with respect to $\boldsymbol{\rho}$ and exhaust all rational subspaces; taken together, they make up the set $\tilde{\mathcal{A}}$. According to Theorem 4, the initial system (1) has n analytic one-parameter families $\mathcal{A}'_j$ of periodic solutions (these are Lyapunov [6] families).

If the eigenvalues $\boldsymbol{\lambda}$ satisfy Condition ω, then by Theorems 6, 5 the component $\mathcal{A}_n = \mathcal{B}$ is also an analytic set.

Let

$$f = g = \langle \boldsymbol{\rho}, \boldsymbol{\lambda} \rangle + \frac{1}{2} \langle \boldsymbol{\rho}, T\boldsymbol{\rho}^* \rangle + \cdots, \tag{15}$$

where T is a symmetric matrix. In the general case, $\det T \neq 0$ and the system of equations in (10) has a one-dimensional solution

$$\boldsymbol{\rho}^* = T^{-1}\boldsymbol{\lambda}^*(a - 1) + o(a - 1), \tag{16}$$

where * means transposition.

If the initial system (1) is real, the coordinates x_j, y_j are connected by the reality relation $\bar{x}_j = -iy_j$. Consequently, $-\arg x_j = \arg y_j - \pi/2$, that is, $\arg(x_j y_j) = \pi/2$.

Thus, purely imaginary ρ_j with $\Im \rho_j \geq 0, j = 1, \ldots, n$ correspond to real values of the initial coordinates. Each set $\mathcal{A}'_j$ has a real part,

$$\Re \mathcal{A}'_j = \{\rho_j : \Re \rho_j = 0, \quad \Im \rho_j \geq 0\},$$

which is a real one-parameter family of periodic solutions. For the real Hamiltonian (15), the matrix T is also real. If the vector $T^{-1}\Im\boldsymbol{\lambda}^*$ has coordinates of different signs, then according to (16), the set $\mathcal{B}$ has only a trivial real part, $\Re \mathcal{B} = 0$.

If all the coordinates of $T^{-1}\boldsymbol{\alpha}^*$ are of the same sign, the real set $\Re \mathcal{B}$ is a one-parameter family of n-dimensional irreducible invariant tori with frequency basis $\alpha_1 a, \ldots, \alpha_n a$. As $a \to 1$, the tori of this family tend toward the fixed point $\mathbf{x} = \mathbf{y} = 0$. For a detailed analysis of various situations, see [2], Part II, §3.

3. A Neighborhood of a Periodic Solution

3.1. *Local coordinates*

Let a real Hamiltonian system with $n + 1$ degrees of freedom have a real 2π-periodic solution $\mathcal{M}$ and the Hamiltonian function is analytic in some neighborhood of it.

According to [4] Chapter II, Section 2.1 near the solution $\mathcal{M}$ we can introduce such real local canonical coordinates $\boldsymbol{\xi} = (\xi_1, \ldots, \xi_n)$, ψ and $\boldsymbol{\eta} = (\eta_1, \ldots, \eta_n)$, ρ, that the solution $\mathcal{M}$ is given by equations

$$\boldsymbol{\xi} = \boldsymbol{\eta} = 0, \quad \rho = 0, \quad \psi = \psi_0 + t \tag{17}$$

and the Hamiltonian has the form

$$\gamma = \Sigma \gamma_{\mathbf{pq}l}(\psi) \boldsymbol{\xi}^{\mathbf{p}} \boldsymbol{\eta}^{\mathbf{q}} \rho^l = \rho + \cdots, \tag{18}$$

where integers $\mathbf{p}, \mathbf{q} \geqslant 0$, integer $l \geqslant 0$, real analytic functions $\gamma_{\mathbf{pq}l}(\psi)$ have in ψ the period 2π and they are expanded in the Fourier series.

Then the Hamiltonian system is

$$\dot{\xi}_j = \frac{\partial \gamma}{\partial \eta_j}, \quad \dot{\eta}_j = -\frac{\partial \gamma}{\partial \xi_j}, \quad j = 1, \ldots, n,$$
$$\dot{\psi} = \frac{\partial \gamma}{\partial \rho}, \quad \dot{\rho} = -\frac{\partial \gamma}{\partial \psi}. \tag{19}$$

3.2. *Normal form*

For $\rho = 0$ and $\psi = t$ quadratic in $\boldsymbol{\xi}, \boldsymbol{\eta}$ part γ_2 of the Hamiltonian (18) defines 2π-periodic linear in $\boldsymbol{\xi}, \boldsymbol{\eta}$ system

$$\dot{\xi}_j = \frac{\partial \gamma_2}{\partial \eta_j}, \quad \dot{\eta}_j = -\frac{\partial \gamma_2}{\partial \xi_j}, \quad j = 1, \ldots, n. \tag{20}$$

Let $\nu_1, \ldots, \nu_{2n}$ be eigenvalues of its monodromy matrix, i.e. matrix of substitution of the fundamental matrix of solutions to the system (20) in the period 2π. Let all $|\nu_j| = 1$ and $\nu_j \neq -1$. We put

$$\alpha_j = \frac{1}{2\pi \mathrm{i}} \ln \nu_j, \quad \alpha_j \in \mathbb{R}, \quad \alpha_j \in \left(-\frac{1}{2}, \frac{1}{2}\right), \quad j = 1, \ldots, 2n. \tag{21}$$

Using correct numeration, one obtains

$$\alpha_{j+n} = -\alpha_j, \; j = 1, \ldots, n. \tag{22}$$

Let $\boldsymbol{\alpha} = (\alpha_1, \ldots, \alpha_n)$.

Theorem 7 ([4], Ch. II; [7]). *There exists a complex formal invertible 2π-periodic in ψ and φ canonical transformation of coordinates in the form of Poisson series*

$$\boldsymbol{\xi}, \psi, \boldsymbol{\eta}, \rho \longleftrightarrow \mathbf{x}, \varphi, \mathbf{y}, r, \tag{23}$$

which reduces the Hamiltonian γ into normal form

$$g(\mathbf{x}, \varphi, \mathbf{y}, r) = r + \mathrm{i} \sum_{j=1}^{n} \alpha_j x_j y_j + \sum g_{\mathbf{pq}lm} \mathbf{x}^{\mathbf{p}} \mathbf{y}^{\mathbf{q}} r^l e^{\mathrm{i}m\varphi}, \tag{24}$$

where $\mathbf{x}, \mathbf{y} \in \mathbb{C}^n$, $0 \leqslant \mathbf{p}, \mathbf{q} \in \mathbb{Z}^n$, $l \geqslant 0$ and m are integers, all terms of the second sum are resonant, i.e.

$$\langle \mathbf{p} - \mathbf{q}, \boldsymbol{\alpha} \rangle + m = 0. \tag{25}$$

Theorem 8. *There exists a canonical transformation*

$$x_j = u_j \exp(-\mathrm{i}\beta_j \varphi), \quad y_j = v_j \exp(\mathrm{i}\beta_j \varphi), \quad j = 1, \ldots, n,$$

$$r = s - \mathrm{i} \sum_{j=1}^{n} \beta_j u_j v_j \tag{26}$$

with rational β_j, which reduces the normal form of the Hamiltonian (24), (25) to an autonomous power series

$$h(\mathbf{u}, \mathbf{v}, s) = s + \mathrm{i} \sum_{j=1}^{n} \gamma_j u_j v_j + \sum h_{\mathbf{pq}l} \mathbf{u}^{\mathbf{p}} \mathbf{v}^{\mathbf{q}} s^l, \tag{27}$$

where in the first sum nonzero γ_j are irrational numbers and in the second sum $0 \leqslant \mathbf{p}, \mathbf{q} \in \mathbb{Z}^n$, $0 \leqslant l \in \mathbb{Z}$, $h_{\mathbf{pq}l} = \mathrm{const} \in \mathbb{C}$ and present only resonant terms with

$$\langle \mathbf{p} - \mathbf{q}, \boldsymbol{\gamma} \rangle = 0, \tag{28}$$

where $\boldsymbol{\gamma} = (\gamma_1, \ldots, \gamma_n)$.

Similar theorem is in [7]. Theorem 8 is a particular case of Theorem 13 that will be given in Section 4 with the proof.

Variable s now is a formal integral of the following system

$$\dot{u}_j = \frac{\partial h}{\partial v_j}, \quad \dot{v}_j = -\frac{\partial h}{\partial u_j}, \; j = 1, \ldots, n, \tag{29}$$

$$\dot{\varphi} = \frac{\partial h}{\partial s}. \tag{30}$$

If the initial Hamiltonian γ is real for real coordinates $\boldsymbol{\xi}, \psi, \boldsymbol{\eta}, \rho$, then in Theorem 7 variables $\mathbf{x}, \mathbf{y}$ are complex but variables ψ, ρ and

φ, r are real. Here according to [4] Chapters I and II variables $\mathbf{x}, \mathbf{y}$ are connected with real variables $\mathbf{X} = (X_1, \ldots, X_n)$, $\mathbf{Y} = (Y_1, \ldots, Y_n)$ by the linear standard transformation

$$x_j = \frac{1}{\sqrt{2\mathrm{i}}} \left(\mathrm{i}X_j - Y_j \right), \quad y_j = \frac{1}{\sqrt{2\mathrm{i}}} \left(\mathrm{i}X_j + Y_j \right), \; j = 1, \ldots, n. \quad (31)$$

3.3. *Convergence of the normalizing transformation*

Condition ω^1
Let $\omega_k = \min |p_0 + \langle \mathbf{p}, \boldsymbol{\alpha} \rangle|$ over $p_0 + \langle \mathbf{p}, \boldsymbol{\alpha} \rangle \neq 0$, $|p_0| + ||\mathbf{p}|| < 2^k$, $p_0 \in \mathbb{Z}$, $\mathbf{p} \in \mathbb{Z}^n$. Then

$$-\sum_{k=1}^{\infty} \frac{\log \omega_k}{2^k} < \infty.$$

Condition $\mathbf{A}^1$
There exists such a Poisson series $a(\mathbf{x}, \varphi, \mathbf{y}, r)$ that in the normal form (24)

$$\frac{\partial g}{\partial y_j} = \mathrm{i}\alpha_j x_j a, \quad \frac{\partial g}{\partial x_j} = \mathrm{i}\alpha_j y_j a, \quad j = 1, \ldots, n,$$
$$\frac{\partial g}{\partial \varphi} = 0, \qquad \frac{\partial g}{\partial r} = a. \tag{32}$$

According to [3] §11 it is equivalent to the condition that Hamiltonian g is a power series of one variable $r + i\sum_{j=1}^{n} \alpha_j x_j y_j$.

Theorem 9 ([3], §11). *The normalizing transformation (23) converges in a neighborhood of our periodic solution if numbers $\boldsymbol{\alpha}$ satisfy Condition ω^1 and the normal form (24), (25) satisfies Condition $\mathbf{A}^1$.*

3.4. *Set $\mathcal{A}^1$*

Let now a be an arbitrary parameter. Then the system of equations (32) defines the set $\mathcal{A}^1$. For the reduced normal form (27),

(29), (30), the set $\mathcal{A}^1$ is defined by the system

$$\frac{\partial h}{\partial v_j} = \mathrm{i}\gamma_j u_j a, \qquad \frac{\partial h}{\partial u_j} = \mathrm{i}\gamma_j v_j a, \quad j = 1, \dots, n, \tag{33}$$

$$\frac{\partial h}{\partial s} = a. \tag{34}$$

Subsystem of equations (33) defines the set of all periodic solutions of the subsystem (29). Equation (30) $\dot\varphi = \partial h/\partial s$ gives dependence of φ from t for each of these solutions.

To each set of increasing indices $I = \{i_1, \dots, i_m\}$, $1 \leqslant i_1$, $i_m \leqslant n$, $1 \leqslant m \leqslant n$, there correspond the *coordinate subspaces*

$$K_I^1 = \{\mathbf{x}, \mathbf{y}, r, \varphi : x_j = y_j = 0 \text{ for all } j \notin I\},$$
$$L_I^1 = \{\mathbf{u}, \mathbf{v}, s, \varphi : u_j = v_j = 0 \text{ for all } j \notin I\}.$$

Now the coordinate r or s corresponds to the frequency 1, so each subspace K_I^1 and L_I^1 corresponds to the set of frequencies $\{1, \alpha_{i_1}, \dots, \alpha_{i_m}\}$, $\{1, \gamma_{i_1}, \dots, \gamma_{i_m}\}$. As in Section 2, we define for it *multiplicity of resonances* $\varkappa_I^1$ and *degree of irrationality* $\sigma_I^1 = m + 1 - \varkappa_I^1$.

As in the transformation (26), all β_j are rational numbers, then these numbers coincide for K_I^1 and L_I^1.

As before, in subspaces K_I^1 and L_I^1 we define *components* $\mathcal{A}_I^1$ of *the set* $\mathcal{A}^1$, as parts of intersection of set $\mathcal{A}^1$ with the subspace, excluding points lying at the smaller coordinate subspaces K_J^1 and L_J^1 correspondingly. Sets $\mathcal{A}_I^1$ in K_I^1 and in L_I^1 are the same and connected by the transformation (26). If $I = \{1, \dots, n\}$, then K_I^1 and L_I^1 are the whole space $\mathbb{C}^{2(n+1)}$, where we denote $K_n^1 = L_n^1$. For components $\mathcal{A}_I^1$, Theorem 3 about families of invariant tori is true.

Let $\tilde{I}^1$ be a set of all indices j with rational α_j or zero γ_j. Define $\mathcal{A}_{\tilde{I}}^1 = \mathcal{A}^1 \cap L_{\tilde{I}^1}^1$. Similarly to Theorem 4, there exists a converge transformation, normalizing at the set $\mathcal{A}_{\tilde{I}}^1$, and the set is analytic. It contains all families of periodic solutions adjoining the initial periodic solution $\mathcal{M}$. In it equations (33) take the form

$$\frac{\partial h}{\partial v_j} = 0 = \frac{\partial h}{\partial u_j}, \quad j \in \tilde{I}^1. \tag{35}$$

This system defines a set of stationary points, and equation $\dot\varphi = \frac{\partial h}{\partial s} = a$ gives dependence of t for them. The set $\mathcal{A}_{\tilde{I}}^1$ contains the set

$\mathcal{A}_0^1$, where all $u_j = v_j = 0$, $j = 1, \ldots, n$, i.e. it is not empty. Hence, each periodic solution of a Hamiltonian system belongs to a family of periodic solutions (see [4] Ch. II, §2).

3.5. *Set* $\mathcal{B}^1$

Let $\Gamma = \{\gamma_1, \ldots, \gamma_n\}$ be a diagonal matrix. At the set $\mathcal{A}^1$ in coordinates $\mathbf{u}, \mathbf{v}, s$ we consider the square $(2n + 1) \times (2n + 1)$ matrix

$$B_1 = \begin{pmatrix} \dfrac{\partial^2 h}{\partial \mathbf{v} \partial \mathbf{u}} - i\Gamma a & \dfrac{\partial^2 h}{\partial \mathbf{v} \partial \mathbf{v}} & \dfrac{\partial^2 h}{\partial \mathbf{v} \partial s} \\ \dfrac{\partial^2 h}{\partial s \partial \mathbf{u}} & \dfrac{\partial^2 h}{\partial s \partial \mathbf{v}} & \dfrac{\partial^2 h}{\partial s \partial s} \\ -\dfrac{\partial^2 h}{\partial \mathbf{u} \partial \mathbf{u}} & -\dfrac{\partial^2 h}{\partial \mathbf{v} \partial \mathbf{u}} + i\Gamma a & -\dfrac{\partial^2 h}{\partial \mathbf{u} \partial s} \end{pmatrix}, \tag{36}$$

where a is the same parameter as in equation (32).

The *set* $\mathcal{B}^1$ is such a subset of the set $\mathcal{A}^1$ on which the matrix B_1 is nilpotent, i.e. $B_1^{2n+1} = 0$.

Theorem 10. *Generically,*

$$\mathcal{B}^1 = \mathcal{A}_n^1$$

and $\Re \mathcal{B}^1$ *is a one-parameter (along a) family of invariant tori of dimension* σ_n^1.

Theorem 11 ([2], Part II, §3). *Under Condition ω^1 there exists such analytic canonical transformation* (20) *that reduces the initial system to the normal form at the set* $\mathcal{B}^1$ *and the set is analytic.*

If no resonant relation $p_0 + \langle \boldsymbol{\alpha}, \mathbf{p} \rangle = 0$ exists for numbers $1, \boldsymbol{\alpha}$, where $p_0 \in \mathbb{Z}$, $\mathbf{p} \in \mathbb{Z}^n$, then generically the set $\Re \mathcal{B}^1$ is a one-parameter family of irreducible invariant tori of dimension $n + 1$.

4. A Neighborhood of an Invariant Torus

4.1. *Local coordinates*

Let a real analytic Hamiltonian system with $k + n$ degrees of freedom have a k-dimensional invariant torus $\mathcal{T}^k$. The torus is named *regular,* if in some canonical analytical coordinates $\boldsymbol{\xi}$, $\boldsymbol{\psi}$, $\boldsymbol{\eta}$, $\boldsymbol{\rho}$ ($\boldsymbol{\psi}, \boldsymbol{\rho} \in \mathbb{R}^k$, $\boldsymbol{\xi}, \boldsymbol{\eta} \in \mathbb{R}^n$):

(1) the torus $\mathcal{T}^k$ is defined by equations

$$\boldsymbol{\xi} = \boldsymbol{\eta} = 0, \quad \boldsymbol{\rho} = 0; \tag{37}$$

(2) coordinates $\boldsymbol{\psi}$ are cyclic (angular) $\mod 2\pi$ and in the torus $\mathcal{T}^k$ they satisfy the system of equations

$$\dot{\psi}_i = \Omega_i = \mathrm{const} \in \mathbb{R}, \quad i = 1, \ldots, k; \tag{38}$$

(3) in a neighborhood of the torus $\mathcal{T}^k$ the system is Hamiltonian

$$
\begin{aligned}
\dot{\xi}_j &= \frac{\partial \gamma}{\partial \eta_j}, \quad \dot{\eta}_j = -\frac{\partial \gamma}{\partial \xi_j}, \quad j = 1, \ldots, n, \\
\dot{\psi}_i &= \frac{\partial \gamma}{\partial \rho_i}, \quad \dot{\rho}_i = -\frac{\partial \gamma}{\partial \psi_i}, \quad i = 1, \ldots, k,
\end{aligned}
\tag{39}
$$

with analytic Hamiltonian function $\gamma(\boldsymbol{\xi}, \boldsymbol{\psi}, \boldsymbol{\eta}, \boldsymbol{\rho})$, which is expanded into the convergent Poisson series

$$\gamma = \sum \gamma_{\mathbf{p,l,q,m}} \boldsymbol{\xi}^{\mathbf{p}} \boldsymbol{\eta}^{\mathbf{q}} \boldsymbol{\rho}^{\mathbf{m}} \exp \mathrm{i} \langle \mathbf{l}, \boldsymbol{\psi} \rangle, \tag{40}$$

where $0 \leqslant \mathbf{p}, \mathbf{q} \in \mathbb{Z}^n$, $0 \leqslant \mathbf{m} \in \mathbb{Z}^k$, $\mathbf{l} \in \mathbb{Z}^k$;

(4) the variational system is reducible, i.e. the square in $\boldsymbol{\xi}, \boldsymbol{\eta}$ part $\gamma_2(\boldsymbol{\xi}, \boldsymbol{\psi}, \boldsymbol{\eta}, \boldsymbol{\rho})$ of the Hamiltonian γ for $\boldsymbol{\rho} = 0$ does not depend on $\boldsymbol{\psi}$:

$$\gamma_2 = \frac{1}{2} \langle \boldsymbol{\zeta}, G\boldsymbol{\zeta} \rangle, \tag{41}$$

where $\boldsymbol{\zeta} = (\boldsymbol{\xi}, \boldsymbol{\eta})$ and G is a constant symmetric $2n \times 2n$ matrix. Let $\lambda_1, \ldots, \lambda_{2n}$ be the eigenvalues of the matrix $A = JG$, where $J = \begin{pmatrix} 0 & E_n \\ -E_n & 0 \end{pmatrix}$ and E_n is the identity $n \times n$ matrix. Applying the correct numbering, one has

$$\lambda_{j+n} = -\lambda_j, \quad j = 1, \ldots, n.$$

We denote $\boldsymbol{\lambda} = (\lambda_1, \ldots, \lambda_n)$ and $\boldsymbol{\Omega} = (\Omega_1, \ldots, \Omega_k)$.

4.2. Normal form

Theorem 12. *There exists a canonical invertible formal transformation of coordinates*

$$\boldsymbol{\xi}, \boldsymbol{\psi}, \boldsymbol{\eta}, \boldsymbol{\rho} \longleftrightarrow \mathbf{x}, \boldsymbol{\varphi}, \mathbf{y}, \mathbf{r} \tag{42}$$

in the form of Poisson series of type (40), *which near a regular invariant torus* $\mathcal{T}^k$ *reduces the Hamiltonian* (40) *to its normal form*

$$g = \sum_{j=1}^{n} \lambda_j x_j y_j + \sum_{i=1}^{k} \Omega_i r_i + \sum g_{\mathbf{plqm}} \mathbf{x}^{\mathbf{p}} \mathbf{y}^{\mathbf{q}} \mathbf{r}^{\mathbf{m}} \exp \mathrm{i} \langle \mathbf{l}, \boldsymbol{\varphi} \rangle, \tag{43}$$

where the third sum contains only resonant terms, for which

$$\langle \mathbf{p} - \mathbf{q}, \boldsymbol{\lambda} \rangle + \mathrm{i} \langle \mathbf{l}, \boldsymbol{\Omega} \rangle = 0. \tag{44}$$

Its proof is similar to the proof of Theorem 3.1 in [4] Ch. II. In what follows in this section, we will consider the case where all eigenvalues $\boldsymbol{\lambda}$ are pure imaginary: $\lambda_j = \mathrm{i}\alpha_j$, $\alpha_j \in \mathbb{R}$, and all $\lambda_j \neq 0$. Let $\boldsymbol{\alpha} = (\alpha_1, \ldots, \alpha_n)$.

Theorem 13. *If equation* $\langle \mathbf{l}, \boldsymbol{\Omega} \rangle = 0$ *in* $\mathbf{l} \in \mathbb{Z}^k$ *has only zero solution* $\mathbf{l} = 0$, *then there exists a canonical transformation*

$$x_j = u_j \exp\left(-\mathrm{i} \langle \mathbf{A}_j, \boldsymbol{\varphi} \rangle\right), \quad y_j = v_j \exp \mathrm{i} \langle \mathbf{A}_j, \boldsymbol{\varphi} \rangle, \; j = 1, \ldots, n,$$
$$\mathbf{r}^* = \mathbf{s}^* - B\mathbf{w}^*,$$
$$\tag{45}$$

where $\mathbf{w} = (u_1 v_1, \ldots, u_n v_n)$, B *is a* $k \times n$ *matrix with rational elements, and matrix* $A = \begin{pmatrix} \mathbf{A}_1 \\ \vdots \\ \mathbf{A}_n \end{pmatrix} = B^*$, *which transforms the Hamiltonian normal form* (43), (44) *to the autonomous series*

$$h(\mathbf{u}, \mathbf{v}, \mathbf{s}) = \mathrm{i} \sum_{j=1}^{n} \gamma_j u_j v_j + \sum_{i=1}^{k} \Omega_i s_i + \sum h_{\mathbf{pqm}} \mathbf{u}^{\mathbf{p}} \mathbf{v}^{\mathbf{q}} \mathbf{s}^{\mathbf{m}}, \tag{46}$$

which does not depend on angles $\boldsymbol{\varphi}$ *and contains only resonant terms with*

$$\langle \mathbf{p} - \mathbf{q}, \boldsymbol{\gamma} \rangle = 0, \tag{47}$$

where $\boldsymbol{\gamma} = (\gamma_1, \ldots, \gamma_n)$.

Here asterisk * means transposition.

Proof. In assumption of theorem; equation (44) takes the form

$$\langle \mathbf{p} - \mathbf{q}, \boldsymbol{\alpha} \rangle + \langle \mathbf{l}, \boldsymbol{\Omega} \rangle = 0. \tag{48}$$

Let equation

$$\langle \mathbf{p}, \boldsymbol{\alpha} \rangle + \langle \mathbf{l}, \boldsymbol{\Omega} \rangle = 0 \tag{49}$$

have μ linearly independent integer solutions $(\mathbf{p}, \mathbf{l})$. Then integer solutions $(\mathbf{p}, \mathbf{l})$ of equation (49) form a lattice in $\mathbb{R}^{n+k}$ and as a consequence of linear independence $\boldsymbol{\Omega}$ we have $0 \leqslant \mu \leqslant n$. Let the set of vectors

$$(\mathbf{p}_1, \mathbf{l}_1), \ldots, (\mathbf{p}_\mu, \mathbf{l}_\mu) \tag{50}$$

form a basis of the lattice. Then the $\mu \times n$ matrix $C = \begin{pmatrix} \mathbf{p}_1 \\ \vdots \\ \mathbf{p}_\mu \end{pmatrix}$ has such μ columns that their determinant is different from zero. For simplicity of notations we assume that they are columns with numbers $1, \ldots, \mu$. Now all vectors $\mathbf{p} = (p_1, \ldots, p_n)$ we divide into two parts $\mathbf{p}' = (p_1, \ldots, p_\mu)$ and $\mathbf{p}'' = (p_{\mu+1}, \ldots, p_n)$. Similarly we divide the matrix $C = (C', C'')$, where matrix C' has dimension $\mu \times \mu$, and matrix C'' has dimension $\mu \times (n - \mu)$. For vectors of the basis (50), from equation (49) we obtain the system of equations

$$C'\boldsymbol{\alpha}'^* + C''\boldsymbol{\alpha}''^* + L\boldsymbol{\Omega}^* = 0,$$

where L is a matrix with integral elements.

Let us solve the last system for $\boldsymbol{\alpha}'^*$:

$$\boldsymbol{\alpha}'^* = -C'^{-1}C''\boldsymbol{\alpha}''^* - C'^{-1}L\boldsymbol{\Omega}^* \stackrel{\text{def}}{=} -D\boldsymbol{\alpha}''^* - \tilde{A}\boldsymbol{\Omega}^*, \tag{51}$$

where matrices $D = C'^{-1}C''$ and $\tilde{A} = C'^{-1}L$ have rational elements and dimensions $\mu \times (n - \mu)$ and $\mu \times k$, respectively. Here numbers $\boldsymbol{\alpha}''$ and $\boldsymbol{\Omega}$ are linearly independent over integral numbers, i.e. the equation

$$\langle \mathbf{p}'', \boldsymbol{\alpha}'' \rangle + \langle \mathbf{l}, \boldsymbol{\Omega} \rangle = 0$$

in integer $(\mathbf{p}'', \mathbf{l})$ has only zero solution $\mathbf{p}'' = 0$, $\mathbf{l} = 0$. Let us define the $n \times k$ matrix as follows:

$$A = \begin{pmatrix} \tilde{A} \\ 0 \end{pmatrix}. \tag{52}$$

Taking into account (51), now we consider integer solutions $(\mathbf{p}, \mathbf{q}, \mathbf{l})$ to the equation (48)

$$\langle (\mathbf{p} - \mathbf{q})', \boldsymbol{\alpha}' \rangle + \langle (\mathbf{p} - \mathbf{q})'', \boldsymbol{\alpha}'' \rangle + \langle \mathbf{l}, \boldsymbol{\Omega} \rangle$$
$$= \left\langle (\mathbf{p} - \mathbf{q})', -D\boldsymbol{\alpha}''^* - \tilde{A}\boldsymbol{\Omega}^* \right\rangle + \langle (\mathbf{p} - \mathbf{q})'', \boldsymbol{\alpha}'' \rangle + \langle \mathbf{l}, \boldsymbol{\Omega} \rangle$$
$$= \left\langle (\mathbf{p} - \mathbf{q})', -\tilde{A}\boldsymbol{\Omega}^* \right\rangle + \langle \mathbf{l}, \boldsymbol{\Omega} \rangle - \left\langle (\mathbf{p} - \mathbf{q})', D\boldsymbol{\alpha}''^* \right\rangle + \langle (\mathbf{p} - \mathbf{q})'', \boldsymbol{\alpha}'' \rangle$$
$$= \left\langle \mathbf{l} - \tilde{A}^*(\mathbf{p} - \mathbf{q})', \boldsymbol{\Omega} \right\rangle + \left\langle (\mathbf{p} - \mathbf{q})'' - D^*(\mathbf{p} - \mathbf{q})', \boldsymbol{\alpha}'' \right\rangle = 0.$$

As $\boldsymbol{\alpha}''$ and $\boldsymbol{\Omega}$ are linearly independent over integral numbers, then the last equality gives equations

$$\mathbf{l} = \tilde{A}^* (\mathbf{p} - \mathbf{q})', \tag{53}$$
$$(\mathbf{p} - \mathbf{q})'' = D^* (\mathbf{p} - \mathbf{q})'.$$

Now let us compute dependence from $\boldsymbol{\varphi}$ for a resonant monomial after substitution (45)

$$\mathbf{x}^{\mathbf{p}} \mathbf{y}^{\mathbf{q}} \exp i \langle \mathbf{l}, \boldsymbol{\varphi} \rangle = \mathbf{u}^{\mathbf{p}} \mathbf{v}^{\mathbf{q}} \exp i \left(\langle \mathbf{q} - \mathbf{p}, A\boldsymbol{\varphi}^* \rangle + \langle \mathbf{l}, \boldsymbol{\varphi} \rangle \right),$$

where matrix A was defined in (52). So here dependence of $\boldsymbol{\varphi}$ is

$$\left\langle (\mathbf{q} - \mathbf{p})', \tilde{A}\boldsymbol{\varphi}^* \right\rangle + \langle \mathbf{l}, \boldsymbol{\varphi} \rangle = \left\langle \tilde{A}^*(\mathbf{q} - \mathbf{p})' + \mathbf{l}, \boldsymbol{\varphi} \right\rangle = \langle 0, \boldsymbol{\varphi} \rangle = 0$$

according to (53).

Other statements of theorem are easily verified. Proof is finished.

$\square$

Remark 1. In system

$$\dot{u}_j = \frac{\partial h}{\partial v_j}, \quad \dot{v}_j = -\frac{\partial h}{\partial u_j}, \quad j = 1, \ldots, n, \tag{54}$$

$$\dot{\varphi}_i = \frac{\partial h}{\partial s_i}, \quad \dot{s}_i = -\frac{\partial h}{\partial \varphi_i}, \quad i = 1, \ldots, k,$$

corresponding to the reduced normal form of Hamiltonian (46), all coordinates $\mathbf{s}$ are parameters. For the stationary points $(\mathbf{u}^0, \mathbf{v}^0, \mathbf{s}^0)$ of the subsystem (54), which satisfy the "algebraic" system of equations

$$\frac{\partial h}{\partial v_j} = \frac{\partial h}{\partial u_j} = 0, \quad j = 1, \ldots, n,$$

equations

$$\dot{\varphi}_i = \left.\frac{\partial h}{\partial s_i}\right|_{\mathbf{u}^0, \mathbf{v}^0, \mathbf{s}^0}, \quad i = 1, \ldots, k,$$

define frequencies on the corresponding invariant tori.

Remark 2. If frequencies $\Omega_1, \ldots, \Omega_k$ are linearly independent over integral numbers, i.e. equation $\langle \mathbf{l}, \boldsymbol{\Omega} \rangle = 0$ in $\mathbf{l} \in \mathbb{Z}^k$ has only zero solution $\mathbf{l} = 0$, then the initial torus $\mathcal{T}^k$ is irreducible.

4.3. *Convergence of the normalizing transformation*

Condition ω^k

Let $\omega_m = \min |\langle \mathbf{p}, \boldsymbol{\alpha} \rangle + \langle \mathbf{l}, \boldsymbol{\Omega} \rangle| = 0$ for $\langle \mathbf{p}, \boldsymbol{\alpha} \rangle + \langle \mathbf{l}, \boldsymbol{\Omega} \rangle \neq 0$, $||\mathbf{p}|| + ||\mathbf{l}|| < 2^m$, $\mathbf{p} \in \mathbb{Z}^n$, $\mathbf{l} \in \mathbb{Z}^k$. Then

$$-\sum_{m=1}^{\infty} \frac{\log \omega_m}{2^m} < 0.$$

Condition $\mathbf{A}^k$

There exists such a Poisson series $a(\mathbf{x}, \boldsymbol{\varphi}, \mathbf{y}, \mathbf{r})$ that in the normal form (43)

$$\frac{\partial g}{\partial y_j} = i\alpha_j x_j a, \quad \frac{\partial g}{\partial x_j} = i\alpha_j y_j a, \quad j = 1, \ldots, n,$$

$$\frac{\partial g}{\partial \varphi_i} = 0, \quad \frac{\partial g}{\partial r_i} = \Omega_i a, \quad i = 1, \ldots, k. \tag{55}$$

It is equivalent to the condition that Hamiltonian g is a power series of one variable

$$\sum_{j=1}^{n} \lambda_j x_j y_j + \sum_{i=1}^{k} \Omega_i r_i.$$

Theorem 14. *The normalizing transformation* (42) *converges in a neighborhood of our invariant torus* $\mathcal{T}^k$ *if numbers* $\boldsymbol{\alpha}$ *and* $\boldsymbol{\Omega}$ *satisfy Condition* ω^k *and the normal form* (43), (44) *satisfies Condition* $\mathbf{A}^k$.

4.4. Set $\mathcal{A}^k$

Let now a be an arbitrary parameter. Then the system of equations (55) defines the formal set $\mathcal{A}^k$. For the reduced normal form (46), (47) the set $\mathcal{A}^k$ is defined by the system of equations

$$\frac{\partial h}{\partial v_j} = i\gamma_j u_j a, \quad \frac{\partial h}{\partial u_j} = i\gamma_j v_j a, \quad j = 1, \ldots, n,$$

$$\frac{\partial h}{\partial s_i} = \Omega_i a, \quad i = 1, \ldots, k. \tag{56}$$

To each set of increasing indices $I = \{i_1, \ldots, , i_m\}$, $1 \leqslant i_1, i_m \leqslant n$, $1 \leqslant m \leqslant n$, there correspond the coordinate subspaces

$$K_I^k = \{\mathbf{x}, \mathbf{y}, \mathbf{r}, \boldsymbol{\varphi} : x_j = y_j = 0 \text{ for all } j \notin I\},$$

$$L_I^k = \{\mathbf{u}, \mathbf{v}, \mathbf{s}, \boldsymbol{\varphi} : u_j = v_j = 0 \text{ for all } j \notin I\}.$$

If $I = \{1, \ldots, n\}$, then $K_I^k = L_I^k = \mathbb{C}^{2(n+k)} \overset{\text{def}}{=} L_n^k$.

Now coordinates $\mathbf{r}, \boldsymbol{\varphi}, \mathbf{s}$ correspond to frequencies $\boldsymbol{\Omega}$. So the coordinate subspaces K_I^k and L_I^k correspond to the set of frequencies $\{\boldsymbol{\Omega}, \alpha_{i_1}, \ldots, \alpha_{i_m}\}$ and $\{\boldsymbol{\Omega}, \gamma_{i_1}, \ldots, \gamma_{i_m}\}$. For each K_I^k and L_I^k, we denote $m_I = m$ and:

(1) *half-dimension* $m_I^k = k + m_I$;
(2) *multiplicity of resonances* $\varkappa_I^k$ as the number of linearly independent integer relations

$$\sum_{j\in I} p_j \alpha_j + \langle \mathbf{l}, \boldsymbol{\Omega} \rangle = 0 \quad \text{and} \quad \sum_{j\in I} p_j \gamma_j + \langle \mathbf{l}, \boldsymbol{\Omega} \rangle = 0$$

with integer p_j and $\mathbf{l} \in \mathbb{Z}^k$;
(3) *degree of irrationality* $\sigma_I^k = k + m_I^k - \varkappa_I^k$;
(4) *subset of frequencies* $\boldsymbol{\Omega}, \boldsymbol{\alpha}_I$ and $\boldsymbol{\Omega}, \boldsymbol{\gamma}_I$.

Numbers $m_I^k, \varkappa_I^k, \sigma_I^k$ coincide for K_I^k and L_I^k, because in transformation (45) all elements of matrices A and B are rational numbers.

As before in subspaces K_I^k and L_I^k we define *components* $\mathcal{A}_I^k$ of the set $\mathcal{A}^k$ as parts of intersections of the set $\mathcal{A}^k$ with the subspaces excluding points belonging to smaller subspaces K_J^k and L_J^k. Sets $\mathcal{A}_I^k$ in K_I^k and in L_I^k are the same and connected by transformation (45). So, there is the component $\mathcal{A}_n^k$.

Theorem 15. *If the normalized Hamiltonian* (43), (44) *is analytic, then each component* $\Re\mathcal{A}_I^k$ *is a family of irreducible invariant tori of dimension* σ_I^k *with frequencies* $\boldsymbol{\Omega}a$, $\boldsymbol{\gamma}_I a$. *In the generic case, these families have* $\varkappa_I^k$ *parameters.*

4.5.   *Set* $\mathcal{B}^k$

Let $\Gamma = \{\gamma_1, \ldots, \gamma_n\}$ be a diagonal matrix. In the set $\mathcal{A}^k$ in coordinates $\mathbf{u}, \mathbf{v}, \mathbf{s}$ we consider the $(2n + k) \times (2n + k)$ matrix

$$B_k = \begin{pmatrix} \dfrac{\partial^2 h}{\partial\mathbf{v}\partial\mathbf{u}} - \mathrm{i}\Gamma a & \dfrac{\partial^2 h}{\partial\mathbf{v}\partial\mathbf{v}} & \dfrac{\partial^2 h}{\partial\mathbf{v}\partial\mathbf{s}} \\[2mm] \dfrac{\partial^2 h}{\partial\mathbf{s}\partial\mathbf{u}} & \dfrac{\partial^2 h}{\partial\mathbf{s}\partial\mathbf{v}} & \dfrac{\partial^2 h}{\partial\mathbf{s}\partial\mathbf{s}} \\[2mm] -\dfrac{\partial^2 h}{\partial\mathbf{u}\partial\mathbf{u}} & -\dfrac{\partial^2 h}{\partial\mathbf{u}\partial\mathbf{v}} + \mathrm{i}\Gamma a & -\dfrac{\partial^2 h}{\partial\mathbf{u}\partial\mathbf{s}} \end{pmatrix},$$

where a is the same parameter as in equations (56). The set $\mathcal{B}^k$ is such a subset of the set $\mathcal{A}^k$, where the matrix B_k is nilpotent, i.e. $B_k^{2n+k} = 0$.

Theorem 16. *Under Condition ω^k, there exists such an analytic canonical transformation* (42), *which reduces the initial Hamiltonian to the normal form in the set $\mathcal{B}^k$ and that set is analytic.*

Theorem 17. *Generically,* $\mathcal{B}^k = \mathcal{A}_n^k$.

Hence, the set $\Re\mathcal{B}^k$ consists of tori of dimension k, if all γ_j are zero, i.e. all α_j are linear combinations of frequencies Ω with rational coefficients. It means that all α_j are commensurable with frequencies. Such set $\Re\mathcal{B}^k$ forms a family with $k+1$ parameters.

5. Remarks

Neighborhood of the n-dimensional invariant torus in system with n degrees of freedom was studied in [4] Ch. II, §3. It was shown that such irreducible torus lie at one-parameter family of irreducible invariant tori of dimension n.

If the variational system near a stationary point or near an invariant torus $\mathcal{T}^k$ or near a periodic solution has eigenvalues λ_j with $\Re\lambda_j \neq 0$, then all stated above relate to their central manifolds and all Theorems are true. So, generically in real analytic Hamiltonian system with n degrees of freedom and without parameters:

(a) periodic solutions form one-parameter families,

(b) n-dimensional regular tori $\mathcal{T}^n$ form one-parameter families,

(c) k-dimensional irreducible regular tori $\mathcal{T}^k$ with $k < n$ form $(l+1)$-parameter families, if exactly $2l$ their eigenvalues have zero real parts, and all their imaginary parts are commensurable with frequencies $(0 \leqslant l \leqslant n-k)$.

Hence, in the system with d parameters periodic solutions and n-dimensional invariant tori $\mathcal{T}^n$ form $(d+1)$-parameter families, but k-dimensional invariant tori $\mathcal{T}^k$ with $k < n$ with mentioned properties form $(d+l+1)$-parameter families.

References

1. A. N. Kolmogorov, On conservation of conditionally-periodic motions under small perturbations of the Hamiltonian function. *Dokl. Akad. Nauk SSSR*, **98**(4), (1954) 527–530 (in Russian).
2. A. D. Bruno, *Local Methods in Nonlinear Differential Equations.* Springer-Verlag, Berlin Heidelberg, New York, London, Paris, Tokyo, 1989.
3. A. D. Bruno, Analytical form of differential equations (II). *Trans. Moscow Math. Soc.*, **26**(1972) 199–239; *Trudy Moskov. Mat. Obsc.*, **25**(1971) 119–262 (in Russian).
4. A. D. Bruno, *The Restricted 3–body Problem: Plane Periodic Orbits*, Walter de Gruyter, Berlin, 1994; Nauka, Moscow, 1990, p. 296 (in Russian).
5. A. D. Bruno, Analytical form of differential equations (I). *Trans. Moscow Math. Soc.*, **25** (1971) 131–288; *Trudy Moskov. Mat. Obsc.*, **25**(1971) 119–262 (in Russian).
6. A. M. Lyapunov, Problème général de la stabilité du mouvement. *Ann. Fac. Sci. Toulouse Math.* **9**(2), 204–474; *Stability of Motion.* Academic Press, New York, London, (1966) [English].
7. A. D. Bruno, Normalization of the periodic Hamiltonian system. *Programming and Computer Software*, **46**(2), (2020) 76–83, doi:10.1134/S0361768820020048.

© 2022 World Scientific Publishing Europe Ltd.
https://doi.org/10.1142/9781800611368_0002

Chapter 2

The Algebraic Potentials Having Tri-Hamiltonian Structures of the Reflection Groups of Types D_4, F_4, H_4

Jiro Sekiguchi

Department of Mathematics
Tokyo University of Agriculture and Technology
Tokyo, Japan
sekiguti@cc.tuat.ac.jp

The purpose of this chapter is to study the relationship between the algebraic potential having tri-Hamiltonian structure of the reflection group of type D_4 (resp., F_4, H_4) and the polynomial potential of type A_3 (resp., B_3, H_3). The present work is based on the paper by S. Romano [12] on algebraic Frobenius manifolds having tri-Hamiltonian structure. We will construct a 4-parameter family of algebraic potentials having tri-hamiltonian structure in the D_4 case (resp., F_4 case) and define an ordinary differential equation from a member of the 4-parameter family and show the coincidence of the differential equation in question and an ordinary differential equation constructed from the polynomial potential of type A_3 (resp., B_3). We will also treat the correspondence between H_4 case and H_3 case and show a partial answer to the case (H_4, H_3).

1. Introduction

We start this introduction with recalling the definition of the WDVV equation. The WDVV equation is the system of nonlinear partial differential equations formulated by Witten and Dijkgraaf–Verlinde–Verlinde. We review it briefly. For the details, refer Dubrovin [3], Sabbah [13]. The following formulation is due to [9]:

Let $F = F_0 + F_1$ be a function of $(x_1, x_2, \ldots, x_n)$. Here, F_0 is the polynomial defined by

$$F_0 = \begin{cases} \dfrac{1}{2} x_1 x_n^2 + \displaystyle\sum_{j=2}^{n/2} x_j x_{n-j+1} x_n & (n : \text{even}) \\ \dfrac{1}{2} x_1 x_n^2 + \displaystyle\sum_{j=2}^{m-1} x_j x_{2m-j} x_n + \dfrac{1}{2} x_m{}^2 x_n & (n : \text{odd}, \ m = (n+1)/2) \end{cases}$$

and F_1 is a function of $(x_1, x_2, \ldots, x_{n-1})$, independent of x_n.

We assume that F is weighted homogeneous. Namely there are non-zero constants $w, w_1, w_2, \ldots, w_n$ such that $EF = wF$, where $E = \frac{1}{w_n} \sum_{j=1}^{n} w_j x_j \partial_{x_j}$ is the Euler vector field. The constants $w_1, w_2, \ldots, w_n$ are assumed to be rational numbers and $0 < w_1 \le w_2 \le \cdots \le w_n = 1$. We define the vector-valued function P by

$$P = (\partial_{x_n} F, \partial_{x_{n-1}} F, \ldots, \partial_{x_1} F). \tag{1}$$

From the definition, P is the gradient vector of F. Moreover, we put

$$C = {}^t(\partial_{x_1} P, \partial_{x_2} P, \ldots, \partial_{x_n} P), \tag{2}$$

which is the Hessian of F. We need the partial differentiations of C which are denoted by

$$\tilde{B}_j = \partial_{x_j} C \quad (j = 1, 2, \ldots, n).$$

It is clear that $\tilde{B}_n = I_n$, the identity matrix. Moreover, we put

$$B_\infty^{(n)} = \operatorname{diag}(r + w_1, \ldots, r + w_n),$$

where r is an arbitrary constant.

Definition 1. The system of partial differential equations

$$\begin{cases} EF = wF \\ [\tilde{B}^{(j)}, \tilde{B}^{(k)}] = O \quad (j, k = 1, 2, \ldots, n) \end{cases} \tag{3}$$

for F is called the WDVV equation. If F is its solution, F is called a potential. Moreover, the coordinate $(x_1, \ldots, x_n)$ is called a flat coordinate.

A polynomial (resp., algebraic) solution to the WDVV equation with respect to the coordinate $(x_1, \ldots, x_n)$ is called a polynomial (resp., algebraic) potential.

Let F be a potential, namely we assume F is a solution to (3). Using F, we have introduced an $n \times n$ matrix C. Then

$$T = EC \tag{4}$$

is an $n \times n$ matrix whose entries coincide with the entries of C up to constant factors. Then the system of differential equations

$$\partial_{x_j} Y = -T^{-1} \tilde{B}^{(j)} B_\infty^{(n)} Y \quad (j = 1, 2, \ldots, n) \tag{5}$$

for $Y = {}^t(\mathbf{y}_1, \ldots, \mathbf{y}_n)$ is integrable (cf. [6,9]). Paying attention to one of the equations of the system (5), we have

$$\partial_{x_n} Y = -T^{-1} B_\infty^{(n)} Y, \tag{6}$$

which is regarded as an ordinary differential equation with respect to the variable x_n. It is clear from the definition that $\det(T)$ is a monic of degree n as a polynomial of x_n. Let $u_1, \ldots, u_n$ be solutions of $\det(T) = 0$. Then (6) has singularities at the points $u_1, \ldots, u_n$.

We consider the condition

(TH) $n = 2k$ and $0 < 2\mu + 1 = w_1 = \cdots = w_k < w_{k+1} = \cdots = w_{2k} = 1$.

If (TH) holds, then the Frobenius manifold defined by F has tri-Hamiltonian structure (cf. [12]). Romano [12] proved a complete description of semisimple tri-Hamiltonian Frobenus manifolds in the lowest non-trivial dimension $n = 4$. His approach relies on the interpretation of semisimple Frobenius manifolds as isomonodromic deformation spaces of certain Fuchsian linear systems (similar to a special case of (6)) which are equivalent to the Painlevé VIμ equation

$$\frac{d^2 y}{dt^2} = \frac{1}{2} \left(\frac{1}{y} + \frac{1}{y-1} + \frac{1}{y-t} \right) \left(\frac{dy}{dt} \right)^2 - \left(\frac{1}{t} + \frac{1}{t-1} + \frac{1}{y-t} \right) \left(\frac{dy}{dt} \right)$$

$$+ \frac{y(y-1)(y-t)}{2t^2(t-1)^2} \left((2\mu - 1)^2 + \frac{t(t-1)}{(y-t)^2} \right). \tag{7}$$

The above special case of the Painlevé VI equation appeared in [4]. It is shown in [4] that the solutions to (7) parametrize semisimple Frobenius manifolds of dimension 3. This implies the existence of a map of three-dimensional Frobenius manifolds to four-dimensional tri-Hamiltonian ones, since both classes are parametrized by the same PVI transcendents.

As is already shown by Dubrovin (cf. [3]), there are three polynomial potentials in $n = 3$ case corresponding to the reflection groups of types A_3, B_3, H_3. It is interesting to apply the above result by Romano to these three cases. As to the Frobenius manifolds having tri-Hamiltonian structure, there are at least two algebraic potentials constructed by Pavlyk [11] and Dinar [1]. The former is denoted by $D_4(1)$ in the main text expected to correspond to the potential of type A_3 and the latter is denoted by $F_4(1)$ and is to the potential of type B_3. The remaining one denoted by $H_4(4)$ is to that of type H_3.

Remark 1. It is better to explain the background of the notation $D_4(1)$, $F_4(1)$, $H_4(4)$. The subregular nilpotent orbits of the simple Lie algebra of type D_4 (resp., F_4) are denoted by $D_4(1)$ (resp., $F_4(1)$). The algebraic potential constructed by Pyvlyk (resp., Y. Dinar) corresponds to the nilpotent orbit $D_4(1)$ (resp., $F_4(1)$). The notation follows those of nilpotent orbits. In spite of the fact that there is no simple Lie algebra whose Weyl group is the reflection group of type H_4, it is possible to define primitive conjugacy classes in the case of the reflection group of type H_4 and the notation $H_4(4)$ is proposed by Douvropoulos [14] for a primitive conjugacy class in this group.

Our effort is focused on the confirmation of the correspondence mentioned above. Namely we pay attention to showing the correspondence between

(C-A3) the polynomial potential of type A_3 and the algebraic potential of type $D_4(1)$,

(C-B3) the polynomial potential of type B_3 and the algebraic potential of type $F_4(1)$,

(C-H3) the polynomial potential of type H_3 and the algebraic potential of type $H_4(4)$.

We recall the polynomial potentials of three variables:

$$\begin{cases} F = \dfrac{x_1 x_3^2 + x_2^2 x_3}{2} - \dfrac{3 x_1^2 x_2^2}{2} + \dfrac{4 x_1^5}{135} & (A_3 \text{ case}), \\[3mm] F = \dfrac{x_1 x_3^2 + x_2^2 x_3}{2} - \dfrac{3 x_1 x_2^3}{2} + \dfrac{27 x_1^3 x_2^2}{2} + \dfrac{2187 x_1^7}{70} & (B_3 \text{ case}), \\[3mm] F = \dfrac{x_1 x_3^2 + x_2^2 x_3}{2} + \dfrac{x_1^2 x_2^3}{6} + \dfrac{x_1^5 x_2^2}{20} + \dfrac{x_1^{11}}{3960} & (H_3 \text{ case}). \end{cases} \tag{8}$$

We easily see that the weights w_1, w_2, w_3 in these cases are as follows:

	w_1	w_2	w_3
A_3	$1/2$	$3/4$	1
B_3	$1/3$	$2/3$	1
H_3	$1/5$	$3/5$	1

Let F be one of the potentials given in (8). We consider the case $r = -1$ of the differential equation (6). Then we obtain

$$\frac{d}{dx}\hat{Y} = \left(\frac{N_1}{x - z_1} + \frac{N_2}{x - z_2} + \frac{N_3}{x - z_3} \right) \hat{Y}, \tag{9}$$

where $\hat{Y} = {}^t(\mathbf{y}_1, \mathbf{y}_2)$ and N_1, N_2, N_3 are 2×2 matrices depending on x_1, x_2. Then putting $Z = (x-u)^{-2\alpha}\hat{Y}$, where α is given by $\alpha = 1 - w_2$ and

$$\xi = L(x) = \frac{(z_1 + z_2 - u)x - z_1 z_2}{x - u},$$

we find that the equation (9) for Y turns out to be the equation

$$\frac{d}{d\xi}Z = \left(\frac{N_1}{\xi - z_1} + \frac{N_2}{\xi - z_2} + \frac{N_3}{\xi - z_{3a}} + \frac{N_4}{\xi - z_{4a}} \right) Z \tag{10}$$

for Z, where $N_4 = \begin{pmatrix} 0 & 0 \\ 0 & \alpha \end{pmatrix}$ and $z_{3a} = L(z_3)$, $z_{4a} = L(\infty)$.

The matrices N_1, N_2, N_3, N_4 at least satisfy the conditions

(D1) $N_1 + N_2 + N_3 + N_4 = 2\alpha I_2$,
(D2) $\det(N_j) = 0, \quad \mathrm{Tr}(N_j) = \alpha \quad (j = 1, 2, 3, 4)$.

The matrix entries of N_1, N_2, N_3 satisfy algebraic relations in addition to (D1), (D2) for the reason that (9) implies a solution to the Painlevé VI equation. The algebraic relations will be given in Lemma 1 for the case A_3 and Lemma 3 for the case B_3. It is underlined here that from the matrices N_1, N_2, N_3 given in the main text, it is possible to reconstruct polynomial potentials in (8) for each of the cases A_3, B_3. Since it is complicated to show the algebraic relations for H_3, we don't discuss them in the case H_3.

We turn to the algebraic potentials of the four-dimensional tri-Hamiltonian Frobenius manifolds denoted by $D_4(1)$, $F_4(1)$, $H_4(4)$. We need some preparation to formulate the definition of the algebraic potential whose type is one of $D_4(1)$, $F_4(1)$, $H_4(4)$. Let α be the same as the number corresponding to rank three reflection group. Similarly, we let $\beta = w_1$ for simplicity.

	α	β
$D_4(1)$	$1/4$	$1/2$
$F_4(1)$	$1/3$	$1/3$
$H_4(4)$	$2/5$	$1/5$

Let x_1, x_2, x_3, x_4 be variables and their weights are $\beta, \beta, 1, 1$, respectively. Let z_0 be an algebraic function of x_1, x_2, x_3, which is a solution of the algebraic equation

$$x_3 - \sum_{j=0}^{1/\beta-1} \gamma_j(x_1, x_2)\zeta^j - \zeta^{1/\beta} = 0 \tag{11}$$

for ζ, where each $\gamma_j(x_1, x_2)$ is a polynomial of x_1, x_2 and

$$x_3 - \sum_{j=0}^{1/\beta-1} \gamma_j(x_1, x_2) - \zeta^{1/\beta}$$

is weighted homogeneous. This condition implies that β is the weight of z_0. Let $\tilde{F}(x_1, x_2, x_3, x_4, \zeta)$ be a weighted homogeneous polynomial of $(x_1, x_2, x_3, x_4, \zeta)$, which takes the form

$$\tilde{F}(x_1, x_2, x_3, x_4, \zeta) = \frac{1}{2}x_1 x_4^2 + x_2 x_3 x_4 + \sum_{k=0}^{1+2/\beta} \sigma_k(x_1, x_2)\zeta^k.$$

In particular, $\sigma_{1+2/\beta}$ is a constant.

We let

$$F(x_1, x_2, x_3, x_4) = \tilde{F}(x_1, x_2, x_3, x_4, z_0)(= \tilde{F}(x_1, x_2, x_3, x_4, \zeta)|_{\zeta=z_0}).$$

Then $F(x_1, x_2, x_3, x_4)$ is an algebraic function. If $F(x_1, x_2, x_3, x_4)$ is also a solution to the WDVV equation, it is an algebraic potential. The potentials constructed by Pavlyk and Dinar are regarded as algebraic potentials in this sense. In the same manner, it is possible to construct an algebraic potential when $\beta = \frac{1}{5}$. This is the case for type $H_4(4)$ in this chapter.

Let F be the algebraic potential whose type is one of $D_4(1)$, $F_4(1)$, $H_4(4)$. Then the matrix T and $\det(T)$ are defined (cf. (4)). Since

$$x_3 - \sum_{j=0}^{1/\beta-1} \gamma_j(x_1, x_2) z_0^j - z_0^{1/\beta} = 0,$$

we eliminate x_3 in the definition of T by the relation

$$x_3 = \sum_{j=0}^{1/\beta-1} \gamma_j(x_1, x_2) z_0^j + z_0^{1/\beta}.$$

As a consequence, all the matrix entries of $T - x_4 I_4$ are polynomials of x_1, x_2, z_0. Similarly, $\det(T)$ is regarded as a monic of the variable x_4 of weight four whose coefficients are polynomials of x_1, x_2, z_0.

We focus our attention on the differential equation (cf. (6))

$$\partial_{x_4} Y = -T^{-1} B_\infty^{(4)} Y. \tag{12}$$

Noting that $B_\infty^{(4)} = \mathrm{diag}(r + \beta, r + \beta, r + 1, r + 1)$, we treat (12) in the case $r = -1$. Then we obtain the differential equation

$$\frac{d}{dx}\hat{Y} = \left(\frac{R_1}{x - u_1} + \frac{R_2}{x - u_2} + \frac{R_3}{x - u_3} + \frac{R_4}{x - u_4} \right) \hat{Y}, \tag{13}$$

where $\hat{Y} = {}^t(y_1, y_2)$, u_1, u_2, u_3, u_4 are solutions to $\det(T) = 0$ as a quartic equation for x_4 and R_1, R_2, R_3, R_4 are 2×2 matrices depending on x_1, x_2, z_0.

The matrices R_1, R_2, R_3, R_4 at least satisfy the conditions

(D'1) $R_1 + R_2 + R_3 + R_4 = 2\alpha I_2$,
(D'2) $\det(R_j) = 0$, $\mathrm{Tr}(R_j) = \alpha$ $(j = 1, 2, 3, 4)$.

The main purpose of this chapter is to show that (13) is equivalent to (10) in each of the cases $D_4(1), F_4(1)$, which implies the correspondences (C-A3), (C-B3).

We are in a position to explain the idea employed in this chapter to accomplish the purpose. As the first step, we find a 2×2 matrix G and the condition among variables such that

$$GR_jG^{-1} = N_j \quad (j = 1, 2, 3, 4).$$

To explain the second step, we define the linear transformation

$$S(x_4) = \frac{(z_1 - z_2)x_4 + u_1z_2 - u_2z_1}{u_1 - u_2}.$$

It is clear from the definition that

$$S(u_1) = z_1, \quad S(u_2) = z_2, \quad S(\infty) = \infty.$$

The second step is to show that u is defined so that

$$S(u_3) = L(z_3), \quad S(u_4) = L(\infty)$$

are valid.

The correspondences (C-A3), (C-B3), (C-H3) would follow from the confirmation of these steps. Even though the idea is simple, there are many difficulties to overcome in the course of reaching the purpose. One of the difficulties stems from the inability to specifically display the solutions u_1, u_2, u_3, u_4. For this reason, the intermediate calculation results remain complicated and do not become simple expressions. Another difficulty is more serious. We explain it in some detail. For each of the cases $D_4(1)$, $F_4(1)$, $H_4(4)$, it is possible to construct a 1-parameter family of algebraic potentials. To the contrary of our expectation, the members of the family seem not to be isomorphic to each other. At this point, it is worthwhile to mention a suggestion by Dubrovin mentioned by Diner. It says that there is an "equivalence" among potentials and there is an equivalent class consisting of algebraic potentials that corresponds to the polynomial potential of a reflection group of rank three. Related topics shall be discussed in Section 3.4.

We present the results of this chapter. In each of the two cases (C-A3), (C-B3), a detailed calculation was performed to confirm the correspondence. In the case (C-H3), we abandoned to confirm the idea because the calculations were beyond the capabilities of the computer. We only confirm the correspondence under the restriction of variables where the solutions u_1, u_2, u_3, u_4 satisfy the condition $u_1 + u_2 = 0$, $u_3 + u_4 = 0$.

We are now explain the contents of this chapter. We treat the three cases (C-A3), (C-B3), (C-H3) separately. In Section 2, we treat the correspondence between the potential of type A_3 and the algebraic potential of type $D_4(1)$. We first treat the case A_3 in Section 2.1. Introducing the potential in this case and showing the matrices C, T, we compute the ordinary differential equation (21) which is the same as (10). In this manner, we obtain four 2×2 matrices N_j $(j = 1, 2, 3, 4)$ containing three variables. In Section 2.2, we first construct a 4-parameter family of algebraic potentials of variables (x_1, x_2, x_3, x_4) with weights $1/2, 1/2, 1, 1$ and introduce a group consisting of automorphisms of the family. Then we recall the definition of the algebraic potentials of type D_4 constructed by Pavlyk [11] and Dinar [2] and identify them with the members of the family. After the preparation on the computation for the case of A_3 in Sections 2.1 and 2.2, we treat the case $D_4(1)$ in Section 2.3. The algebraic potential of type $D_4(1)$ given Section 2.3 is contained in the same orbit with those constructed by Pavlyk and Dinar under the group action introduced in Section 2.2. As in this introduction, let x_1, x_2, x_3, x_4 be the variables and let z_0 be an algebraic function of x_1, x_2, x_3 defined by (31). Let u_1, u_2, u_3, u_4 be the solutions to $\det(T) = 0$ as an equation of x_4. We determine the ordinary differential equation (34) which contains four 2×2 matrices R_{ja} $(j = 1, 2, 3, 4)$ as residue matrices. To continue the argument, we take a matrix M_4 such that

$$M_4 R_{4a} M_4{}^{-1} = \begin{pmatrix} 0 & 0 \\ 0 & \frac{1}{4} \end{pmatrix}.$$

Then we define $R_{jb} = M_4 R_{ja} M_4{}^{-1}$ $(j = 1, 2, 3, 4)$. The most difficult part of showing the correspondence is to identify the 4-tuples (N_1, N_2, N_3, N_4) (cf. (17), (22)) and $(R_{1b}, R_{2b}, R_{3b}, R_{4b})$ (cf. (35)). Since $N_4 = R_{4b}$ is diagonal, the problem is reduced to finding diagonal matrices P, Q such that the (2,1)-entries of $P R_{3b} P^{-1}$, $Q N_3 Q^{-1}$ equal to 1. Under this normalization, we show the condition such that $P R_{jb} P^{-1} = Q N_j Q^{-1}$ $(j = 1, 2, 3, 4)$. To confirm these equalities, we are deeply indebted to the software Mathematica. In the last part of this section, we investigate the case where the solutions u_1, u_2, u_3, u_4 satisfy the condition $u_1 + u_2 = 0$, $u_3 + u_4 = 0$. In Section 3, we treat the correspondence between the potential of type B_3 and the algebraic potential of type $F_4(1)$. The argument in this section is similar to that in Section 2. The main difference from the case $D_4(1)$ is that in $F_4(1)$ case, $\det(T)$ is the product of quadratic polynomials of x_4. This property simplifies the calculations at each stage compared with $D_4(1)$ case. We treat the B_3 case in Section 3.1. In Section 3.2, we shall construct a 4-parameter family $\mathcal{F}$ of algebraic potentials of variables (x_1, x_2, x_3, x_4) with the weights $1/3, 1/3, 1, 1$ and introduce a group G of automorphisms of the family. We treat an algebraic potential of type $F_4(1)$ which is *not* transformed to that obtained by Dinar and show that it corresponds to the polynomial potential of type B_3 by an argument similar to that in Section 2.3. The argument in Section 3.3 goes well if we treat the algebraic potential which is isomorphic to that obtained by Dinar. The reason why we don't treat Dinar's is that the computation becomes complicated compared with ours (see Remark 9). In Section 3.4, we shall introduce a 1-parameter subfamily $\mathcal{F}'$ of $\mathcal{F}$ with the following conditions: (1) Any member of $\mathcal{F}$ is conjugate to one of $\mathcal{F}'$ under G-action. (2) By G-action, any two of $\mathcal{F}'$ are not conjugate to each other. We define an "equivalence" relation among potentials of type $F_4(1)$ and all the members of $\mathcal{F}'$ are "equivalent" in this sense and discuss an evidence that the equivalence in this sense is the same as the equivalence suggested by Dubrovin. In Section 4, we treat the correspondence between the potential of type H_3 and the algebraic potential of type $H_4(4)$. It is easy to formulate the idea explained above to this case, but it is hard to reach the purpose because the computations at various stages are beyond the capability of computers. For this reason, we abandon investigating the general case and restrict our attention to the case where u_1, u_2, u_3, u_4 satisfy the condition $u_1 + u_2 = 0$, $u_3 + u_4 = 0$. Under this condition, we confirm the identification of (13) and (10).

2. A_3 and $D_4(1)$

In Section 2.1, we first collect the polynomial potential of type A_3, the matrices C and T (see (2), (4) for the definition of C and T). Then we introduce an ordinary differential equation (15) of variable t_3 whose singular points come from the solutions of $\det(T) = 0$ as a polynomial of t_3, its modification (16) and 2×2 matrices N_1, N_2, N_3 (cf. (17)). In Section 2.2, we introduce a 4-parameter family $\mathcal{F}$ of algebraic potentials of the variables x_1, x_2, x_3, x_4 with weights $1/2, 1/2, 1, 1$ and define a group G consisting of automorphisms of the family $\mathcal{F}$. Pavlyk [11] and Dinar [2] constructed algebraic potentials of type $D_4(1)$. We confirm that both the potentials of Pavlyk and Dinar are regarded as members of $\mathcal{F}$. More precisely, we do that they are contained in the same G-orbit in $\mathcal{F}$. In Section 2.3, we take an algebraic potential F which is contained in the G-orbit of Pavlyk's potential in $\mathcal{F}$. Then we construct an ordinary differential equation from F and show the coincidence of it and (21). This follows from Theorems 2–4.

2.1. A_3 case

We start by introducing the polynomial potential which is obtained by Dubrovin. We introduce here the following potential which is the same as the one obtained by Dubrovin [3] by a coordinate transformation.

$$F = \frac{y_1 y_3{}^2 + y_2{}^2 y_3}{2} - \frac{1}{3} y_1{}^2 y_2{}^2 + \frac{4}{135} y_1{}^5,$$

$$C = \begin{pmatrix} y_3 & -\frac{4y_1 y_2}{3} & \frac{2}{27}\left(8y_1{}^3 - 9y_2{}^2\right) \\ y_2 & \frac{1}{3}\left(3y_3 - 2y_1{}^2\right) & -\frac{4y_1 y_2}{3} \\ y_1 & y_2 & y_3 \end{pmatrix},$$

$$T = \begin{pmatrix} y_3 & -\frac{5y_1 y_2}{3} & \frac{1}{9}\left(8y_1{}^3 - 9y_2{}^2\right) \\ \frac{3y_2}{4} & \frac{1}{3}\left(3y_3 - 2y_1{}^2\right) & -\frac{5y_1 y_2}{3} \\ \frac{y_1}{2} & \frac{3y_2}{4} & y_3 \end{pmatrix}.$$

We change the variables y_1, y_2, y_3 by t_1, t_2, t_3 so that they satisfy the relation

$$y_1 = (-t_1{}^2 - t_1 t_2 - t_2{}^2)/2,$$

$$y_2 = 2(t_1 - t_2)(2t_1 + t_2)(t_1 + 2t_2)/9,$$

$$y_3 = (-5t_1{}^4 - 10t_1{}^3 t_2 + 3t_1{}^2 t_2{}^2 - 10t_1 t_2{}^3 - 5t_2{}^4 + 18t_3)/18.$$

Then we have

$$\det(T) = (t_3 - z_1)(t_3 - z_2)(t_3 - z_3),$$

where

$$z_1 = 0, \ z_2 = t_1{}^3(t_1 + 2t_2), \ z_3 = t_2{}^3(2t_1 + t_2). \tag{14}$$

Introducing the matrix $B_\infty^{(3)}$ depending on r by

$$B_\infty^{(3)} = \begin{pmatrix} r + \frac{1}{2} & 0 & 0 \\ 0 & r + \frac{3}{4} & 0 \\ 0 & 0 & r + 1 \end{pmatrix},$$

we consider the ordinary differential equation

$$\frac{d}{dt_3}\tilde{Y} = -T^{-1}B_\infty^{(3)}\tilde{Y}, \tag{15}$$

where $\tilde{Y} = {}^t(\tilde{\mathbf{y}}_1, \tilde{\mathbf{y}}_2, \tilde{\mathbf{y}}_3)$. We focus our attention to the case $r = -1$ of the equation (15). Then we obtain the following differential equation:

$$\frac{d}{dt_3}\hat{Y} = \left(\frac{N_1}{t_3 - z_1} + \frac{N_2}{t_3 - z_2} + \frac{N_3}{t_3 - z_3} \right) \hat{Y}, \tag{16}$$

where $\hat{Y} = {}^t(\tilde{\mathbf{y}}_1, \tilde{\mathbf{y}}_2)$ and

$$\begin{cases} N_1 = \begin{pmatrix} \frac{t_1{}^2 + 7t_2t_1 + t_2{}^2}{36t_1t_2} & \frac{(t_1 - t_2)(t_1{}^2 + 7t_2t_1 + t_2{}^2)}{108t_1t_2} \\ -\frac{t_1 - t_2}{12t_1t_2} & -\frac{(t_1 - t_2)^2}{36t_1t_2} \end{pmatrix}, \\[2em] N_2 = \begin{pmatrix} \frac{5t_1{}^2 + 5t_2t_1 - t_2{}^2}{36t_1(t_1 + t_2)} & -\frac{(2t_1 + t_2)(5t_1{}^2 + 5t_2t_1 - t_2{}^2)}{108t_1(t_1 + t_2)} \\ -\frac{2t_1 + t_2}{12t_1(t_1 + t_2)} & \frac{(2t_1 + t_2)^2}{36t_1(t_1 + t_2)} \end{pmatrix}, \\[2em] N_3 = \begin{pmatrix} \frac{-t_1{}^2 + 5t_2t_1 + 5t_2{}^2}{36t_2(t_1 + t_2)} & \frac{(t_1 + 2t_2)(-t_1{}^2 + 5t_2t_1 + 5t_2{}^2)}{108t_2(t_1 + t_2)} \\ \frac{t_1 + 2t_2}{12t_2(t_1 + t_2)} & \frac{(t_1 + 2t_2)^2}{36t_2(t_1 + t_2)} \end{pmatrix}. \end{cases} \tag{17}$$

It is easy to show that

$$N_1 + N_2 + N_3 = \begin{pmatrix} \frac{1}{2} & 0 \\ 0 & \frac{1}{4} \end{pmatrix}.$$

Letting $Y = (t_3 - u)^{-1/2}\hat{Y}$, we obtain the differential equation

$$\frac{d}{dt_3}Y = \left(\frac{N_1}{t_3 - z_1} + \frac{N_2}{t_3 - z_2} + \frac{N_3}{t_3 - z_3} + \frac{-\frac{1}{2}}{t_3 - u} \right) Y, \tag{18}$$

for Y, where u is a function of t_1, t_2 different from all of z_1, z_2, z_3.

We define the linear fractional transformation $L(t_3)$ of t_3 by

$$\xi = L(t_3) = \frac{(z_1 + z_2 - u)t_3 - z_1 z_2}{t_3 - u}.$$

It follows from the definition that

$$L(z_1) = z_1, \ L(z_2) = z_2, \ L(u) = \infty.$$

We define z_{3a}, z_{4a} by

$$z_{3a} = L(z_3), \quad z_{4a} = L(\infty). \tag{19}$$

Then it is straightforward to show

$$\begin{cases} z_{3a} = \dfrac{z_1 z_2 - (z_1 + z_2 - u)z_3}{u - z_3}, \\ z_{4a} = z_1 + z_2 - u. \end{cases} \tag{20}$$

The differential equation (18) turns out to be

$$\frac{d}{d\xi}Y = \left(\frac{N_1}{\xi - z_1} + \frac{N_2}{\xi - z_2} + \frac{N_3}{\xi - z_{3a}} + \frac{N_4}{\xi - z_{4a}} \right) Y, \tag{21}$$

where

$$N_4 = \begin{pmatrix} 0 & 0 \\ 0 & \frac{1}{4} \end{pmatrix}. \tag{22}$$

It is easy to show that $0, \frac{1}{4}$ are the eigenvalues of each of the matrices N_1, N_2, N_3, N_4 and

$$N_1 + N_2 + N_3 + N_4 = \frac{1}{2} I_2.$$

2.2. *A 4-parameter family of algebraic potentials of type $D_4(1)$*

The purpose of this section is to construct a family of algebraic potentials of four variables (x_1, x_2, x_3, x_4) with weights $1/2, 1/2, 1, 1$ which contain the algebraic potentials constructed by Pavlyk [11] and Dinar [2].

Let z_0 be an algebraic function of (x_1, x_2, x_3) defined by the equation

$$z_0{}^2 - x_3 + d_1 x_1^2 + d_2 x_1 x_2 + d_3 x_2^2 = 0, \tag{23}$$

where d_1, d_2, d_3 are constants to be determined. In particular, z is weighted homogeneous of degree $1/2$. Using (x_1, x_2, x_3, x_4) and z_0, we introduce a function $F = F(x_1, x_2, x_3, x_4)$ satisfying the following conditions:

(a) $F = F_0 + F_1$.
(b) $F_0 = \frac{1}{2} x_1 x_4^2 + x_2 x_3 x_4$.
(c) F_1 is a function of (x_1, x_2, x_3) satisfying the following conditions:

- There is a homogeneous polynomial $G = G(x_1, x_2, Z)$ of degree 5 such that $F_1 = G(x_1, x_2, z_0)$.
- $\frac{\partial^5}{\partial Z^5} G \neq 0$.

We consider the 4×4 matrix C defined by

$$C = \begin{pmatrix} \partial_{x_1}\partial_{x_4}F & \partial_{x_1}\partial_{x_3}F & \partial_{x_1}\partial_{x_2}F & \partial_{x_1}\partial_{x_1}F \\ \partial_{x_2}\partial_{x_4}F & \partial_{x_2}\partial_{x_3}F & \partial_{x_2}\partial_{x_2}F & \partial_{x_2}\partial_{x_1}F \\ \partial_{x_3}\partial_{x_4}F & \partial_{x_3}\partial_{x_3}F & \partial_{x_3}\partial_{x_2}F & \partial_{x_3}\partial_{x_1}F \\ \partial_{x_4}\partial_{x_4}F & \partial_{x_4}\partial_{x_3}F & \partial_{x_4}\partial_{x_2}F & \partial_{x_4}\partial_{x_1}F \end{pmatrix}.$$

It follows from the definition that $\partial_{x_4} C$ is the identity matrix.

Definition 2. If F satisfies $[\partial_{x_j} C, \partial_{x_k} C] = O \ (\forall j, k)$, F is called an algebraic potential of type $D_4(1)$.

Remark 2. The definition of algebraic potential above is not a precise one. Our main interest is to seek functions F satisfying the commutativity of the matrices $\partial_{x_j} C \ (j = 1, 2, 3, 4)$. To formulate precisely, we need to define semisimplicity, irreducibility, etc. of algebraic Frobenius manifolds in its definition.

Theorem 1. *Let z be an algebraic function of (x_1, x_2, x_3) defined by the equation*

$$-\left(p_1{}^2 p_2{}^2 + p_2 q_1{}^2 + q_1 q_2\right) x_1{}^2 + p_2 x_2{}^2 + q_2 x_1 x_2 - x_3 + z_0{}^2 = 0 \tag{24}$$

and let $F = F_{(p_1,p_2,q_1,q_2)}(x_1, x_2, x_3, x_4)$ *be the algebraic function of* (x_1, x_2, x_3, x_4) *defined by*

$$F = 2x_1 x_2{}^2 x_3 \left(2p_1{}^2 p_2{}^2 + 2p_2 q_1{}^2 + q_1 q_2\right)$$

$$- 2x_1{}^2 x_2 x_3 \left(2p_1{}^2 p_2{}^2 q_1 - p_1{}^2 p_2 q_2 + 2p_2 q_1{}^3 + q_1{}^2 q_2\right)$$

$$- \frac{1}{12} x_1 x_2{}^4 \left(12p_1{}^2 p_2{}^3 + 12p_2{}^2 q_1{}^2 - 4p_2 q_1 q_2 - 5q_2{}^2\right)$$

$$- \frac{1}{3} x_1{}^2 x_2{}^3 \left(4p_1{}^2 p_2{}^3 q_1 + 8p_1{}^2 p_2{}^2 q_2 + 4p_2{}^2 q_1{}^3 + 12p_2 q_1{}^2 q_2 + 5q_1 q_2{}^2\right)$$

$$- \frac{1}{2} x_1 x_3{}^2 \left(p_1{}^2 p_2 + q_1{}^2\right)$$

$$+ \frac{2}{3} x_1{}^3 x_3 \left(2p_1{}^4 p_2{}^3 + 4p_1{}^2 p_2{}^2 q_1{}^2 + p_1{}^2 p_2 q_1 q_2 + p_1{}^2 q_2{}^2 + 2p_2 q_1{}^4 + q_1{}^3 q_2\right)$$

$$+ \frac{1}{6} x_1{}^3 x_2{}^2 \left(28p_1{}^4 p_2{}^4 + 56p_1{}^2 p_2{}^3 q_1{}^2 + 44p_1{}^2 p_2{}^2 q_1 q_2 - p_1{}^2 p_2 q_2{}^2 + 28p_2{}^2 q_1{}^4\right.$$

$$\left. + 44p_2 q_1{}^3 q_2 + 15q_1{}^2 q_2{}^2\right)$$

$$- \frac{1}{3} x_1{}^4 x_2 \left(12p_1{}^4 p_2{}^4 q_1 - 8p_1{}^4 p_2{}^3 q_2 + 24p_1{}^2 p_2{}^3 q_1{}^3 + 8p_1{}^2 p_2{}^2 q_1{}^2 q_2\right.$$

$$\left. - 7p_1{}^2 p_2 q_1 q_2{}^2 - p_1{}^2 q_2{}^3 + 12p_2{}^2 q_1{}^5 + 16p_2 q_1{}^4 q_2 + 5q_1{}^3 q_2{}^2\right)$$

$$- \frac{1}{60} x_1{}^5 \left(76p_1{}^6 p_2{}^5 + 84p_1{}^4 p_2{}^4 q_1{}^2 + 204p_1{}^4 p_2{}^3 q_1 q_2 + 11p_1{}^4 p_2{}^2 q_2{}^2\right.$$

$$- 60p_1{}^2 p_2{}^3 q_1{}^4 + 120p_1{}^2 p_2{}^2 q_1{}^3 q_2$$

$$\left. + 130p_1{}^2 p_2 q_1{}^2 q_2{}^2 + 20p_1{}^2 q_1 q_2{}^3 - 68p_2{}^2 q_1{}^6 - 84p_2 q_1{}^5 q_2 - 25q_1{}^4 q_2{}^2\right)$$

$$+ \frac{8p_1}{15} z_0{}^5 + \frac{4}{15} p_2 x_2{}^5 (2p_2 q_1 + q_2) - \frac{2}{3} x_2{}^3 x_3 (2p_2 q_1 + q_2) + q_1 x_2 x_3{}^2$$

$$+ \frac{x_1 x_4{}^2}{2} + x_2 x_3 x_4,$$

then F *is an algebraic potential of type* $D_4(1)$. *Here,* (p_1, p_2, q_1, q_2) *is a 4-tuple of constants.*

The theorem follows by direct computation.

We let $R_F (= R_{F_{(p_1,p_2,q_1,q_2)}}) = -(p_1{}^2 p_2{}^2 + p_2 q_1{}^2 + q_1 q_2)x_1{}^2 + p_2 x_2{}^2 + q_2 x_1 x_2 - x_3 + z_0{}^2$. It is better to treat the pair $\mathbf{F}_{(p_1,p_2,q_1,q_2)} = (F, R_R)$ instead of the polynomial F since the relation $R_F = 0$ plays a basic role in defining the algebraic potential F. In the following argument, we consider a group action on the 4-tuple (p_1, p_2, q_1, q_2).

For this reason, we let $\pi(\mathbf{F}_{(p_1,p_2,q_1,q_2)}) = (p_1,p_2,q_1,q_2)$ and identify $\mathbf{C}^4$ with the totality of 4-tuples (p_1,p_2,q_1,q_2). Moreover, we put $\mathcal{F} = \{\mathbf{F}_{(p_1,p_2,q_1,q_2)} \mid (p_1,p_2,q_1,q_2) \in \mathbf{C}^4\}$.

We consider linear transformations of the coordinate system (x_1,x_2,x_3,x_4) to another system (y_1,y_2,y_3,y_4) of the form

$$x_1 = a_1 y_1 + a_2 y_2, \quad x_2 = a_3 y_1 + a_4 y_2,$$

$$x_3 = b_1 y_3 + b_2 y_4, \quad x_4 = b_3 y_3 + b_4 y_4 \tag{25}$$

$(a_i, b_i$ are constants). By this transformation, the weights of the variables are left fixed, namely the weights of y_1, y_2, y_3, y_4 are $1/2, 1/2, 1, 1$, respectively. Conversely, any linear transformation between (x_1,x_2,x_3,x_4) and (y_1,y_2,y_3,y_4) leaving the weights fixed is of the form (25). By the substitution of (25) to (24), z is regarded as an algebraic function of y_1, y_2, y_3, y_4. Similarly, F is regarded as a function of y_1, y_2, y_3, y_4 which is written by $\tilde{H} = \tilde{H}(y_1, y_2, y_3, y_4)$ to avoid confusion. Then $C_{\tilde{H}}$ that follows plays the role of C:

$$C_{\tilde{H}} = \begin{pmatrix} \partial_{y_1}\partial_{y_4}\tilde{H} & \partial_{y_1}\partial_{y_3}\tilde{H} & \partial_{y_1}\partial_{y_2}\tilde{H} & \partial_{y_1}\partial_{y_1}\tilde{H} \\ \partial_{y_2}\partial_{y_4}\tilde{H} & \partial_{y_2}\partial_{y_3}\tilde{H} & \partial_{y_2}\partial_{y_2}\tilde{H} & \partial_{y_2}\partial_{y_1}\tilde{H} \\ \partial_{y_3}\partial_{y_4}\tilde{H} & \partial_{y_3}\partial_{y_3}\tilde{H} & \partial_{y_3}\partial_{y_2}\tilde{H} & \partial_{y_3}\partial_{y_1}\tilde{H} \\ \partial_{y_4}\partial_{y_4}\tilde{H} & \partial_{y_4}\partial_{y_3}\tilde{H} & \partial_{y_4}\partial_{y_2}\tilde{H} & \partial_{y_4}\partial_{y_1}\tilde{H} \end{pmatrix}.$$

We first postulate the following condition (d) for the linear transformation (25) so that $\tilde{H}$ satisfy the conditions (a), (b), (c) given in the first part of this paragraph.

Condition (d): $\frac{\partial}{\partial x_4} C_{\tilde{H}}$ is a non-zero scalar matrix.

Then we obtain the following relations among the constants a_i, b_i:

$$b_2 = 0, \quad a_2 = 0, \quad b_3 = -a_3 b_1 / a_1, \quad b_1 = a_1 b_4 / a_4. \tag{26}$$

Letting $z = kw$ for a non-zero constant k, we normalize the relation for w obtained from R_F by the substitution (25) similar to the equation (23). As a result, we find

$$a_1 = a_4 k^2 / b_4 \tag{27}$$

and $R_F|_{x_1=a_1 y_1+a_2 y_2,\, x_2=a_3 y_1+a_4 y_2,\, x_3=b_1 y_3+b_2 y_4,\, x_4=b_3 y_3+b_4 y_4}$ coincides with the polynomial of the form

$$\tilde{R} = w^2 - y_3 + d_1' y_1^2 + d_2' y_1 x_2 + d_3' y_2^2$$

$(d_1', d_2', d_3'$ are constants) up to a non-zero constant factor. As a consequence, we obtain the algebraic function $H(= \tilde{H}/(a_4 b_4 k^2))$ and the polynomial $R_H = \tilde{R}$. It is straightforward to show that H satisfies the conditions (a), (b), (c).

coef. of x_2^2 in R_F	p_2	coef. of y_2^2 in R_H	$\dfrac{a_4^2 p_2}{k^2}$
coef. of $x_1 x_2$ in R_F	$\frac{1}{2} q_2$	coef. of $y_1 y_2$ in R_H	$\dfrac{a_3 a_4 p_2}{k^2} + \dfrac{a_4^2 q_2}{2 b_4}$
coef. of z^5 in F	$\frac{8}{15} p_1$	coef. of w^5 in H	$\dfrac{8 k^3 p_1}{15 a_4 b_4}$
coef. of $x_2 x_3^2$ in F	q_1	coef. of $y_2 y_3^2$ in H	$-\dfrac{a_3}{a_4} + \dfrac{k^2 q_1}{b_4}$

For a moment, let $(\tilde{F}, \tilde{R})$ be the pair by substituting (p_1, p_2, q_1, q_2) with $\left(\frac{k^3 p_1}{a_4 b_4}, \frac{a_4^2 p_2}{k^2}, -\frac{a_3}{a_4} + \frac{k^2 q_1}{b_4}, \frac{2 a_3 a_4 p_2}{k^2} + \frac{a_4^2 q_2}{b_4} \right)$ in $(F,\ R_F)$. Then it can be shown that $(\tilde{F},\ \tilde{R})$ coincides with $(H,\ R_H)$ by replacing $(x_1, x_2, x_3, x_4, z_0)$ with $(y_1, y_2, y_3, y_4, w_0)$.

We have thus obtained the following:

Proposition 1. *The action*

$$(p_1, p_2, q_1, q_2) \rightarrow \left(\frac{k^3 p_1}{a_4 b_4}, \frac{a_4^2 p_2}{k^2}, -\frac{a_3}{a_4} + \frac{k^2 q_1}{b_4}, \frac{2 a_3 a_4 p_2}{k^2} + \frac{a_4^2 q_2}{b_4} \right)$$

is an automorphism of the parameter space $\mathbf{C}^4$ *and induces an automorphism of* $\mathcal{F}$.

Let $G = \{(a_3, a_4, b_4, k) \in \mathbf{C}^4 |\ a_4 b_4 k \neq 0\}$ and define

$$(a_3, a_4, b_4, k)(a_3', a_4', b_4', k') = \left(\frac{a_4'(a_3 a_4' k'^2 + a_3' b_4')}{b_4'},\ a_4 a_4',\ b_4 b_4',\ kk' \right).$$

Then G is a group with this product. This product matches with the action of G on $\mathcal{F}$ and that on $\pi(\mathcal{F}) = \mathbf{C}^4$.

We now recall the definition of the matrix C. We define

$$T = \left(\frac{1}{2}(x_1 \partial_{x_1} + x_2 \partial_{x_2}) + x_3 \partial_{x_3} + x_4 \partial_{x_4} \right) C.$$

It can be shown that if $p_1 = 0$, then T is not semisimple. We focus our attention on such an algebraic potential that the corresponding matrix T is semisimple.

Proposition 2. (i) *Let* $\mathcal{A} = \{(p_1, p_2, q_1, q_2) \in \mathbf{C}^4 |\ p_1 \neq 0\}$ *be the subspace of the parameter space of* $\mathcal{F}$. *Then by the action of* G, $\mathcal{A}$ *is decomposed into the following orbits.*

(O.1)$_r$ $G \cdot (1, 1, 0, r)$, *where* $r \in \mathbf{C}$.
(O.2) $G \cdot (1, 0, 0, 1)$.
(O.3) $G \cdot (1, 0, 0, 0)$.
(ii) *If* $G \cdot (1, 1, 0, r) = G \cdot (1, 1, 0, r')$, *then* $r = \pm r'$.
(iii) *In the cases* (O.1)$_r$ *and* (O.2), T *is semisimple. On the other hand, in the case* (O.3), T *is not semisimple.*

Proof. Take $P = (p_1, p_2, q_1, q_2) \in \mathcal{A}$. Then $p_1 \neq 0$. Taking $g = (1, p_1, 1, 1) \in G$, we find that $g \cdot P = (1, p_1^2 p_2, (p_1 q_1 - 1)/p_1, p_1(2p_2 + p_1 q_2))$. This shows that P is transformed to $P' = (1, p_2', q_1', q_2') \in \mathcal{A}$.

We first treat the case $p_2' \neq 0$. Take $g' = (1, 1/\sqrt{p_2'}, \sqrt{p_2'}, 1) \in G$. Then $g' \cdot P' = (1, 1, (q_1' - p_2'^{3/2})/p_2', (2p_2'^2 + q_2')/p_2'^{3/2})$. We write $g' \cdot P' = P'' = (1, 1, q_1'', q_2'')$ for simplicity. Take $g'' = (q_1'', 1, 1, 1)$. Then $g'' \cdot P'' = (1, 1, 0, 2q_1'' + q_2'')$. As a consequence, P is transformed to $Q = (1, 1, 0, r)$ for some $r \in \mathbf{C}$. If $G \cdot (1, 1, 0, r) = G \cdot (1, 1, 0, r')$, then it is easy to show that $r = \pm r$.

We next treat the case $p_2' = 0$. Take $h = (q_1', 1, 1, 1)$. Then $h \cdot P' = P_4 = (1, 0, 0, 2q_1' + q_2')$. (Note that $p_2' = 0$.) We treat the action of G on $(1, 0, 0, r)$ $(r \in \mathbf{C})$. Since $(0, k^3, 1, k) \cdot (1, 0, 0, r) = (1, 0, 0, k^6 r)$, $(1, 0, 0, r)$ is transformed to $(1, 0, 0, 1)$ in case $r \neq 0$, whereas is done to $(1, 0, 0, 0)$.

It is straightforward to show that if (p_1, p_2, q_1, q_2) coincides with either $(1, 1, 0, r)$ or $(1, 0, 0, 1)$, T is semisimple, whereas if $(p_1, p_2, q_1, q_2) = (1, 0, 0, 0)$, then T is not semisimple. $\square$

We now identify the algebraic potentials constructed by O. Pavlyk and Y. Dinar with algebraic potentials in $\mathcal{F}$.

We first treat the algebraic potential obtained by Pavlyk [11]. It reads

$$F_{\mathrm{PVLK}} = n_2 n_3 n_4 + \frac{1}{2} n_1 n_4^2 + \frac{z^5}{2^5 \cdot 3^4 \cdot 5} + \frac{n_1 n_3^2}{6} - \frac{n_3 n_1^3}{108}$$

$$+ \frac{n_1 n_2^2 n_3}{12} + \frac{19 n_1^5}{2^8 \cdot 3^4 \cdot 5} + \frac{7 n_1^3 n_2^2}{2^7 \cdot 3^3} + \frac{n_1 n_2^4}{2^8 \cdot 3}, \qquad (28)$$

where n_1, n_2, n_3, n_4 are variables of weights $1/2, 1/2, 1, 1$, respectively, and z is an algebraic function of the variables n_1, n_2, n_3 defined by the equation

$$z^2 - (n_1^2 + 3n_2^2 + 48n_3) = 0. \qquad (29)$$

By change of variables

$$n_1 = x_1, \quad n_2 = \frac{x_2}{3^{1/2}}, \quad n_3 = \frac{x_3}{48}, \quad n_4 = \frac{x_4}{48 \cdot 3^{1/2}}, \quad z = z_0,$$

we find that F_{PVLK} turns out to be $F_{(1,-1,0,0)}$ and the equation for z does to

$$-x_1^2 - x_2^2 - x_3 + z^2 = 0.$$

We next treat the algebraic potential constructed by Dinar [2]. It reads

$$\begin{aligned}
F_{\mathrm{DNR}} = {} & \frac{1}{2880}(35s_1{}^5 - 48\sqrt{3}s_1{}^3 s_2 + 510s_1{}^3 s_4{}^2 + 1128s_1 s_2{}^2 \\
& - 1840\sqrt{3}s_1 s_2 s_4{}^2 + 360s_1 s_3{}^2 + 775s_1 s_4{}^4 - 720\sqrt{5}s_2 s_3 s_4) \\
& + \frac{1}{180}z(-\sqrt{5}s_1{}^4 + 8\sqrt{15}s_1{}^2 s_2 - 10\sqrt{5}s_1{}^2 s_4{}^2 - 48\sqrt{5}s_2{}^2 \\
& + 40\sqrt{15}s_2 s_4{}^2 - 25\sqrt{5}s_4{}^4),
\end{aligned} \tag{30}$$

where z is an algebraic function of $s_1 s_2, s_4$ defined by the equation

$$\frac{3s_1{}^2}{20} - \frac{2s_1 z}{\sqrt{5}} + \frac{\sqrt{3}s_2}{5} - \frac{s_4{}^2}{4} + z^2 = 0.$$

By the coordinate change

$$s_1 = -2\sqrt{5}x_1, \ s_4 = -2x_2, \ s_2 = -\frac{5}{\sqrt{3}}x_3, \ s_3 = \frac{5}{\sqrt{3}}x_4, \ z = z_0 - 2x_1,$$

we find that F_{DNR} turns out to be $F_{(1,-1,0,0)}$.

Since it is easy to see that $(1, -1, 0, 0) \in G \cdot (1, 1, 0, 0)$, both of the algebraic potentials constructed by Pavlyk and Dinar are contained in $G \cdot (1, 1, 0, 0)$.

Remark 3. We define

$$I(p_1, p_2, q_1, q_2) = \frac{2p_2 q_1 + q_2}{p_1 p_2^{3/2}}.$$

Then it is easy to show that $I(p_1, p_2, q_1, q_2)^2$ is G-invariant and that if $(p_1, p_2, q_1, q_2) \in G(1, 1, 0, r)$, then $r = \pm I(p_1, p_2, q_1, q_2)$.

2.3. $D_4(1)$ *case*

Let x_1, x_2, x_3, x_4 be the variables with weights $1/2,1/2,1,1$, respectively, and let z_0 be an algebraic function of x_1, x_2, x_3 defined by the equation

$$-x_1{}^2 + x_2{}^2 + x_3 + z_0{}^2 = 0. \tag{31}$$

We introduce the algebraic function F by

$$F = \frac{8z_0{}^5}{15} - \frac{19x_1{}^5}{15} + \frac{14x_1{}^3 x_2{}^2}{3} - \frac{4x_1{}^3 x_3}{3} - x_1 x_2{}^4$$
$$- 4x_1 x_2{}^2 x_3 - \frac{x_1 x_3{}^2}{2} + \frac{x_1 x_4{}^2}{2} + x_2 x_3 x_4.$$

It is easy to see that F is weighted homogeneous in the sense that if

$$E = \frac{1}{2}(x_2 \partial_{x_1} + x_2 \partial_{x_2}) + x_3 \partial_{x_3} + x_4 \partial_{x_4},$$

then $EF = \frac{5}{2}F$. It can be shown that F is a solution to the WDVV equation, or equivalently, F is an algebraic potential. Moreover, it is easy to show that F is transformed to the potential $F_{(1,-1,0,0)}$ by a linear transformation of coordinates.

As is explained in the Introduction, it is possible to construct the 4×4 matrices C and T. Using the matrix T, we define $f_0 = \det(T)$. Then f_0 is regarded as a polynomial of x_1, x_2, x_4, z_0 by eliminating x_3 by the relation (31) and its concrete form is given by

$$f_0 = 3600x_1{}^6 x_2{}^2 - 2880x_1{}^5 x_2{}^2 z_0 - 768x_1{}^5 z_0{}^3 - 1360x_1{}^4 x_2{}^4 - 912x_1{}^4 x_2{}^2 z_0{}^2$$
$$+ 192x_1{}^4 z_0{}^4 + 1088x_1{}^3 x_2{}^4 z_0 + 960x_1{}^3 x_2{}^2 z_0{}^3 + 240x_1{}^3 z_0{}^5 + 720x_1{}^2 x_2{}^6$$
$$+ 624x_1{}^2 x_2{}^4 z_0{}^2 + 72x_1{}^2 x_2{}^2 z_0{}^4 - 12x_1{}^2 z_0{}^6 - 576x_1 x_2{}^6 z_0 - 544x_1 x_2{}^4 z_0{}^3$$
$$- 192x_1 x_2{}^2 z_0{}^5 - 24x_1 z_0{}^7 - 16x_2{}^4 z_0{}^4 - 12x_2{}^2 z_0{}^6 - 3z_0{}^8$$
$$- 8x_2 x_4 \left(90x_1{}^5 - 36x_1{}^4 z_0 - 92x_1{}^3 x_2{}^2 - 24x_1{}^3 z_0{}^2 + 60x_1{}^2 x_2{}^2 z_0 + 18x_1{}^2 z_0{}^3 \right.$$
$$\left. - 6x_1 x_2{}^4 + 12x_2{}^4 z_0 + 10x_2{}^2 z_0{}^3 + 3z_0{}^5\right)$$
$$+ 6x_4{}^2 \left(6x_1{}^4 - 2x_1{}^2 z_0{}^2 - 8x_1 x_2{}^2 z_0 - 4x_1 z_0{}^3 - 2x_2{}^4 - 2x_2{}^2 z_0{}^2 + z_0{}^4\right)$$
$$- 8x_2 x_4{}^3 (2x_1 - z_0) + x_4{}^4.$$

In the sequel, we regard f_0 as a polynomial of x_4 and write $f_0(x_4)$ when we stress the variable x_4. Let $\mathcal{Q} = \mathbf{Q}(x_1, x_2, z_0)$ be the rational function field over $\mathbf{Q}$. Then f_0 is an element of $\mathcal{Q}$. Let $\mathcal{K}$ be the field extension of $\mathcal{Q}$ attached by all the roots u_1, u_2, u_3, u_4 of the algebraic equation $f_0(x_4) = 0$.

Introducing the matrix

$$
B_\infty^{(4)} = \begin{pmatrix} r + \frac{1}{2} & 0 & 0 & 0 \\ 0 & r + \frac{1}{2} & 0 & 0 \\ 0 & 0 & r + 1 & 0 \\ 0 & 0 & 0 & r + 1 \end{pmatrix},
$$

we consider the ordinary differential equation

$$
\frac{d}{dx_4} \tilde{Z} = -T^{-1} B_\infty^{(4)} \tilde{Z}, \tag{32}
$$

where $\tilde{Z} = {}^t(\tilde{\mathbf{z}}_1, \tilde{\mathbf{z}}_2, \tilde{\mathbf{z}}_3, \tilde{\mathbf{z}}_4)$. We focus our attention on the case $r = -1$ of the equation (32). We let $\tilde{R} = (-T^{-1} B_\infty^{(4)})_{r=-1}$. Let $\tilde{R}(i,j)$ be the (i,j)-entry of $\tilde{R}$ and define the 2×2 matrix R by $R = \begin{pmatrix} \tilde{R}(1,1) & \tilde{R}(1,2) \\ \tilde{R}(2,1) & \tilde{R}(2,2) \end{pmatrix}$. Since it follows from the definition of $B_\infty^{(4)}$ that $\tilde{R}(i,j) = 0$ for $i = 1, 2, 3, 4$, $j = 3, 4$, we conclude that

$$
\frac{d}{dx_4} Z = RZ \tag{33}
$$

is well-defined, where $Z = {}^t(\tilde{\mathbf{z}}_1, \tilde{\mathbf{z}}_2)$. Letting

$$
R_{ja} = \lim_{x_4 \to u_j} (x_4 - u_j) R \quad (j = 1, 2, 3, 4),
$$

we obtain the differential equation

$$
\frac{d}{dx_4} Z = \left(\frac{R_{1a}}{x_4 - u_1} + \frac{R_{2a}}{x_4 - u_2} + \frac{R_{3a}}{x_4 - u_3} + \frac{R_{4a}}{x_4 - u_4} \right) Z, \tag{34}
$$

which is the same as (33). It is possible to show the following properties of $R_{1a}, R_{2a}, R_{3a}, R_{4a}$:

(P0) $R = \sum_{j=1}^4 \frac{R_{ja}}{x_4 - u_j}$.

(P1) $0, \frac{1}{4}$ are the eigenvalues of each of $R_{1a}, R_{2a}, R_{3a}, R_{4a}$.

(P2) $R_{1a} + R_{2a} + R_{3a} + R_{4a} = \frac{1}{2} I_2$.

We are going to write down the concrete forms of $R_{1a}, R_{2a}, R_{3a}, R_{4a}$. For this purpose, we start with computing the explicit form of the matrix R. It reads

$$R = \frac{1}{f_0} \begin{pmatrix} a_{11} & a_{12} \\ a_{21} & a_{22} \end{pmatrix},$$

where

$$\begin{aligned}
a_{11} = \frac{1}{2}(& -60x_2x_1{}^5 + 6x_4x_1{}^4 + 20x_2z_0x_1{}^4 + 160x_2{}^3x_1{}^3 + 32x_2z_0{}^2x_1{}^3 + 2x_4z_0x_1{}^3 \\
& -30x_2z_0{}^3x_1{}^2 + 44x_2{}^2x_4x_1{}^2 - 132x_2{}^3z_0x_1{}^2 + 60x_2{}^5x_1 + 8x_2z_0{}^4x_1 \\
& -18x_4z_0{}^3x_1 - 16x_2x_4{}^2x_1 + 48x_2{}^3z_0{}^2x_1 - 50x_2{}^2x_4z_0x_1 - 10x_2z_0{}^5 \\
& +5x_4z_0{}^4 + x_4{}^3 - 38x_2{}^3z_0{}^3 - 4x_2{}^2x_4z_0{}^2 - 6x_2{}^4x_4 - 48x_2{}^5z_0 + 8x_2x_4{}^2z_0),
\end{aligned}$$

$$\begin{aligned}
a_{12} = \frac{1}{2}(& 3z_0{}^6 + 18x_1z_0{}^5 - 3x_1{}^2z_0{}^4 + 11x_2{}^2z_0{}^4 - 114x_1{}^3z_0{}^3 + 86x_1x_2{}^2z_0{}^3 \\
& +10x_2x_4z_0{}^3 + 12x_2{}^4z_0{}^2 - 96x_1{}^2x_2{}^2z_0{}^2 - x_4{}^2z_0{}^2 + 12x_1x_2x_4z_0{}^2 + 96x_1{}^5z_0 \\
& +132x_1x_2{}^4z_0 - 244x_1{}^3x_2{}^2z_0 + 4x_1x_4{}^2z_0 + 18x_2{}^3x_4z_0 + 30x_1{}^2x_2x_4z_0 \\
& -204x_1{}^2x_2{}^4 + 332x_1{}^4x_2{}^2 + 5x_1{}^2x_4{}^2 + 3x_2{}^2x_4{}^2 - 80x_1{}^3x_2x_4),
\end{aligned}$$

$$\begin{aligned}
a_{21} = \frac{1}{2}(& z_0{}^6 - 2x_1z_0{}^5 - 21x_1{}^2z_0{}^4 + 5x_2{}^2z_0{}^4 + 22x_1{}^3z_0{}^3 - 2x_1x_2{}^2z_0{}^3 + 2x_2x_4z_0{}^3 \\
& +40x_1{}^4z_0{}^2 + 4x_2{}^4z_0{}^2 - 24x_1{}^2x_2{}^2z_0{}^2 + x_4{}^2z_0{}^2 + 4x_1x_2x_4z_0{}^2 + 20x_1x_2{}^4z_0 \\
& +140x_1{}^3x_2{}^2z_0 + 10x_2{}^3x_4z_0 - 10x_1{}^2x_2x_4z_0 - 20x_1{}^2x_2{}^4 - 140x_1{}^4x_2{}^2 \\
& -x_1{}^2x_4{}^2 + x_2{}^2x_4{}^2 - 8x_1x_2{}^3x_4 + 24x_1{}^3x_2x_4),
\end{aligned}$$

$$\begin{aligned}
a_{22} = \frac{1}{2}(& -300x_2x_1{}^5 + 30x_4x_1{}^4 + 124x_2z_0x_1{}^4 + 208x_2{}^3x_1{}^3 + 64x_2z_0{}^2x_1{}^3 \\
& -2x_4z_0x_1{}^3 - 42x_2z_0{}^3x_1{}^2 - 12x_4z_0{}^2x_1{}^2 - 44x_2{}^2x_4x_1{}^2 - 108x_2{}^3z_0x_1{}^2 \\
& -36x_2{}^5x_1 - 8x_2z_0{}^4x_1 - 6x_4z_0{}^3x_1 - 8x_2x_4{}^2x_1 - 48x_2{}^3z_0{}^2x_1 + 2x_2{}^2x_4z_0x_1 \\
& -2x_2z_0{}^5 + x_4z_0{}^4 + x_4{}^3 - 2x_2{}^3z_0{}^3 - 8x_2{}^2x_4z_0{}^2 - 6x_2{}^4x_4 + 4x_2x_4{}^2z_0).
\end{aligned}$$

By direct computation, we find that $\det \begin{pmatrix} a_{11} & a_{12} \\ a_{21} & a_{22} \end{pmatrix}$ is divisible by f_0.

We let $a_{ij}^{(k)} = a_{ij}|_{x_4=u_k}$. Then

$$R_{ka} = \lim_{x_4 \to u_k} ((x_4 - u_k)R) = \frac{1}{f_0'(u_k)} \begin{pmatrix} a_{11}^{(k)} & a_{12}^{(k)} \\ a_{21}^{(k)} & a_{22}^{(k)} \end{pmatrix}.$$

It is easy to show that

$$f_0'(u_1) = (u_1 - u_2)(u_1 - u_3)(u_1 - u_4).$$

We now introduce

$$\delta_k = f_0'(u_k), \quad \varphi_k = 2a_{11}^{(k)}, \quad \psi_k = 2a_{21}^{(k)}.$$

Then it follows from direct computation that

$$a_{12}^{(k)} = \varphi_k \left(-\varphi_k + \frac{1}{2}\delta_k\right) \Big/ (2\psi_k), \quad a_{22}^{(k)} = \frac{1}{2}\left(-\varphi_k + \frac{1}{2}\delta_k\right)$$

and that

$$R_{ka} = M_k^{-1} \begin{pmatrix} 0 & 0 \\ 0 & \frac{1}{4} \end{pmatrix} M_k = \frac{1}{2\delta_k} \begin{pmatrix} \varphi_k & \varphi_k(\frac{1}{2}\delta_k - \varphi_k)/\psi_k \\ \psi_k & \frac{1}{2}\delta_k - \varphi_k \end{pmatrix},$$

where

$$M_k = \begin{pmatrix} -\psi_k & \varphi_k \\ \psi_k & -\varphi_k + \frac{1}{2}\delta_k \end{pmatrix}.$$

Letting $R_{kb} = M_4 R_{ka} M_4^{-1}$ $(k = 1, 2, 3, 4)$, and $\hat{Z} = M_4 Z$, we find that the equation (34) for Z turns out to be the equation for $\hat{Z}$ given by

$$\frac{d}{dx_4}\hat{Z} = \left(\frac{R_{1b}}{x_4 - u_1} + \frac{R_{2b}}{x_4 - u_2} + \frac{R_{3b}}{x_4 - u_3} + \frac{R_{4b}}{x_4 - u_4}\right)\hat{Z}. \qquad (35)$$

It is underlined here that $R_{4b} = \begin{pmatrix} 0 & 0 \\ 0 & \frac{1}{4} \end{pmatrix}$. Writing

$$R_{kb} = \frac{1}{\delta_k \delta_4 \psi_k \psi_4} \begin{pmatrix} r_{11}^{(k)} & r_{12}^{(k)} \\ r_{21}^{(k)} & r_{22}^{(k)} \end{pmatrix},$$

we have

$$r_{11}^{(k)} = (-\varphi_4\psi_k + \varphi_k\psi_4)\left(-\varphi_4\psi_k + \varphi_k\psi_4 + \frac{1}{2}(\delta_4\psi_k - \delta_k\psi_4)\right),$$

$$r_{12}^{(k)} = (-\varphi_4\psi_k + \varphi_k\psi_4)\left(-\varphi_4\psi_k + \varphi_k\psi_4 - \frac{1}{2}\delta_k\psi_4\right),$$

$$r_{21}^{(k)} = -\left(-\varphi_4\psi_k + \varphi_k\psi_4 + \frac{1}{2}\delta_4\psi_k\right)\left(-\varphi_4\psi_k + \varphi_k\psi_4 + \frac{1}{2}(\delta_4\psi_k - \delta_k\psi_4)\right),$$

$$r_{22}^{(k)} = -\left(-\varphi_4\psi_k + \varphi_k\psi_4 + \frac{1}{2}\delta_4\psi_k\right)\left(-\varphi_4\psi_k + \varphi_k\psi_4 - \frac{1}{2}\delta_k\psi_4\right).$$

We will now find out the condition satisfied by the two 4-tuples of matrices $(R_{1b}, R_{2b}, R_{3b}, R_{4b})$ and (N_1, N_2, N_3, N_4) so that (35) is identified with (21). We recall the definition of matrices N_1, N_2, N_3, N_4 of

                                J. Sekiguchi

the case of the reflection group of type A_3. We write them down by using t_1 and $\nu = \frac{t_2}{t_1}$. Denoting $P = \begin{pmatrix} 1 & 0 \\ 0 & h_0 \end{pmatrix}$, where $h_0 = \frac{1+2\nu}{12\nu(1+\nu)t_1}$ and $N_j{}' = P^{-1}N_j P$ $(j = 1, 2, 3, 4)$, we find that

$$
\begin{cases}
N_1{}' = \begin{pmatrix} \frac{1+7\nu+\nu^2}{36\nu} & \frac{(1-\nu)(1+2\nu)\left(1+7\nu+\nu^2\right)}{1296\nu^2(1+\nu)} \\ -\frac{(1+\nu)(1-\nu)}{(1+2\nu)} & -\frac{(1-\nu)^2}{36\nu} \end{pmatrix}, \\[2ex]
N_2{}' = \begin{pmatrix} \frac{5+5\nu-\nu^2}{36(1+\nu)} & -\frac{(2+\nu)(1+2\nu)\left(5+5\nu-\nu^2\right)}{1296\nu(1+\nu)^2} \\ -\frac{\nu(2+\nu)}{(1+2\nu)} & \frac{(2+\nu)^2}{36(1+\nu)} \end{pmatrix}, \\[2ex]
N_3{}' = \begin{pmatrix} \frac{-1+5\nu+5\nu^2}{36\nu(1+\nu)} & \frac{(1+2\nu)^2\left(-1+5\nu+5\nu^2\right)}{1296\nu^2(1+\nu)^2} \\ 1 & \frac{(1+2\nu)^2}{36\nu(1+\nu)} \end{pmatrix}, \\[2ex]
N_4{}' = \begin{pmatrix} 0 & 0 \\ 0 & \frac{1}{4} \end{pmatrix}.
\end{cases}
\tag{36}
$$

To continue the argument, we need some preparation. Let

$$
L_j = \begin{pmatrix} c_j & c_j\left(\frac{1}{4}-c_j\right)/d_j \\ d_j & \frac{1}{4}-c_j \end{pmatrix} \quad (j = 1, 2),
$$

$$
L_3 = \begin{pmatrix} c_3 & c_3\left(\frac{1}{4}-c_3\right) \\ 1 & \frac{1}{4}-c_3 \end{pmatrix}, \quad L_4 = \begin{pmatrix} 0 & 0 \\ 0 & \frac{1}{4} \end{pmatrix}
$$

be 2×2 matrices satisfying

$$
L_1 + L_2 + L_3 + L_4 = \frac{1}{2}I_2,
\tag{37}
$$

which is equivalent to

$$
\begin{cases}
c_1 + c_2 + c_3 = \frac{1}{2}, \\
d_1 + d_2 + 1 = 0, \\
\frac{c_1}{d_1}\left(\frac{1}{4}-c_1\right) + \frac{c_2}{d_2}\left(\frac{1}{4}-c_2\right) + c_3\left(\frac{1}{4}-c_3\right) = 0.
\end{cases}
\tag{38}
$$

Lemma 1. *We assume*

$$
d_j{}^2(1-8c_j) + (1+d_j)(1-4c_j)(5-36c_j) = 0 \quad (j = 1, 2)
\tag{39}
$$

$$
d_1 d_2\left(-72c_1 d_1 - 72c_1 + 2d_1{}^2 + 17d_1 + 20\right) \neq 0
\tag{40}
$$

in addition to (38). *Then there is a constant ν such that $L_j = N_j{}'$ $(j = 1, 2, 3, 4)$.*

Proof. We first eliminate c_3 and d_2 by the first and the second equation of (38). Then c_2 is written as a rational function of c_1, d_1. It reads

$$c_2 = \frac{\left(\begin{array}{c} 144c_1{}^2d_1{}^2 + 288c_1{}^2d_1 + 144c_1{}^2 \\ -108c_1d_1{}^2 - 144c_1d_1 - 36c_1 + d_1{}^3 + 15d_1{}^2 + 19d_1 \end{array}\right)}{4d_1\left(-72c_1d_1 - 72c_1 + 2d_1{}^2 + 17d_1 + 20\right)}. \quad (41)$$

We note that the denominator of the right-hand side of this equation does not vanish under the condition (40). Substituting (41) for c_2 in the equation (38), we obtain

$$144c_1{}^2d_1 + 144c_1{}^2 - 8c_1d_1{}^2 - 56c_1d_1 - 56c_1 + d_1{}^2 + 5d_1 + 5 = 0. \quad (42)$$

We now put $c_1 = \frac{1+7\nu+\nu^2}{36\nu}$ for some constant ν. Then (42) turns out to be

$$\left(2d_1\nu + d_1 - \nu^2 + 1\right)\left(d_1\nu^2 + 2d_1\nu + \nu^2 - 1\right) = 0. \quad (43)$$

Since both the definition of c_1 and (43) are invariant by the change $\nu \to 1/\nu$, we may choose ν so that $2d_1\nu + d_1 - \nu^2 + 1 = 0$, or equivalently, $d_1 = \frac{\nu^2-1}{2\nu+1}$. Then c_2, c_3, d_2 are determined uniquely and the lemma follows. $\quad\square$

Theorem 2. *There is $g_0 \in \mathcal{K}$ such that*

$$\begin{pmatrix} 1 & 0 \\ 0 & g_0 \end{pmatrix} R_{kb} \begin{pmatrix} 1 & 0 \\ 0 & g_0 \end{pmatrix}^{-1} = N_k \quad (k = 1, 2, 3, 4). \quad (44)$$

Proof. It is sufficient to show the existence of $g_0' \in \mathcal{K}$ such that

$$\begin{pmatrix} 1 & 0 \\ 0 & g_0' \end{pmatrix} R_{kb} \begin{pmatrix} 1 & 0 \\ 0 & g_0' \end{pmatrix}^{-1} = N_k' \quad (k = 1, 2, 3, 4), \quad (45)$$

which we will now prove.

We let $R_{jc} = Q^{-1} R_{jb} Q$, where $Q = \begin{pmatrix} 1 & 0 \\ 0 & (R_{3b})_{21} \end{pmatrix}$. For simplicity, we denote

$$R_{jc} = \begin{pmatrix} c_j' & c_j'(\frac{1}{4} - c_j')/d_j' \\ d_j' & \frac{1}{4} - c_j' \end{pmatrix} \quad (j = 1, 2),$$

$$R_{3c} = \begin{pmatrix} c_3' & c_3'(\frac{1}{4} - c_3') \\ 1 & \frac{1}{4} - c_3' \end{pmatrix}, \quad R_{4c} = \begin{pmatrix} 0 & 0 \\ 0 & \frac{1}{4} \end{pmatrix}$$

similarly to the definition of L_1, L_2, L_3, L_4. Since the 4-tuples $(R_{1a}, R_{2a}, R_{3a}, R_{4a})$ and $(R_{1c}, R_{2c}, R_{3c}, R_{4c})$ are conjugate, it follows from the property (P2) for $(R_{1a}, R_{2a}, R_{3a}, R_{4a})$ that

$$R_{1c} + R_{2c} + R_{3c} + R_{4c} = \frac{1}{2} I_2. \tag{46}$$

On the other hand, by direct computation we find that

$$d_j'^2(1 - 8c_j') + (1 + d_j')(1 - 4c_j')(5 - 36c_j') = 0 \quad (j = 1, 2) \tag{47}$$

and

$$d_1' d_2' \left(-72 c_1' d_1' - 72 c_1' + 2 d_1'^2 + 17 d_1' + 20 \right) \neq 0. \tag{48}$$

It is underlined here that to prove (47), (48), we are deeply indebted to the software Mathematica. The proof of (48) is simpler than that of (47), since it is enough to check (48) for special values of x_1, x_2, z_0. As a consequence, we have shown the assumptions of Lemma 1 for R_{jc} $(j = 1, 2, 3, 4)$. Then there exists ν such that

$$R_{jc} = N_j' \quad (j = 1, 2, 3, 4)$$

and the theorem follows. $\qquad\qquad\square$

Concerning the dependence of ν on x_1, x_2, z_0, we have the following:

Theorem 3. *If* (44) *holds, then*

$$\nu = \frac{\delta_3 \psi_3 (6 r_{11}^{(2)} - \delta_2 \delta_4 \psi_2 \psi_4)}{\delta_2 \psi_2 (6 r_{11}^{(3)} - \delta_3 \delta_4 \psi_3 \psi_4)}. \tag{49}$$

Proof. Paying attention on the (1,1)-entries of both sides of (44), we find that

$$\frac{1}{\delta_4 \psi_4} \left(\frac{r_{11}^{(1)}}{\delta_1 \psi_1}, \frac{r_{11}^{(2)}}{\delta_2 \psi_2}, \frac{r_{11}^{(3)}}{\delta_3 \psi_3} \right)$$

$$= \left(\frac{1 + 7\nu + \nu^2}{36\nu}, \frac{5 + 5\nu - \nu^2}{36(1 + \nu)}, \frac{-1 + 5\nu + 5\nu^2}{36\nu(1 + \nu)} \right). \tag{50}$$

Multiplying 6 and subtracting $(1,1,1)$, we find that (50) turns out to be

$$\frac{6}{\delta_4\psi_4}\left(\frac{r_{11}^{(1)}}{\delta_1\psi_1}, \frac{r_{11}^{(2)}}{\delta_2\psi_2}, \frac{r_{11}^{(3)}}{\delta_3\psi_3}\right) - (1,1,1)$$

$$= (1+\nu+\nu^2)\left(\frac{1}{\nu}, -\frac{1}{1+\nu}, -\frac{1}{\nu(1+\nu)}\right). \tag{51}$$

Then comparing the second and third entries of both sides, we obtain the formula (49). $\qquad\square$

We are going to determine u. For this purpose, we define the linear transformation

$$S(x_4) = \frac{(z_1 - z_2)x_4 + u_1 z_2 - u_2 z_1}{u_1 - u_2},$$

where z_1, z_2 are defined in (14). Then it is clear that

$$S(u_1) = z_1, \quad S(u_2) = z_2, \quad S(\infty) = \infty.$$

Lemma 2. *If ν is defined by (49), then*

$$\frac{(u_1 - u_3)(u_2 - u_4)}{(u_1 - u_2)(u_3 - u_4)} = \frac{\nu^3(2+\nu)}{1+2\nu}. \tag{52}$$

This lemma can be shown with the aid of Mathematica. We recall the definition of z_{3a}, z_{4a} (cf. (19)).

Theorem 4. *If ν is defined by (49) and if $S(u_4) = z_{4a}$, then*

$$u = \frac{t_1^4(1+2\nu)(u_2 - u_4)}{u_2 - u_1} \tag{53}$$

and $S(u_3) = z_{3a}$.

50   *J. Sekiguchi*

Proof. We first recall that $\nu = \frac{t_2}{t_1}$. Then it follows from (14) that

$$z_1 = 0, \quad z_2 = t_1{}^4(2 + \nu), \quad z_3 = t_1{}^4(2 + \nu). \tag{54}$$

The assumption $S(u_4) = z_{4a}$ combined with (20), (54) implies (53). It follows from (53) and (52) that

$$u = \frac{z_2(u_2 - u_4)}{u_2 - u_1} = \frac{z_3(u_3 - u_4)}{u_3 - u_1}.$$

Then

$$u - z_2 = \frac{z_2(u_1 - u_4)}{u_2 - u_4}, \quad u - z_3 = \frac{z_3(u_1 - u_4)}{u_3 - u_1},$$

which combined with (20) implies that

$$z_{3a} = \frac{z_3(u - z_2)}{u - z_3} = \frac{z_2(u_3 - u_1)}{u_2 - u_1} = S(u_3).$$

and the theorem follows. $\qquad\square$

The composition $t_3 = L^{-1}(S(x_4))$ is given by

$$t_3 = \frac{z_2(u_2 - u_4)(x_4 - u_1)}{(u_2 - u_1)(x_4 - u_4)},$$

where u and z_2 are defined by (53) and (54), respectively. Then

x_4	u_1	u_2	u_3	u_4	∞
t_3	0	z_2	z_3	∞	u

Remark 4. On the one hand, the differential equation (21) depends on ξ, t_1, t_2, u; ξ is its variable and t_1, t_2, u are its parameters. By letting $t_2 = \nu t_1$, we regard t_1, ν, u as its parameters. On the other hand, the differential equation (35) depends on x_1, x_2, x_4, z_0; x_4 is its variable and x_1, x_2, z_0 are regarded as its parameters. Moreover, since the singular points u_1, u_2, u_3, u_4 of (35) are solutions of $f_0(x_4) = 0$, where $f_0(x_4)$ is a quartic polynomial of x_4, it follows that u_1, u_2, u_3, u_4 are algebraic functions of x_1, x_2, z_0. As consequences of Theorems 3 and 4, we find that the differential equation (35) is reduced to the differential equation (21) by defining ν and $u/t_1{}^4$ as functions of x_1, x_2, z_0 by (49) and (53).

Unfortunately, the formulas (49) and (53) are very complicated to write down explicitly.

It is worth mentioning the difference between the procedure employed by Romano [12] and the argument in this chapter. The method by Romano is to construct the potential obtained by Pavlyk from the polynomial potential of type A_3 constructed by Dubrovin. His idea is natural but (49) and (53) suggest that it seems hard to accomplish the task of obtaining the expressions of the variables x_1, x_2, x_4, z_0 by the variables t_1, t_2, t_3. On the contrary, the idea of this chapter is purely algebraic, namely to check the comparison among the matrix entries of N_1, N_2, N_3 and those of $R_{1a}, R_{2a}, R_{3a}, R_{4a}$.

Remark 5. Since the formulas (49) and (53) are complicated to write down, it is valuable to treat a special case to compute their expressions. We consider the case $x_2 = 0$. Then

$$f_0(x_4) = -3z_0{}^3(2x_1 - z_0)^2(4x_1 + z_0)^3$$
$$+ 6\left(6x_1{}^4 - 2x_1{}^2z_0{}^2 - 4x_1z_0{}^3 + z_0{}^4\right)x_4{}^2 + x_4{}^4.$$

We may take the solutions u_1, u_2, u_3, u_4 of $f_0(x_4) = 0$ by

$$u_1 = \sqrt{A + 2\sqrt{3}B},$$

$$u_2 = -\sqrt{A + 2\sqrt{3}B},$$

$$u_3 = \sqrt{A - 2\sqrt{3}B},$$

$$u_4 = -\sqrt{A - 2\sqrt{3}B},$$

where

$$A = -18x_1{}^4 + 6x_1{}^2z_0{}^2 + 12x_1z_0{}^3 - 3z_0{}^4,$$
$$B = (x_1 + z_0)^2(3x_1{}^2 - 2x_1z_0 + z_0{}^2)^3.$$

Since

$$A^2 - 12B = -3(2x_1 - z_0)^2z_0{}^3(4x_1 + z_0)^3,$$

it follows that

$$u_1u_3 = \sqrt{-3(2x_1 - z_0)^2z_0{}^3(4x_1 + z_0)^3}.$$

By a direct computation, we find that (49) is reduced to

$$\nu = \frac{u_1 u_3 - z_0^2 (4x_1 + z_0)^2}{2z_0^2 (4x_1 + z_0)^2}$$

and (53) is reduced to

$$u = \frac{t_1^4 \{u_1(u_1 + u_3) + 6(x_1^4 - 2x_1^2 z_0^2 - 4x_1 z_0^3 + z_0^4)\}}{2z_0^2 (4x_1 + z_0)^2}.$$

3. B_3 and F_4

The construction of this section is similar to Section 2, but the argument is simpler than that because $\det(T)$ is factored into two quadratic polynomials for the variable x_4. In Section 3.1, we first collect the polynomial potential of type B_3, the matrices C and T (see (2), (4) for the definition of C and T). Then we introduce an ordinary differential equation (58) of variable t_3 whose singular points are the solutions of $\det(T) = 0$ as a polynomial of t_3, its modification (61) and 2×2 matrices N_1, N_2, N_3 (cf. (59)). In Section 3.2, we introduce a 4-parameter family $\mathcal{F}$ of algebraic potentials of the variables x_1, x_2, x_3, x_4 with attached weights $1/3, 1/3, 1, 1$ and define a group consisting of automorphisms of the family $\mathcal{F}$. Dinar [1] constructed an algebraic potential of type F_4. We confirm that the potential of Dinar is regarded as a member of $\mathcal{F}$. In Section 3.3, we take an algebraic potential F which is a member of $\mathcal{F}$ but is not isomorphic to Dinar's by the isomorphisms defined in Section 3.2. Then we construct an ordinary differential equation from F and show its similarities with (21). This follows from Theorems 5 and 6. The reason why we didn't treat the algebraic potential of Dinar is explained in Remark 9. In Section 3.4, we shall introduce a 1-parameter subfamily $\mathcal{F}'$ of $\mathcal{F}$ with the following conditions: (1) Any member of $\mathcal{F}$ is conjugate to one of $\mathcal{F}'$ under G-action. (2) By G-action, any two of $\mathcal{F}'$ are not conjugate to each other. We define an "equivalence" relation among potentials of type $F_4(1)$ and all the members of the family are "equivalent" in this sense and discuss an evidence that the equivalence in this sense is the same as the equivalence suggested by Dubrovin.

3.1. B_3 *case*

We start by introducing the polynomial potential which is obtained by B. Dubrovin. We introduce here the following potential, which is the same as the one obtained by Dubrovin [3] by a coordinate transformation.

$$F = \frac{2187 t_1{}^7}{70} + \frac{27 t_1{}^3 t_2{}^2}{2} - \frac{3 t_1 t_2{}^3}{2} + \frac{t_1 t_3{}^2}{2} + \frac{t_2{}^2 t_3}{2},$$

$$C = \begin{pmatrix} t_3 & \frac{9}{2}\left(18 t_1{}^2 t_2 - t_2{}^2\right) & \frac{81}{5} t_1 \left(81 t_1{}^4 + 5 t_2{}^2\right) \\ t_2 & 9 t_1 \left(3 t_1{}^2 - t_2\right) + t_3 & -\frac{9}{2}\left(t_2{}^2 - 18 t_1{}^2 t_2\right) \\ t_1 & t_2 & t_3 \end{pmatrix},$$

$$T = \begin{pmatrix} t_3 & 6\left(18 t_1{}^2 - t_2\right) t_2 & 27 t_1 \left(81 t_1{}^4 + 5 t_2{}^2\right) \\ \frac{2 t_2}{3} & 27 t_1{}^3 - 9 t_2 t_1 + t_3 & 6\left(18 t_1{}^2 - t_2\right) t_2 \\ \frac{t_1}{3} & \frac{2 t_2}{3} & t_3 \end{pmatrix}.$$

In this case,

$$\det(T) = \left(-27 t_1{}^3 - 9 t_1 t_2 - t_3\right)$$

$$\times \left(729 t_1{}^6 - 486 t_1{}^4 t_2 + 27 t_1{}^2 t_2{}^2 + 18 t_1 t_2 t_3 - 8 t_2{}^3 - t_3{}^2\right).$$

We introduce the algebraic function w_0 of (t_1, t_2) by the relation

$$2 t_2 - 9 t_1^2 + w_0^2 = 0. \tag{55}$$

By this relation, we eliminate t_2 in $\det(T)$. Then

$$\det(T) = \frac{1}{8}\left(135 t_1{}^3 - 9 t_1 w_0{}^2 + 2 t_3\right)\left(81 t_1{}^3 - 9 t_1 w_0{}^2 - 2 t_3 - 2 w_0{}^3\right)$$

$$\times \left(81 t_1{}^3 - 9 t_1 w_0{}^2 - 2 t_3 + 2 w_0{}^3\right).$$

Letting

$$\begin{cases} z_1 = \frac{1}{2}\left(81 t_1{}^3 - 9 t_1 w_0{}^2 - 2 w_0{}^3\right), \\ z_2 = \frac{1}{2}\left(81 t_1{}^3 - 9 t_1 w_0{}^2 + 2 w_0{}^3\right), \\ z_3 = -\frac{9}{2}\left(15 t_1{}^3 - t_1 w_0{}^2\right), \end{cases} \tag{56}$$

we find that

$$\det(T) = (t_3 - z_1)(t_3 - z_2)(t_3 - z_3).$$

Introducing the matrix $B_\infty^{(3)}$ depending on r by

$$B_\infty^{(3)} = \begin{pmatrix} r + \frac{1}{3} & 0 & 0 \\ 0 & r + \frac{2}{3} & 0 \\ 0 & 0 & r + 1 \end{pmatrix},$$

we consider the ordinary differential equation

$$\frac{d}{dt_3}\tilde{Y} = -T^{-1}B_\infty^{(3)}\tilde{Y}, \tag{57}$$

where $\tilde{Y} = {}^t(\tilde{\mathbf{y}}_1, \tilde{\mathbf{y}}_2, \tilde{\mathbf{y}}_3)$. We treat the case $r = -1$ of the equation (57). Then we obtain the following differential equation:

$$\frac{d}{dt_3}\hat{Y} = \left(\frac{N_1}{t_3 - z_1} + \frac{N_2}{t_3 - z_2} + \frac{N_3}{t_3 - z_3} \right) \hat{Y}, \tag{58}$$

where $\hat{Y} = {}^t(\tilde{\mathbf{y}}_1, \tilde{\mathbf{y}}_2)$ and

$$\begin{cases} N_1 = \begin{pmatrix} \dfrac{-9t_1{}^2 + 6w_0 t_1 + w_0{}^2}{6w_0(6t_1 + w_0)} & \dfrac{(3t_1 + w_0)\left(9t_1{}^2 - 6w_0 t_1 - w_0{}^2\right)}{4w_0(6t_1 + w_0)} \\[3mm] -\dfrac{3t_1 + w_0}{9w_0(6t_1 + w_0)} & \dfrac{(3t_1 + w_0)^2}{6w_0(6t_1 + w_0)} \end{pmatrix}, \\[10mm] N_2 = \begin{pmatrix} \dfrac{-9t_1{}^2 - 6w_0 t_1 + w_0{}^2}{6w_0(w_0 - 6t_1)} & -\dfrac{(3t_1 - w_0)\left(9t_1{}^2 + 6w_0 t_1 - w_0{}^2\right)}{4(6t_1 - w_0)w_0} \\[3mm] \dfrac{3t_1 - w_0}{9(6t_1 - w_0)w_0} & \dfrac{(w_0 - 3t_1)^2}{6w_0(w_0 - 6t_1)} \end{pmatrix}, \\[10mm] N_3 = \begin{pmatrix} \dfrac{27t_1{}^2 - w_0{}^2}{3(6t_1 - w_0)(6t_1 + w_0)} & \dfrac{3t_1\left(27t_1{}^2 - w_0{}^2\right)}{2(6t_1 - w_0)(6t_1 + w_0)} \\[3mm] \dfrac{2t_1}{3(6t_1 - w_0)(6t_1 + w_0)} & \dfrac{3t_1{}^2}{(6t_1 - w_0)(6t_1 + w_0)} \end{pmatrix}. \end{cases} \tag{59}$$

Remark 6. It is easy to obtain the Schlesinger system from the triplet $\{N_1, N_2, N_3\}$. We let

$$N_j^\sharp = JN_k J^{-1}, \quad (k = 1, 2, 3),$$

where $J = \begin{pmatrix} 1 & 0 \\ 0 & w_0 \end{pmatrix}$ and introduce t by

$$t = \frac{(w_0 - 3t_1)(6t_1 + w_0)^2}{2w_0{}^3}.$$

For a moment, we assume that t_1 is a constant and regard w_0 as a function of t. Then it follows that

$$\begin{cases} \dfrac{d}{dt}N_1{}^\sharp = \dfrac{1}{t}[N_3{}^\sharp,\ N_1{}^\sharp], \\[2ex] \dfrac{d}{dt}N_2{}^\sharp = \dfrac{1}{t-1}[N_3{}^\sharp,\ N_2{}^\sharp], \\[2ex] \dfrac{d}{dt}N_3{}^\sharp = \dfrac{1}{t}[N_1{}^\sharp,\ N_3{}^\sharp] + \dfrac{1}{t-1}[N_2{}^\sharp,\ N_3{}^\sharp], \end{cases}$$

which is the Schlesinger system.

Letting $Y = (t_3 - u)^{-2/3}\hat{Y}$, we obtain the differential equation

$$\frac{d}{dt_3}Y = \left(\frac{N_1}{t_3 - z_1} + \frac{N_2}{t_3 - z_2} + \frac{N_3}{t_3 - z_3} + \frac{-\frac{2}{3}}{t_3 - u}\right)Y \qquad (60)$$

for Y, where u is a function of t_1, t_2 different from all of z_1, z_2, z_3.

We define the linear fractional transformation $L(t_3)$ of t_3 by

$$\xi = L(t_3) = \frac{(z_1 + z_2 - u)t_3 - z_1 z_2}{t_3 - u}.$$

It is easy to see that

$$L(z_1) = z_1, \quad L(z_2) = z_2, \quad L(u) = \infty.$$

We let

$$z_{3a} = L(z_3), \quad z_{4a} = L(\infty).$$

Then the differential equation (60) turns out to be

$$\frac{d}{d\xi}Y = \left(\frac{N_1}{\xi - z_1} + \frac{N_2}{\xi - z_2} + \frac{N_3}{\xi - z_{3a}} + \frac{N_4}{\xi - z_{4a}}\right)Y, \qquad (61)$$

where

$$N_4 = \begin{pmatrix} 0 & 0 \\ 0 & \frac{1}{3} \end{pmatrix}. \qquad (62)$$

It is easy to show that $0, \frac{1}{3}$ are the eigenvalues of each of the matrices N_1, N_2, N_3, N_4 and that $N_1 + N_2 + N_3 + N_4 = \frac{2}{3}I_2$.

3.2.   *A 4-parameter family of algebraic potentials of type $F_4(1)$*

In this paragraph, we present a family of algebraic potentials of four variables (x_1, x_2, x_3, x_4) which contains the algebraic potential constructed by Dinar [1]. The following argument is similar to Section 2.2.

Let x_1, x_2, x_3, x_4 be the variables with weights $1/3, 1/3, 1, 1$, respectively, and let z_0 be the algebraic function of x_1, x_2, x_3, x_4 defined by the equation

$$\frac{1}{36}p_2 x_1{}^2 x_2 \left(-8p_1{}^3 p_4 + 27 p_2 p_3 - 54 p_3{}^2 p_4\right)$$

$$+ \frac{1}{18}x_1 x_2{}^2 \left(-4p_1{}^3 p_4 + 27 p_2 p_3 - 27 p_3{}^2 p_4\right)$$

$$+ \frac{1}{5832}x_1{}^3 \left(64 p_1{}^6 p_4{}^3 - 324 p_1{}^3 p_2{}^2 p_4 + 864 p_1{}^3 p_3{}^2 p_4{}^3 + 729 p_2{}^3 p_3 \right.$$

$$\left. - 2187 p_2{}^2 p_3{}^2 p_4 + 2916 p_3{}^4 p_4{}^3\right)$$

$$+ p_3 x_2{}^3 - x_3 + \frac{1}{4}p_1 z(p_2 x_1 + 2x_2)^2 + z^3 = 0 \tag{63}$$

and let $F = F_{(p_1,p_2,p_3,p_4)}(x_1, x_2, x_3, x_4)$ be the algebraic function of (x_1, x_2, x_3, x_4) defined by

$$
\begin{aligned}
&F \\
&= \frac{1}{2}x_1 x_4{}^2 + x_2 x_3 x_4 - \frac{p_4}{28570268160}(851968 p_4{}^7 p_1{}^{15} + 19316736 p_3{}^2 p_4{}^7 p_1{}^{12} \\
&\quad + 9289728 p_2 p_3 p_4{}^6 p_1{}^{12} - 8515584 p_2{}^2 p_4{}^5 p_1{}^{12} + 133373952 p_3{}^4 p_4{}^7 p_1{}^9 \\
&\quad + 250822656 p_2 p_3{}^3 p_4{}^6 p_1{}^9 - 229920768 p_2{}^2 p_3{}^2 p_4{}^5 p_1{}^9 + 17418240 p_2{}^3 p_3 p_4{}^4 p_1{}^9 \\
&\quad + 1306368 p_2{}^4 p_4{}^3 p_1{}^9 + 40310784 p_3{}^6 p_4{}^7 p_1{}^6 + 2539579392 p_2 p_3{}^5 p_4{}^6 p_1{}^6 \\
&\quad - 2327947776 p_2{}^2 p_3{}^4 p_4{}^5 p_1{}^6 + 352719360 p_2{}^3 p_3{}^3 p_4{}^4 p_1{}^6 \\
&\quad + 61725888 p_2{}^4 p_3{}^2 p_4{}^3 p_1{}^6 - 44089920 p_2{}^5 p_3 p_4{}^2 p_1{}^6 + 38946096 p_2{}^6 p_4 p_1{}^6 \\
&\quad - 2766327552 p_3{}^8 p_4{}^7 p_1{}^3 + 11428107264 p_2 p_3{}^7 p_4{}^6 p_1{}^3 - 10475764992 p_2{}^2 p_3{}^6 p_4{}^5 p_1{}^3 \\
&\quad + 2380855680 p_2{}^3 p_3{}^5 p_4{}^4 p_1{}^3 + 654735312 p_2{}^4 p_3{}^4 p_4{}^3 p_1{}^3 - 595213920 p_2{}^5 p_3{}^3 p_4{}^2 p_1{}^3 \\
&\quad - 85030560 p_2{}^7 p_3 p_1{}^3 + 347208120 p_2{}^6 p_3{}^2 p_4 p_1{}^3 - 7652750400 p_3{}^{10} p_4{}^7 \\
&\quad + 19284931008 p_2 p_3{}^9 p_4{}^6 - 17677853424 p_2{}^2 p_3{}^8 p_4{}^5 + 5356925280 p_2{}^3 p_3{}^7 p_4{}^4 \\
&\quad - 57395628 p_2{}^7 p_3{}^3 + 2008846980 p_2{}^4 p_3{}^6 p_4{}^3 - 2008846980 p_2{}^5 p_3{}^5 p_4{}^2 \\
&\quad + 569173311 p_2{}^6 p_3{}^4 p_4)x_1{}^7 + \frac{p_4}{37791360}(-24576 p_3 p_4{}^6 p_1{}^{12} + 45056 p_2 p_4{}^5 p_1{}^{12} \\
&\quad - 663552 p_3{}^3 p_4{}^6 p_1{}^9 + 1216512 p_2 p_3{}^2 p_4{}^5 p_1{}^9 - 138240 p_2{}^2 p_3 p_4{}^4 p_1{}^9 \\
&\quad - 13824 p_2{}^3 p_4{}^3 p_1{}^9 - 6718464 p_3{}^5 p_4{}^6 p_1{}^6 + 12317184 p_2 p_3{}^4 p_4{}^5 p_1{}^6 \\
&\quad - 2799360 p_2{}^2 p_3{}^3 p_4{}^4 p_1{}^6 - 653184 p_2{}^3 p_3{}^2 p_4{}^3 p_1{}^6 + 583200 p_2{}^4 p_3 p_4{}^2 p_1{}^6 \\
&\quad - 618192 p_2{}^5 p_4 p_1{}^6 - 30233088 p_3{}^7 p_4{}^6 p_1{}^3 + 55427328 p_2 p_3{}^6 p_4{}^5 p_1{}^3 \\
&\quad - 18895680 p_2{}^2 p_3{}^5 p_4{}^4 p_1{}^3 - 6928416 p_2{}^3 p_3{}^4 p_4{}^3 p_1{}^3 + 7873200 p_2{}^4 p_3{}^3 p_4{}^2 p_1{}^3 \\
&\quad + 1574640 p_2{}^6 p_3 p_1{}^3 - 5511240 p_2{}^5 p_3{}^2 p_4 p_1{}^3
\end{aligned}
$$

$$-51018336p_3{}^9p_4{}^6 + 93533616p_2p_3{}^8p_4{}^5 - 42515280p_2{}^2p_3{}^7p_4{}^4$$

$$+1062882p_2{}^6p_3{}^3 - 21257640p_2{}^3p_3{}^6p_4{}^3$$

$$+26572050p_2{}^4p_3{}^5p_4{}^2 - 9034497p_2{}^5p_3{}^4p_4)x_2x_1{}^6 + \frac{p_4}{37791360}(45056p_4{}^5p_1{}^{12}$$

$$+1216512p_3{}^2p_4{}^5p_1{}^9 - 276480p_2p_3p_4{}^4p_1{}^9 - 41472p_2{}^2p_4{}^3p_1{}^9 + 12317184p_3{}^4p_4{}^5p_1{}^6$$

$$-5598720p_2p_3{}^3p_4{}^4p_1{}^6 - 1959552p_2{}^2p_3{}^2p_4{}^3p_1{}^6 + 2332800p_2{}^3p_3p_4{}^2p_1{}^6$$

$$-3090960p_2{}^4p_4p_1{}^6 + 55427328p_3{}^6p_4{}^5p_1{}^3 - 37791360p_2p_3{}^5p_4{}^4p_1{}^3$$

$$-20785248p_2{}^2p_3{}^4p_4{}^3p_1{}^3 + 31492800p_2{}^3p_3{}^3p_4{}^2p_1{}^3 + 9447840p_2{}^5p_3p_1{}^3$$

$$-27556200p_2{}^4p_3{}^2p_4p_1{}^3 + 93533616p_3{}^8p_4{}^5 - 85030560p_2p_3{}^7p_4{}^4 + 6377292p_2{}^5p_3{}^3$$

$$-63772920p_2{}^2p_3{}^6p_4{}^3 + 106288200p_2{}^3p_3{}^5p_4{}^2 - 45172485p_2{}^4p_3{}^4p_4)x_2{}^2x_1{}^5$$

$$-\frac{p_4}{524880}(2560p_3p_4{}^4p_1{}^9 + 768p_2p_4{}^3p_1{}^9 + 51840p_3{}^3p_4{}^4p_1{}^6 + 36288p_2p_3{}^2p_4{}^3p_1{}^6$$

$$-64800p_2{}^2p_3p_4{}^2p_1{}^6 + 114480p_2{}^3p_4p_1{}^6 + 349920p_3{}^5p_4{}^4p_1{}^3 + 384912p_2p_3{}^4p_4{}^3p_1{}^3$$

$$-874800p_2{}^2p_3{}^3p_4{}^2p_1{}^3 - 437400p_2{}^4p_3p_1{}^3 + 1020600p_2{}^3p_3{}^2p_4p_1{}^3 + 787320p_3{}^7p_4{}^4$$

$$-295245p_2{}^4p_3{}^3 + 1180980p_2p_3{}^6p_4{}^3 - 2952450p_2{}^2p_3{}^5p_4{}^2 + 1673055p_2{}^3p_3{}^4p_4)x_2{}^3x_1{}^4$$

$$+\frac{p_4}{349920}(-256p_1{}^9p_4{}^3 - 114480p_1{}^6p_2{}^2p_4 + 43200p_1{}^6p_2p_3p_4{}^2 - 12096p_1{}^6p_3{}^2p_4{}^3$$

$$+583200p_1{}^3p_2{}^3p_3 - 1020600p_1{}^3p_2{}^2p_3{}^2p_4 + 583200p_1{}^3p_2p_3{}^3p_4{}^2 - 128304p_1{}^3p_3{}^4p_4{}^3$$

$$+393660p_2{}^3p_3{}^3 - 1673055p_2{}^2p_3{}^4p_4 + 1968300p_2p_3{}^5p_4{}^2 - 393660p_3{}^6p_4{}^3)x_2{}^4x_1{}^3$$

$$+\frac{p_4}{3240}(-848p_1{}^6p_2p_4 + 160p_1{}^6p_3p_4{}^2 + 6480p_1{}^3p_2{}^2p_3 - 7560p_1{}^3p_2p_3{}^2p_4$$

$$+2160p_1{}^3p_3{}^3p_4{}^2 + 4374p_2{}^2p_3{}^3 - 12393p_2p_3{}^4p_4 + 7290p_3{}^5p_4{}^2)x_2{}^5x_1{}^2$$

$$-\frac{p_4}{9720}(848p_4p_1{}^6 - 12960p_2p_3p_1{}^3 + 7560p_3{}^2p_4p_1{}^3 - 8748p_2p_3{}^3 + 12393p_3{}^4p_4)x_2{}^6x_1$$

$$+\frac{1}{105}p_3(40p_1{}^3 + 27p_3{}^2)p_4x_2{}^7 - \frac{p_4}{2099520}\{(5120p_4{}^4p_1{}^9 + 103680p_3{}^2p_4{}^4p_1{}^6$$

$$-69120p_2p_3p_4{}^3p_1{}^6 + 51840p_2{}^2p_4{}^2p_1{}^6 + 49572p_2{}^4p_1{}^3 + 699840p_3{}^4p_4{}^4p_1{}^3$$

$$-933120p_2p_3{}^3p_4{}^3p_1{}^3 + 699840p_2{}^2p_3{}^2p_4{}^2p_1{}^3 - 116640p_2{}^3p_3p_4p_1{}^3$$

$$+1574640p_3{}^6p_4{}^4 - 3149280p_2p_3{}^5p_4{}^3 + 98415p_2{}^4p_3{}^2 + 2361960p_2{}^2p_3{}^4p_4{}^2$$

$$-787320p_2{}^3p_3{}^3p_4)x_1{}^4 + 216(-640p_3p_4{}^3p_1{}^6 + 960p_2p_4{}^2p_1{}^6 + 1836p_2{}^3p_1{}^3$$

$$-8640p_3{}^3p_4{}^3p_1{}^3 + 12960p_2p_3{}^2p_4{}^2p_1{}^3 - 3240p_2{}^2p_3p_4p_1{}^3 - 29160p_3{}^5p_4{}^3$$

$$+3645p_2{}^3p_3{}^2 + 43740p_2p_3{}^4p_4{}^2 - 21870p_2{}^2p_3{}^3p_4)x_2x_1{}^3 + 648(320p_4{}^2p_1{}^6$$

$$+1836p_2{}^2p_1{}^3 + 4320p_3{}^2p_4{}^2p_1{}^3 - 2160p_2p_3p_4p_1{}^3 + 3645p_2{}^2p_3{}^2 + 14580p_3{}^4p_4{}^2$$

$$-14580p_2p_3{}^3p_4)x_2{}^2x_1{}^2 + 23328(68p_2p_1{}^3 - 40p_3p_4p_1{}^3 + 135p_2p_3{}^2 - 270p_3{}^3p_4)x_2{}^3x_1$$

$$+11664(68p_1{}^3 + 135p_3{}^2)x_2{}^4\}x_3 - \frac{1}{216}\{(16p_4{}^2p_1{}^3 + 27p_2{}^2 + 108p_3{}^2p_4{}^2$$

$$-108p_2p_3p_4)x_1 + 108(p_2 - 2p_3p_4)x_2\}x_3{}^2 + \frac{17}{2880}p_1{}^4p_4(p_2x_1 + 2x_2)^6z$$

$$-\frac{p_1{}^2p_4}{38880}(p_2x_1 + 2x_2)^2\{162p_2x_1{}^2x_2(-8p_1{}^3p_4 + 27p_2p_3 - 54p_3{}^2p_4)$$

$$+324x_1x_2{}^2(-4p_1{}^3p_4 + 27p_2p_3 - 27p_3{}^2p_4) + x_1{}^3(64p_1{}^6p_4{}^3 - 324p_1{}^3p_2{}^2p_4$$

$$+864p_1{}^3p_3{}^2p_4{}^3 + 729p_2{}^3p_3 - 2187p_2{}^2p_3{}^2p_4 + 2916p_3{}^4p_4{}^3) + 5832p_3x_2{}^3$$

$$-5832x_3\}z^2 + \frac{3}{7}p_1p_4z^7.$$

Then F is an algebraic potential. Namely if we define the matrix C (cf. (2)) by using F, we have

$$[\partial_{x_j} C,\ \partial_{x_k} C] = O \quad (\forall j, k).$$

We denote by R_F the left-hand side of equation ((63)). We define the pair $\mathbf{F}_{(p_1, p_2, p_4, p_4)} = (F,\ R_R)$ instead of the polynomial F itself since the relation $R_F = 0$ plays a basic role in defining the algebraic potential F. In the following argument, we consider a group action on the 4-tuple (p_1, p_2, p_3, p_4). For this reason, we put $\pi(\mathbf{F}_{(p_1, p_2, p_3, p_4)}) = (p_1, p_2, p_3, p_4)$ and identify $\mathbf{C}^4$ with the totality of 4-tuples (p_1, p_2, p_3, p_4). Moreover, we let $\mathcal{F} = \{\mathbf{F}_{(p_1, p_2, p_3, p_4)} \mid (p_1, p_2, p_3, p_4) \in \mathbf{C}^4,\ p_1 p_3 p_4 \neq 0\}$.

We consider a linear transformation of the coordinate system (x_1, x_2, x_3, x_4) of the form (25). Then we find an automorphism of $\mathbf{C}^4$ by

$$(p_1, p_2, p_3, p_4) \to \left(a_1{}^2 p_1,\ a_2 p_2 + a_3,\ a_1{}^3 p_3,\ \frac{a_2}{a_1{}^3} p_4 \right) \qquad (64)$$

and this induces an automorphism of the family $\mathcal{F}$. Therefore, $G = \{(a_1, a_2, a_3) \in \mathbf{C}^4 \mid a_1 a_2 \neq 0\}$ is regarded as an automorphism group whose action on the family $\mathcal{F}$ is defined by (64).

Remark 7. Substituting $p_1 = -3q_1^2$ and eliminating x_3 by using the equation ((63)), we find that $\det(T)$ turns out to be the product of two quadratic polynomials of the variable x_4.

We are going to identify the algebraic potential obtained by Dinar [1] with a member of $\mathcal{F}$. For this purpose, we recall the algebraic potential by Dinar. It reads

$$
\begin{aligned}
F_{DNR2} \\
&= \frac{11443 s_1{}^7}{174182400} - \frac{29459 s_1{}^6 s_2}{580608000} + \frac{6089 s_1{}^5 s_2{}^2}{276480000} - \frac{254609 s_1{}^4 s_2{}^3}{34836480000} + \frac{17 s_1{}^4 s_3}{44800} + \frac{152263 s_1{}^3 s_2{}^4}{116121600000} \\
&\quad - \frac{2647 s_1{}^3 s_2 s_3}{336000} - \frac{300457 s_1{}^2 s_2{}^5}{5806080000000} + \frac{6059 s_1{}^2 s_2{}^2 s_3}{2240000} \\
&\quad + z^2 \left(\frac{9 s_1{}^5}{44800} + \frac{3 s_1{}^4 s_2}{89600} - \frac{3 s_1{}^3 s_2{}^2}{89600} - \frac{3 s_1{}^2 s_2{}^3}{640000} + \frac{81 s_1{}^2 s_3}{2800} + \frac{153 s_1 s_2{}^4}{89600000} + \frac{243 s_1 s_2 s_3}{14000} \right. \\
&\qquad \left. + \frac{1107 s_2{}^5}{4480000000} + \frac{729 s_2{}^2 s_3}{280000} \right) \\
&\quad + z \left(\frac{409 s_1{}^6}{2419200} - \frac{191 s_1{}^5 s_2}{1344000} + \frac{187 s_1{}^4 s_2{}^2}{5376000} + \frac{67 s_1{}^3 s_2{}^3}{13440000} + \frac{27 s_1{}^3 s_3}{560} - \frac{319 s_1{}^2 s_2{}^4}{179200000} \right. \\
&\qquad \left. - \frac{117 s_1{}^2 s_2 s_3}{5600} + \frac{529 s_1 s_2{}^5}{13440000000} + \frac{9 s_1 s_2{}^2 s_3}{56000} + \frac{1247 s_2{}^6}{38400000000} + \frac{369 s_2{}^3 s_3}{560000} + \frac{243 s_3{}^2}{70} \right) \\
&\quad - \frac{1973651 s_1 s_2{}^6}{174182400000000} - \frac{18223 s_1 s_2{}^3 s_3}{33600000} + \frac{s_1 s_3{}^2}{20} - 2 s_1 s_3 s_4 + 2 s_1 s_4{}^2 + \frac{292289 s_2{}^7}{193536000000000} \\
&\quad + \frac{60131 s_2{}^4 s_3}{1344000000} + \frac{3 s_2 s_3{}^2}{200} + \frac{3 s_2 s_3 s_4}{5},
\end{aligned}
$$

$$(65)$$

where s_1, s_2, s_3, s_4 are variables of weights $1/3, 1/3, 1, 1$, respectively, and z is an algebraic function of s_1, s_2, s_3 defined by the equation

$$-\frac{s_1{}^3}{96} - z\left(\frac{s_1{}^2}{48} + \frac{s_1 s_2}{80} + \frac{3 s_2{}^2}{1600}\right) + \frac{13 s_1{}^2 s_2}{2880} - \frac{s_1 s_2{}^2}{28800}$$

$$-\frac{41 s_2{}^3}{288000} - \frac{3 s_3}{2} + z^3 = 0. \tag{66}$$

By the change of variables

$$s_1 = x_1, \quad s_2 = \frac{10}{3}(x_1 + 3x_2), \quad s_3 = \frac{2x_3}{3}, \quad s_4 = x_4 \quad (z = z_0),$$

we find that F_{DNR2} turns out to be $F_{(-3/16, 4/3, -41/288, -24/5)}$.

3.3. $F_4(1)$ *case I*

In this section, we treat the algebraic potential introduced in the previous paragraph for the case $(p_1, p_2, p_3, p_4) = (-3, 0, -14/9, 81/4)$. The main reason why we treat this algebraic potential contained in $\mathcal{F}$ is that since only rational numbers appear in each step of the computations, we can avoid the technical difficulties caused by the appearance of irrational numbers.

We start the argument by introducing an algebraic potential related with the reflection group of type F_4. Let x_1, x_2, x_3, x_4 be the variables with weights $1/3, 1/3, 1, 1$, respectively, and let z be the algebraic function of x_1, x_2, x_3, x_4 defined by the relationship

$$10368 x_1{}^3 + 48 x_1 x_2{}^2 - \frac{14}{9} x_2{}^3 - 3 x_2{}^2 z - x_3 + z^3 = 0. \tag{67}$$

We introduce the algebraic function F by

$$F = \frac{x_1 x_4{}^2}{2} + x_2 x_3 x_4 + \frac{19190513074176 x_1{}^7}{35} - \frac{27088846848 x_1{}^6 x_2}{5}$$

$$+ \frac{4514807808 x_1{}^5 x_2{}^2}{5}$$

$$- 151538688 x_1{}^4 x_2{}^3 + 1327104 x_1{}^3 x_2{}^4 + 96768 x_1{}^2 x_2{}^5$$

$$+ \frac{3152 x_1 x_2{}^6}{5} - \frac{1618 x_2{}^7}{45}$$

$$
\begin{aligned}
&+ z\left(-50388480 x_1{}^4 x_2{}^2 + 7838208 x_1{}^3 x_2{}^3 + 186624 x_1{}^2 x_2{}^4 \right.\\
&\left.\quad + 3024 x_1 x_2{}^5 - \frac{735 x_2{}^6}{4}\right) + z^3\left(16796160 x_1{}^4 - 2612736 x_1{}^3 x_2\right.\\
&\left.\quad - 62208 x_1{}^2 x_2{}^2 - 1008 x_1 x_2{}^3 - \frac{241 x_2{}^4}{4}\right)\\
&+ z^2\left(2916 x_1 x_2{}^4 - \frac{567 x_2{}^5}{2}\right) + z^4\left(189 x_2{}^3 - 1944 x_1 x_2{}^2\right)\\
&+ \frac{2187 x_2{}^2 z^5}{20} + z^6\left(324 x_1 - \frac{63 x_2}{2}\right) - \frac{729 z^7}{28}.
\end{aligned}
\tag{68}
$$

It is easy to see that F is weighted homogeneous, in the sense that if

$$
E = \frac{1}{3}(x_1 \partial_{x_1} + x_2 \partial_{x_2}) + x_3 \partial_{x_3} + x_4 \partial_{x_4},
$$

then $EF = \frac{7}{3}F$. It can be shown that F is a solution to the WDVV equation, or equivalently, F is an algebraic potential.

By (67), we eliminate x_3 and regard each matrix entry of T as a polynomial of (x_1, x_2, x_4, z_0). As a consequence, $\det(T)$ is a polynomial of (x_1, x_2, x_4, z). It is better to change the variables so that the computations become simpler. We introduce y_0, y_1, y_2, y_4 by the following relations

$$
\begin{cases}
x_1 = \frac{1}{216}(-y_0 - y_1 - 3y_2),\\
x_2 = \frac{1}{12}(y_0 + 4y_1 - 6y_2),\\
z = \frac{1}{4}(2y_2 - y_0),\\
x_4 = \frac{1}{4}\left(-y_0{}^3 + 4y_0{}^2 y_1 - 6y_0{}^2 y_2 - 12 y_0 y_1 y_2 + 12 y_1{}^3 + 24 y_1{}^2 y_2 \right.\\
\qquad\quad \left. -36 y_1 y_2{}^2 - 72 y_2{}^3\right) + y_4.
\end{cases}
\tag{69}
$$

It is easy to write the explicit form of $\det(T)$ by using the variables y_0, y_1, y_2, y_4. Actually, introducing the polynomials

$$
\begin{cases}
h_1 = y_0{}^2 - 3y_1{}^2 + 6y_1 y_2 + 9y_2{}^2,\\
h_2 = y_0{}^2 + 3y_1{}^2 + 6y_1 y_2 - 9y_2{}^2,\\
h_3 = -9y_0\left(y_1{}^2 + 3y_2{}^2\right),
\end{cases}
\tag{70}
$$

we define

$$\varphi_1 = \left(y_4 + \frac{1}{2}h_3\right)^2 - h_1{}^3,$$

$$\varphi_2 = \left(y_4 - \frac{1}{2}h_3\right)^2 - h_2{}^3.$$

Then

$$\det(T) = \varphi_1\varphi_2.$$

Let $\mathcal{Q} = \mathbf{Q}(y_1, y_2, y_0)$ be the function field over $\mathbf{Q}$. We regard φ_1, φ_2 as quadratic polynomials of y_4 over $\mathcal{Q}$. So we write $\varphi_1(y_4)$, $\varphi_2(y_4)$ instead of φ_1, φ_2, respectively, when we stress the variable y_4. Denoting $k_1 = h_1{}^{1/2}$, $k_2 = h_2{}^{1/2}$, we find that u_1, u_2 (resp., v_1, v_2) are solutions of the equation $\varphi_1(y_4) = 0$ (resp., $\varphi_2(y_4) = 0$), where

$$\begin{cases} u_1 = -\frac{1}{2}h_3 + k_1{}^3, \\ u_2 = -\frac{1}{2}h_3 - k_1{}^3, \\ v_1 = \frac{1}{2}h_3 + k_2{}^3, \\ v_2 = \frac{1}{2}h_3 - k_2{}^3. \end{cases} \tag{71}$$

Let $\mathcal{K} = \mathcal{Q}u_1v_1 + \mathcal{Q}u_1 + \mathcal{Q}v_1 + \mathcal{Q}$ be the module over $\mathcal{Q}$. Then $\mathcal{K}$ is regarded as the quotient field of $\mathcal{Q}[u_1, v_1]$.

We let

$$B_\infty^{(4)} = \begin{pmatrix} r + \frac{1}{3} & 0 & 0 & 0 \\ 0 & r + \frac{1}{3} & 0 & 0 \\ 0 & 0 & r + 1 & 0 \\ 0 & 0 & 0 & r + 1 \end{pmatrix}$$

and as is explained in the Introduction, using $B_\infty^{(4)}$, we consider the ordinary differential equation

$$\frac{d}{dy_4}\tilde{Z} = -T^{-1}B_\infty^{(4)}\tilde{Z}, \tag{72}$$

where $\tilde{Z} = {}^t(\tilde{\mathbf{z}}_1, \tilde{\mathbf{z}}_2, \tilde{\mathbf{z}}_3, \tilde{\mathbf{z}}_4)$. We treat the case $r = -1$ of the equation (72). We let $\tilde{R} = (-T^{-1}B_\infty^{(4)})_{r=-1}$. Let $\tilde{R}(i, j)$ be the

(i, j)-entry of $\tilde{R}$ and define the 2×2 matrix R by

$$R = \begin{pmatrix} \tilde{R}(1,1) & \tilde{R}(1,2) \\ \tilde{R}(2,1) & \tilde{R}(2,2) \end{pmatrix}.$$

Since it follows from the definition of B_∞ that $\tilde{R}(i,j) = 0$ for $i = 3, 4, j = 1, 2, 3, 4$, we obtain the differential equation

$$\frac{d}{dy_4} Z = RZ, \tag{73}$$

where $Z = {}^t(\tilde{\mathbf{z}}_1, \tilde{\mathbf{z}}_2)$. Letting

$$R_{U,ja} = \lim_{y_4 \to u_j} (y_4 - u_j) R \ \ (j = 1, 2),$$

$$R_{V,ja} = \lim_{y_4 \to v_j} (y_4 - v_j) R \ \ (j = 1, 2),$$

we obtain the differential equation

$$\frac{d}{dy_4} Z = \left(\frac{R_{U,1a}}{y_4 - u_1} + \frac{R_{U,2a}}{y_4 - u_2} + \frac{R_{V,1a}}{y_4 - v_1} + \frac{R_{V,2a}}{y_4 - v_2} \right) Z, \tag{74}$$

which is the same as (73). We note that all of $R_{U,1a}, R_{U,2a}$, $R_{V,1a}, R_{V,2a}$ are 2×2 matrices whose matrix entries are contained in $\mathcal{K}$. In spite of the fact that it is complicated to reproduce here the concrete forms of $R_{U,1a}, R_{U,2a}, R_{V,1a}, R_{V,2a}$, it is possible to show their following properties:

(P1) $0, \frac{1}{3}$ are the eigenvalues of each of $R_{U,1a}, R_{U,2a}, R_{V,1a}, R_{V,2a}$.
(P2) $R_{U,1a} + R_{U,2a} + R_{V,1a} + R_{V,2a} = \frac{2}{3} I_2$.

Our purpose is to reduce (73) to (61) by obtaining the expressions of t_1, w_0, t_3 in terms of y_0, y_1, y_2, y_4.

We focus our efforts on writing down the concrete forms of $R_{U,1a}, R_{U,2a}, R_{V,1a}, R_{V,2a}$. For this purpose, we start with computing the explicit form of the 2×2 matrix R. Let a_{ij} be the

(i,j)-entry of the matrix $\varphi_1(y_4)\varphi_2(y_4)R$. Then

$$R = \frac{1}{\varphi_1(y_4)\varphi_2(y_4)} \begin{pmatrix} a_{11} & a_{12} \\ a_{21} & a_{22} \end{pmatrix}.$$

In particular,

a_{11}

$$= \frac{1}{24}(-4y_0{}^9 - 96y_0{}^7y_1y_2 - 84y_0{}^6y_1{}^3 - 24y_0{}^6y_1{}^2y_2 + 36y_0{}^6y_1y_2{}^2 + 504y_0{}^6y_2{}^3$$

$$+ 735y_0{}^5y_1{}^4 - 630y_0{}^5y_1{}^2y_2{}^2 - 5049y_0{}^5y_2{}^4 - 228y_0{}^4y_1{}^5 - 2250y_0{}^4y_1{}^4y_2$$

$$- 792y_0{}^4y_1{}^3y_2{}^2 + 1188y_0{}^4y_1{}^2y_2{}^3 + 13500y_0{}^4y_1y_2{}^4 + 3078y_0{}^4y_2{}^5 + 8172y_0{}^3y_1{}^5y_2$$

$$- 4536y_0{}^3y_1{}^3y_2{}^3 - 66420y_0{}^3y_1y_2{}^5 + 108y_0{}^2y_1{}^7 - 1224y_0{}^2y_1{}^6y_2 - 13068y_0{}^2y_1{}^5y_2{}^2$$

$$- 32184y_0{}^2y_1{}^4y_2{}^3 + 48276y_0{}^2y_1{}^3y_2{}^4 + 78408y_0{}^2y_1{}^2y_2{}^5 + 16524y_0{}^2y_1y_2{}^6$$

$$- 5832y_0{}^2y_2{}^7 + 2376y_0y_1{}^8 + 12528y_0y_1{}^6y_2{}^2 - 2592y_0y_1{}^4y_2{}^4 - 97200y_0y_1{}^2y_2{}^6$$

$$- 122472y_0y_2{}^8 - 864y_1{}^9 + 3888y_1{}^8y_2 + 19008y_1{}^7y_2{}^2 - 54432y_1{}^6y_2{}^3 - 134784y_1{}^5y_2{}^4$$

$$+ 202176y_1{}^4y_2{}^5 + 326592y_1{}^3y_2{}^6 - 256608y_1{}^2y_2{}^7 - 209952y_1y_2{}^8 + 104976y_2{}^9)$$

$$+ \frac{1}{6}(-4y_0{}^6 + 4y_0{}^5y_1 + 6y_0{}^5y_2 - 72y_0{}^4y_1y_2 + 30y_0{}^3y_1{}^3 + 60y_0{}^3y_1{}^2y_2 + 90y_0{}^3y_1y_2{}^2$$

$$+ 180y_0{}^3y_2{}^3 - 219y_0{}^2y_1{}^4 - 270y_0{}^2y_1{}^2y_2{}^2 - 1431y_0{}^2y_2{}^4 + 72y_0y_1{}^5 + 180y_0y_1{}^4y_2$$

$$+ 1080y_0y_1y_2{}^4 + 972y_0y_2{}^5 - 720y_1{}^5y_2 + 3024y_1{}^3y_2{}^3 - 5184y_1y_2{}^5)y_4$$

$$+ \frac{1}{6}(y_0{}^3 - 4y_0{}^2y_1 + 6y_0{}^2y_2 + 12y_0y_1y_2 - 12y_1{}^3 - 24y_1{}^2y_2 + 36y_1y_2{}^2 + 72y_2{}^3)y_4^2$$

$$+ \frac{2}{3}y_4{}^3,$$

a_{21}

$$= \frac{1}{432}(-4y_0{}^9 - 96y_0{}^7y_1y_2 + 42y_0{}^6y_1{}^3 - 6y_0{}^6y_1{}^2y_2 - 18y_0{}^6y_1y_2{}^2 + 126y_0{}^6y_2{}^3$$

$$+ 87y_0{}^5y_1{}^4 - 630y_0{}^5y_1{}^2y_2{}^2 + 783y_0{}^5y_2{}^4 - 57y_0{}^4y_1{}^5 + 1125y_0{}^4y_1{}^4y_2$$

$$- 198y_0{}^4y_1{}^3y_2{}^2 - 594y_0{}^4y_1{}^2y_2{}^3 + 3375y_0{}^4y_1y_2{}^4 - 1539y_0{}^4y_2{}^5 + 396y_0{}^3y_1{}^5y_2$$

$$- 4536y_0{}^3y_1{}^3y_2{}^3 + 3564y_0{}^3y_1y_2{}^5 - 54y_0{}^2y_1{}^7 - 306y_0{}^2y_1{}^6y_2 + 6534y_0{}^2y_1{}^5y_2{}^2$$

$$- 8046y_0{}^2y_1{}^4y_2{}^3 - 24138y_0{}^2y_1{}^3y_2{}^4 + 19602y_0{}^2y_1{}^2y_2{}^5 - 8262y_0{}^2y_1y_2{}^6$$

$$- 1458y_0{}^2y_2{}^7 + 432y_0y_1{}^8 + 864y_0y_1{}^6y_2{}^2 - 2592y_0y_1{}^4y_2{}^4 + 7776y_0y_1{}^2y_2{}^6$$

$$+ 34992y_0y_2{}^8 - 216y_1{}^9 - 1944y_1{}^8y_2 + 4752y_1{}^7y_2{}^2 + 27216y_1{}^6y_2{}^3 - 33696y_1{}^5y_2{}^4$$

$$- 101088{y_1}^4{y_2}^5 + 81648{y_1}^3{y_2}^6 + 128304{y_1}^2{y_2}^7 - 52488y_1{y_2}^8 - 52488{y_2}^9)$$

$$+ \frac{1}{108}({y_0}^5 y_1 - 3{y_0}^5 y_2 - 15{y_0}^3{y_1}^3 + 15{y_0}^3{y_1}^2 y_2 - 45{y_0}^3 y_1{y_2}^2 + 45{y_0}^3{y_2}^3$$

$$- 30{y_0}^2{y_1}^4 + 270{y_0}^2{y_2}^4 + 18y_0{y_1}^5 - 90y_0{y_1}^4 y_2 + 270y_0 y_1{y_2}^4 - 486y_0{y_2}^5$$

$$- 72{y_1}^5 y_2 + 648y_1{y_2}^5)y_4$$

$$+ \frac{1}{108}({y_0}^3 - {y_0}^2 y_1 - 3{y_0}^2 y_2 + 12y_0 y_1 y_2 + 6{y_1}^3 - 6{y_1}^2 y_2 - 18y_1{y_2}^2 + 18{y_2}^3){y_4}^2.$$

We don't write the explicit forms of a_{12} and a_{22} since it suffices to know these matrix elements a_{11} and a_{21} in the following argument.

We note that $\det(R)$ is divisible by $\varphi_1\varphi_2$.

We let $a_{U,ij}^{(k)} = a_{ij}|_{y_4=u_k}$, $a_{V,ij}^{(k)} = a_{ij}|_{y_4=v_k}$ $(j, k = 1, 2)$. Then

$$R_{U,ka} = \frac{1}{\varphi_1'(u_k)\varphi_2(u_k)} \begin{pmatrix} a_{U,11}^{(k)} & a_{U,12}^{(k)} \\ a_{U,21}^{(k)} & a_{U,22}^{(k)} \end{pmatrix},$$

$$R_{V,ka} = \frac{1}{\varphi_1(v_k)\varphi_2'(v_k)} \begin{pmatrix} a_{V,11}^{(k)} & a_{V,12}^{(k)} \\ a_{V,21}^{(k)} & a_{V,22}^{(k)} \end{pmatrix}.$$

We now introduce

$$\delta_{U,k} = \varphi_1'(u_k)\varphi_2(u_k), \quad \delta_{V,k} = \varphi_1(v_k)\varphi_2'(v_k),$$
$$\sigma_{U,k} = a_{U,11}^{(k)}, \quad\quad\quad\quad \sigma_{V,k} = a_{V,11}^{(k)},$$
$$\tau_{U,k} = a_{U,21}^{(k)}, \quad\quad\quad\quad \tau_{V,k} = a_{V,21}^{(k)}.$$

Then it follows from direct computation that

$$a_{U,12}^{(k)} = \sigma_{V,k}\left(-\sigma_{U,k} + \frac{1}{3}\delta_{U,k}\right)/\tau_{U,k}, \quad a_{V,12}^{(k)} = \sigma_{V,k}\left(-\sigma_{V,k} + \frac{1}{3}\delta_{V,k}\right)/\tau_{V,k},$$

$$a_{U,22}^{(k)} = -\sigma_{U,k} + \frac{1}{3}\delta_{U,k}, \quad\quad\quad\quad a_{V,22}^{(k)} = -\sigma_{V,k} + \frac{1}{3}\delta_{V,k}$$

and that

$$R_{U,ka} = M_{U,k}^{-1}\begin{pmatrix} 0 & 0 \\ 0 & \frac{1}{3} \end{pmatrix} M_{U,k} = \frac{1}{\delta_{U,k}}\begin{pmatrix} \sigma_{U,k} & \sigma_{U,k}(\frac{1}{3}\delta_{U,k} - \sigma_{U,k})/\tau_{U,k} \\ \tau_{U,k} & \frac{1}{3}\delta_{U,k} - \sigma_{U,k} \end{pmatrix},$$

$$R_{V,ka} = M_{V,k}^{-1}\begin{pmatrix} 0 & 0 \\ 0 & \frac{1}{3} \end{pmatrix} M_{V,k} = \frac{1}{\delta_{V,k}}\begin{pmatrix} \sigma_{V,k} & \sigma_{V,k}(\frac{1}{3}\delta_{V,k} - \sigma_{V,k})/\tau_{V,k} \\ \tau_{V,k} & \frac{1}{3}\delta_{V,k} - \sigma_{V,k} \end{pmatrix},$$

where

$$M_{U,k} = \begin{pmatrix} -\tau_{U,k} & \sigma_{U,k} \\ \tau_{U,k} & -\sigma_{U,k} + \frac{1}{3}\delta_{U,k} \end{pmatrix},$$

$$M_{V,k} = \begin{pmatrix} -\tau_{V,k} & \sigma_{V,k} \\ \tau_{V,k} & -\sigma_{V,k} + \frac{1}{3}\delta_{V,k} \end{pmatrix}.$$

Letting

$$R_{U,kb} = M_{V,2}R_{U,ka}M_{V,2}^{-1}, \; R_{V,kb} = M_{V,2}R_{V,ka}M_{V,2}^{-1} \; (k = 1,2)$$

and $\hat{Z} = M_{V,2}Z$, we find that the equation (74) for Z turns out to be the equation for $\hat{Z}$ given by

$$\frac{d}{dy_4}\hat{Z} = \left(\frac{R_{U,1b}}{y_4 - u_1} + \frac{R_{U,2b}}{y_4 - u_2} + \frac{R_{V,1b}}{y_4 - v_1} + \frac{R_{V,2b}}{y_4 - v_2} \right) \hat{Z}. \qquad (75)$$

It is underlined here that $R_{V,2b} = \begin{pmatrix} 0 & 0 \\ 0 & \frac{1}{3} \end{pmatrix}$.

We are going to find out the condition satisfied by the two 4-tuples of matrices $(R_{U,1b}, R_{U,2b}, R_{V,1b}, R_{V,2b})$ and (N_1, N_2, N_3, N_4) defined by (59), (62) so that (75) is identified with (61). We first recall the definition of z_1, z_2, z_3 (cf. (56)). Letting $\mu = \frac{w_0}{t_1}$, we rewrite them by using t_1 and μ. Then

$$\begin{cases} z_1 = \frac{1}{2}t_1{}^3(81 - 9\mu^2 - 2\mu^3), \\ z_2 = \frac{1}{2}t_1{}^3(81 - 9\mu^2 + 2\mu^3), \\ z_3 = -\frac{9}{2}t_1{}^3(15 - \mu^2). \end{cases} \qquad (76)$$

We next recall the definition of matrices N_1, N_2, N_3, N_4. Denoting $P = \begin{pmatrix} 1 & 0 \\ 0 & h_0 \end{pmatrix}$, where $h_0 = -\frac{2}{3t_1(\mu-6)(\mu+6))}$ and $N_j' = P^{-1}N_jP$

$(j = 1, 2, 3, 4)$, we find that

$$
\begin{cases}
N_1' = \begin{pmatrix} \frac{\mu^2+6\mu-9}{6\mu(\mu+6)} & \frac{(\mu+3)(\mu^2+6\mu-9)}{6\mu(\mu+6)^2(\mu-6)} \\ \frac{(\mu+3)(\mu-6)}{6\mu} & \frac{(\mu+3)^2}{6\mu(\mu+6)} \end{pmatrix}, \\[2em]
N_2' = \begin{pmatrix} \frac{\mu^2-6\mu-9}{6(\mu-6)\mu} & -\frac{(\mu-3)(\mu^2-6\mu-9)}{6\mu(\mu+6)(\mu-6)^2} \\ -\frac{(\mu-3)(\mu+6)}{6\mu} & \frac{(\mu-3)^2}{6(\mu-6)\mu} \end{pmatrix}, \\[2em]
N_3' = \begin{pmatrix} \frac{\mu^2-27}{3(\mu-6)(\mu+6)} & -\frac{(\mu^2-27)}{(\mu-6)^2(\mu+6)^2} \\ 1 & -\frac{3}{(\mu-6)(\mu+6)} \end{pmatrix}, \\[2em]
N_4' = \begin{pmatrix} 0 & 0 \\ 0 & \frac{1}{3} \end{pmatrix}.
\end{cases}
\tag{77}
$$

To continue the argument, we need some preparation. Let

$$
L_j = \begin{pmatrix} c_j & c_j(\frac{1}{3} - c_j)/d_j \\ d_j & \frac{1}{3} - c_j \end{pmatrix} \quad (j = 1, 2),
$$

$$
L_3 = \begin{pmatrix} c_3 & c_3(\frac{1}{3} - c_3) \\ 1 & \frac{1}{3} - c_3 \end{pmatrix}, \quad L_4 = \begin{pmatrix} 0 & 0 \\ 0 & \frac{1}{3} \end{pmatrix}
$$

be 2×2 matrices satisfying

$$
L_1 + L_2 + L_3 + L_4 = \frac{2}{3}I_2.
\tag{78}
$$

The condition (78) is equivalent to

$$
\begin{cases}
c_1 + c_2 + c_3 = \frac{2}{3}, \\
d_1 + d_2 + 1 = 0, \\
\frac{c_1}{d_1}(\frac{1}{3} - c_1) + \frac{c_2}{d_2}(\frac{1}{3} - c_2) + c_3(\frac{1}{3} - c_3) = 0.
\end{cases}
\tag{79}
$$

Lemma 3. *We assume*

$$
-144c_j{}^2 d_j - 144c_j{}^2 + 12c_j d_j{}^2 + 72c_j d_j + 69c_j - 2d_j{}^2 - 8d_j - 7 = 0
$$

$$
(j = 1, 2)
\tag{80}
$$

in addition to (79). *Then there is a constant* μ *such that* $L_j = N_j{}' \ (j = 1, 2, 3, 4)$.

This lemma is provable by an argument similar to that of Lemma 1.

We introduce the polynomial h_4 by

$$h_4 = y_0{}^2 - 12y_1y_2. \tag{81}$$

It is easy to show the relationship among h_1, h_2, h_3, h_4:

$$h_1^3 + h_2^3 - h_3^2 - 3h_1h_2h_4 + h_4^3 = 0. \tag{82}$$

Using h_1, h_2, h_4 (cf. (70) and (81)), we define

$$c_1' = \frac{h_4{}^2 + 4k_1k_2h_4 - 4k_1{}^2k_2{}^2}{24k_1k_2(h_4 - k_1k_2)},$$

$$d_1' = \frac{(h_4 + k_1k_2)(h_4 - 2k_1k_2)}{2h_4k_1k_2},$$

$$c_2' = -\frac{h_4{}^2 - 4k_1k_2h_4 - 4k_1{}^2k_2{}^2}{24k_1k_2(h_4 + k_1k_2)},$$

$$d_2' = -\frac{(h_4 - k_1k_2)(h_4 + 2k_1k_2)}{2h_4k_1k_2},$$

$$c_3' = \frac{3h_4{}^2 - 4k_1{}^2k_2{}^2}{12(h_4{}^2 - k_1{}^2k_2{}^2)}. \tag{83}$$

Then it is straightforward to show the relationship similar to (80) for c_j', d_j' $(j = 1, 2)$, namely

$$-144c_j'{}^2 d_j' - 144c_j'{}^2 + 12c_j'd_j^2 + 72c_j'd_j' + 69c_j'$$
$$-2d_j'{}^2 - 8d_j' - 7 = 0 \quad (j = 1, 2). \tag{84}$$

Lemma 4. *We define* $R_{U,jc} = G_R^{-1} R_{U,jb} G_R$ $(j = 1, 2)$, $R_{V,jc} = G_R^{-1} R_{V,jb} G_R$ $(j = 1, 2)$, *where* $G_R = \begin{pmatrix} 1 & 0 \\ 0 & R_{V,1b}(2, 1) \end{pmatrix}$ $(R_{V,1b}(2, 1)$ *is the (2,1)-entry of* $R_{V,1b})$. *Then*

$$R_{U,jc} = \begin{pmatrix} c_j' & c_j'(\frac{1}{3} - c_j')/d_j' \\ d_j' & \frac{1}{3} - c_j' \end{pmatrix} \quad (j = 1, 2),$$

$$R_{V,1c} = \begin{pmatrix} c_3' & c_3'(\frac{1}{3} - c_3') \\ 1 & \frac{1}{3} - c_3' \end{pmatrix}, \quad R_{V,2c} = \begin{pmatrix} 0 & 0 \\ 0 & \frac{1}{3} \end{pmatrix}.$$

This lemma is proved by direct computation with the help of the software Mathematica.

Theorem 5. *We let*

$$\mu = -\frac{6k_1 k_2}{h_4}. \tag{85}$$

Then

$$R_{U,jc} = N'_j \ (j = 1, 2), \quad R_{V,jc} = N'_{j+2} \ (j = 1, 2). \tag{86}$$

Proof. By substituting (85) in these formulas, we conclude that

$$R_{U,1c} = N'_1, \quad R_{U,2c} = N'_2, \quad R_{V,1c} = N'_3,$$

where the matrices $R_{U,1c}, R_{U,2c}, R_{V,1c}$ are defined in Lemma 4 and the matrices N'_1, N'_2, N_3 are defined in (77). Then the theorem follows. $\qquad\qquad\square$

Remark 8. There are two possibilities of μ satisfying $R_{V,1c} = N_3{}'$. If we take

$$\mu = \frac{6k_1 k_2}{h_4},$$

then

$$R_{U,1c} = N'_2, \quad R_{U,2c} = N'_1$$

follows.

We are going to determine u. For this purpose, we recall the definition of z_1, z_2, z_3 (cf. (56)), that of u_1, u_2, v_1, v_2 (cf. (71)) and that of μ (cf. (85)). We define the fractional linear transformation

$$W(y_4) = \frac{(u_2 - v_2)(z_2 - z_1)(y_4 - u_1)}{(u_2 - u_1)(y_4 - v_2)} + z_1. \tag{87}$$

It is clear that

$$W(u_1) = z_1, \ W(u_2) = z_2, \ W(v_2) = \infty.$$

Theorem 6. *Assume (85). Denoting*

$$u = \frac{27t_1^3(-3h_4{}^3 + 12h_4 k_1{}^2 k_2{}^2 + 16h_3 k_2{}^3 - 16k_2{}^6)}{2h_4{}^3}, \tag{88}$$

we have

$$W(v_1) = z_3, \quad W(\infty) = u.$$

Proof. The relation (82) among h_1, h_2, h_3, h_4 implies $W(v_1) = z_3$. On the other hand, (88) is equivalent to $W(\infty) = u$. $\qquad\square$

We conclude that by the transformation $\xi = S(y_4)$, the differential equation (75) turns out to be (61).

Remark 9. The author confirmed that the computation in each step in this paragraph goes well and the conclusions similar to Theorems 5 and 6 are established when we take the algebraic potential with the parameter $(p_1, p_2, p_3, p_4) = (-3, 0, -82/9, 27/169)$, which is isomorphic to Dinar's under G-action instead of that with $(p_1, p_2, p_3, p_4) = (-3, 0, -14/9, 81/4)$. But the argument becomes complicated and lengthy because of the appearance of the irrational numbers. For this reason, he abandoned performing this case $(-3, 0, -82/9, 27/169)$.

3.4. $F_4(1)$ *case II*

In this section, we study the relationship between a general member of $\mathcal{F}$ and the potential treated in Section 3.3.

By virtue of the action of G defined by (64), we may take

$$(p_1, p_2, p_3, p_4) = \left(-3,\ 0,\ -\frac{2(m+1)}{m-1},\ \frac{(m-1)^2}{4}\right) \qquad (89)$$

for the number m. It is noted here that the case treated in Section 3.3 corresponds to $m = -8$. It is possible to define the algebraic potential F_m of type $F_4(1)$ defined by (68) for the case (89). The algebraic function z satisfies the equation

$$2m^2\,(1-m)^2\,x_1{}^3 - 6mx_1x_2{}^2 - \frac{2\,(m+1)}{m-1}x_2{}^3 - x_3 - 3x_2{}^2 z + z^3 = 0.$$
$$(90)$$

The argument in Section 3.2 shows the following:

(1) Any algebraic potential of the form (68) is transformed to F_m for some m under the action of G.
(2) The potentials F_m and $F_{m'}$ transform each other under the action of G if and only if $m = m'$.

It is better to introduce n_1 such that $m = n_1{}^3$ and n_2 by the equation

$$3n_2{}^2 - (n_1{}^2 + n_1 + 1) = 0$$

to avoid the complexity in the subsequent argument. Moreover, we let $\mathrm{n} = (n_1, n_2)$ and $F_{[\mathrm{n}]} = F_{n_1{}^3}$.

We employ the flat coordinate $(x_{\mathrm{n}1}, x_{\mathrm{n}2}, x_{\mathrm{n}3}, x_{\mathrm{n}4})$ instead of (x_1, x_2, x_3, x_4) to stress the dependence on n. The defining equation of the algebraic function z_{n} is then

$$2n_1{}^6 \left(1 - n_1{}^3\right)^2 x_{\mathrm{n}1}{}^3 - 6n_1{}^3 x_{\mathrm{n}1} x_{\mathrm{n}2}{}^2 - \frac{2\left(n_1{}^3 + 1\right)}{n_1{}^3 - 1}$$

$$x_{\mathrm{n}2}{}^3 - x_{\mathrm{n}3} - 3x_{\mathrm{n}2}{}^2 z_{\mathrm{n}} + z_{\mathrm{n}}{}^3 = 0. \tag{91}$$

Let $T_{[\mathrm{n}]}$ be the 4×4 matrix corresponding to $F_{[\mathrm{n}]}$, in this case (cf. (4)). In order to write $\det(T_{[\mathrm{n}]})$, we need some preparation. Let $y_{\mathrm{n}0}$, $y_{\mathrm{n}1}$, $y_{\mathrm{n}2}$, $y_{\mathrm{n}4}$ be the variables defined by the following relation

$$\begin{cases}
x_{\mathrm{n}1} = -\frac{y_{\mathrm{n}1}(n_1+n_2)+y_{\mathrm{n}2}(n_1-3n_2+2)-y_{\mathrm{n}0}}{18(n_1-1)n_1{}^2 n_2{}^3}, \\[2mm]
x_{\mathrm{n}2} = \frac{y_{\mathrm{n}2}\left(2n_1{}^2-3n_2+1\right)+y_{\mathrm{n}1}(2n_1 n_2-n_2+1)+(n_1+1)y_{\mathrm{n}0}}{6n_1 n_2}, \\[2mm]
z_{\mathrm{n}} = -\frac{y_{\mathrm{n}2}\left(2n_1{}^2+2n_1+3n_2-1\right)+(2n_1+1)(n_2-1)y_{\mathrm{n}1}+(n_1-1)y_{\mathrm{n}0}}{6n_1 n_2}, \\[2mm]
x_{\mathrm{n}4} = \frac{1}{12n_2}\left\{3(n_1+1)y_{\mathrm{n}0}{}^3 - 3y_{\mathrm{n}0}{}^2 y_{\mathrm{n}1}(2n_1 n_2 - n_2 + 1)\right. \\[1mm]
\qquad - 3y_{\mathrm{n}0}{}^2 y_{\mathrm{n}2}\left(2n_1{}^2 - 3n_2 + 1\right) \\[1mm]
\qquad - 12(n_1 - 1)(n_1 + 1)(2n_2 - 1)y_{\mathrm{n}0}y_{\mathrm{n}1}y_{\mathrm{n}2} \\[1mm]
\qquad + 2(n_1 - 1)(2n_2 - 1)(3n_2 - 2n_1{}^2 + 1)\left(y_{\mathrm{n}1}{}^3 - 3y_{\mathrm{n}1}y_{\mathrm{n}2}{}^2\right) \\[1mm]
\qquad \left. + 6(n_1 - 1)(2n_2 - 1)((2n_1 - 1)n_2 + 1)\left(y_{\mathrm{n}1}{}^2 y_{\mathrm{n}2} - 3y_{\mathrm{n}2}{}^3\right)\right\} + y_{\mathrm{n}4}.
\end{cases} \tag{92}$$

Moreover, we define four polynomials

$$\begin{cases}
h_{\mathrm{n}1} = y_{\mathrm{n}0}{}^2 + (n_1 - 1)(2n_2 - 1)(y_{\mathrm{n}1} - 3y_{\mathrm{n}2})(y_{\mathrm{n}1} + y_{\mathrm{n}2}), \\
h_{\mathrm{n}2} = y_{\mathrm{n}0}{}^2 - (n_1 - 1)(2n_2 - 1)(y_{\mathrm{n}1} - y_{\mathrm{n}2})(y_{\mathrm{n}1} + 3y_{\mathrm{n}2}), \\
h_{\mathrm{n}3} = 3(n_1 - 1)(2n_2 - 1)y_{\mathrm{n}0}\left(y_{\mathrm{n}1}{}^2 + 3y_{\mathrm{n}2}{}^2\right), \\
h_{\mathrm{n}4} = 4(n_1 - 1)(2n_2 - 1)y_{\mathrm{n}1}y_{\mathrm{n}2} + y_{\mathrm{n}0}{}^2.
\end{cases} \tag{93}$$

Then it is straightforward to show

$$h_{\mathrm{n}1}{}^3 + h_{\mathrm{n}2}{}^3 - h_{\mathrm{n}3}{}^2 - 3h_{\mathrm{n}1}h_{\mathrm{n}2}h_{\mathrm{n}4} + h_{\mathrm{n}4}{}^3 = 0.$$

By using the variables y_{n0}, y_{n1}, y_{n2}, y_{n4}, we find that

$$\det(T_{[n]}) = \varphi_{n1}\varphi_{n2},$$

where

$$\varphi_{n1} = \left(y_{n4} + \frac{1}{2}h_{n3}\right)^2 - h_{n1}{}^3,$$

$$\varphi_{n2} = \left(y_{n4} - \frac{1}{2}h_{n3}\right)^2 - h_{n2}{}^3.$$

The argument in Section 3.3 from the definition of $B_\infty^{(4)}$ till the part before Lemma 4 goes the parallel way in the general case, but it is hard to prove the analogue of Lemma 4 directly due to the $=$ limited capacity of computers. For this reason, we abandon the argument parallel to that in Section 3.3.

We note that both the numbers n_1, n_2 are rational, the computations become simpler compared with the general case, namely the case where at least one of n_1, n_2 is irrational, and confirmed the validity of Lemma 4 at least for the case where (n_1, n_2) is one of the pairs

$$(-1/2, \ -1/2), \ (-2, \ \pm 1), \ (11/2, \ \pm 7/2), \ (-13/2, \ \pm 7/2),$$
$$(22, \ \pm 13), \ (-23, \ \pm 13).$$

We are going to discuss the problem of isomorphisms between the discriminants of the form "$\det(T_n)$".

We define a linear transformation between two coordinates $(y_{n0}, y_{n1}, y_{n2}, y_{n4})$ and (y_0, y_1, y_2, y_4) by

$$y_{n0} = y_0, \ y_{n1} = \kappa y_1, \ y_{n2} = \kappa y_2, \ y_{n4} = y_4, \tag{94}$$

where κ is a solution of

$$(n_1 - 1)(2n_2 - 1)\kappa^2 + 3 = 0.$$

It is easy to show that the polynomials $h_{n1}, h_{n2}, h_{n3}, h_{n4}$ (cf. (93)) transform each other to h_1, h_2, h_3, h_4 (cf. (70) and (81)) by (94). As a consequence, it is clear that $\varphi_{n1}\varphi_{n2} = \varphi_1\varphi_2$.

We next consider the transformation between two coordinates $(x_{n1}, x_{n2}, z_n, x_{n4})$ and (x_1, x_2, z, x_4) defined by the successive use of (92), (94), (69) and denote it by

$$(x_{n1}, x_{n2}, z_n, x_{n4}) = H(x_1, x_2, z, x_4). \tag{95}$$

It follows from the definition that H induces a linear transformation between (x_{n1}, x_{n2}, z_n) and (x_1, x_2, z) and that there is a homogeneous polynomial $\eta(x_1, x_2, z)$ such that $x_{n4} = x_4 + \eta(x_1, x_2, z)$. It is observed that by (95), the potential F defined by (68) is not transformed to the algebraic potential. This means that H is not an isomorphism between Frobenius manifolds.

The author constructed the family of algebraic potentials of type $F_4(1)$ with four parameters and a group of linear automorphisms of the family. Contrary to his expectation that all the members of the family transform each other under the group action, the conclusion is different. Namely there are members depending on one parameter which do not transform under the group action. One possibility to explain this phenomenon is that there is a group of automorphisms of the family which is properly larger than the group of linear automorphisms introduced in Section 3.3 and its action on the family $\mathcal{F}$ is transitive. Or there is no such group. The author asked Dinar about the conjecture by Dubrovin by e-mail. In his reply mail, he wrote '*First notice that Dubrovin conjecture is about equivalence between Frobenius manifolds not about isomorphisms. Also, equivalence does not mean the potentials will be identical under the given transformation.* To explain an idea to formulate the equivalence for the case $F_4(1)$, we need to prepare some notation. Let $F(x_1, x_2, x_3, x_4)$ be an algebraic potential of the family $\mathcal{F}$. We consider the matrix T corresponding to F and put $\Delta_F = \det(T)$. Then the algebraic potentials F_A, F_B are Δ-equivalent if Δ_{F_A} and Δ_{F_B} transform each other by a coordinate transformation of the form

$$\begin{cases} y_1 = a_1 x_1 + a_2 x_2 + a_3 z, \\ y_2 = c_1 x_1 + b_2 x_2 + b_3 z, \\ w = c_1 x_1 + c_2 x_2 + c_3 z, \\ y_4 = d_1 x_4 + \varphi(x_1, x_2, z), \end{cases}$$

where φ is a homogeneous polynomial of degree 3 and a_j, b_j, c_j, d_1 are constants. It is underlined here that by the above coordinate transformation, F_A is transformed to a function which is not necessarily a potential. It is not known whether the equivalence in the sense of Dubrovin written in Dinar's e-mail is the same as the Δ-equivalence in the above sense or not. It is not clear for the author whether "Δ-equivalence" is well-defined for arbitrary potentials of four variables. At any rate, it is already shown that all the members of the family $\mathcal{F}$ are Δ-equivalent.

4. H_3 and H_4

The conclusion of this section is incomplete compared with the results in the preceding two cases. The argument of Section 4.1 is similar to those of Sections 2.1 and 3.1. It is possible to construct a 4-parameter family of algebraic potential related with H_4 case, the author abandoned to write it down because it is a bit lengthy. In Section 4.2, we give an algebraic potential with variables x_1, x_2, x_3, x_4 and attached weights $1/5, 1/5, 1, 1$. Then we construct an ordinary differential equation from F. After the computation following the construction of the differential equation, we abandoned to continue the investigation and we restrict our attention to the case $x_2 = -2x_1$. Under this restriction, the computation becomes simpler and the partial answer to our purpose is obtained. The result is given in Theorem 8.

4.1. H_3 *case*

We start by introducing the polynomial potential

$$F = \frac{t_1 t_3{}^2 + t_2{}^2 t_3}{2} + \frac{1}{6} t_1{}^2 t_2{}^3 + \frac{1}{20} t_1{}^5 t_2{}^2 + \frac{1}{3960} t_1{}^9$$

which is the same as the one obtained by Dubrovin [3] by a coordinate transformation. The matrices C and T are

$$C = \begin{pmatrix} t_3 & \frac{t_2 t_1{}^4}{2} + t_2{}^2 t_1 & \frac{t_1{}^9}{36} + t_2{}^2 t_1{}^3 + \frac{t_2{}^3}{3} \\ t_2 & \frac{t_1{}^5}{10} + t_2 t_1{}^2 + t_3 & \frac{t_2 t_1{}^4}{2} + t_2{}^2 t_1 \\ t_1 & t_2 & t_3 \end{pmatrix},$$

$$T = \begin{pmatrix} t_3 & \frac{7}{10} t_1 t_2 \left(t_1{}^3 + 2t_2 \right) & \frac{1}{20} \left(t_1{}^9 + 36 t_2{}^2 t_1{}^3 + 12 t_2{}^3 \right) \\ \frac{3t_2}{5} & \frac{1}{10} \left(t_1{}^5 + 10 t_2 t_1{}^2 + 10 t_3 \right) & \frac{7}{10} t_1 t_2 \left(t_1{}^3 + 2t_2 \right) \\ \frac{t_1}{5} & \frac{3t_2}{5} & t_3 \end{pmatrix}.$$

Letting

$$t_2 = -\frac{\left(s^2 - 4s - 1 \right) \left(s^2 + 4s - 1 \right) \left(5s^2 - 1 \right) t_1{}^3}{\left(3s^2 + 1 \right)^3},$$

74

J. Sekiguchi

we find that

$$\det(T) = (t_3 - z_1)(t_3 - z_2)(t_3 - z_3),$$

where

$$\begin{cases}
z_1 = \dfrac{\left(357s^{10}-6125s^8-20480s^7-5150s^6+4096s^5+1910s^4-215s^2+7\right)t_1{}^5}{10(3s^2+1)^5}, \\[2ex]
z_2 = \dfrac{\left(357s^{10}-6125s^8+20480s^7-5150s^6-4096s^5+1910s^4-215s^2+7\right)t_1{}^5}{10(3s^2+1)^5}, \\[2ex]
z_3 = -\dfrac{\left(507s^{10}-3955s^8-6690s^6+3530s^4-585s^2+25\right)t_1{}^5}{10(3s^2+1)^5}.
\end{cases} \qquad (96)$$

Introducing the matrix $B_\infty^{(3)}$ depending on r by

$$B_\infty^{(3)} = \begin{pmatrix} r+\frac{1}{5} & 0 & 0 \\ 0 & r+\frac{3}{5} & 0 \\ 0 & 0 & r+1 \end{pmatrix},$$

we consider the ordinary differential equation

$$\frac{d}{dt_3}\tilde{Y} = -T^{-1}B_\infty^{(3)}\tilde{Y}, \qquad (97)$$

where $\tilde{Y} = {}^t(\tilde{\mathbf{y}}_1, \tilde{\mathbf{y}}_2, \tilde{\mathbf{y}}_3)$. We treat the case $r = -1$ of the equation (97). Then we obtain the following differential equation

$$\frac{d}{dt_3}\hat{Y} = \left(\frac{N_1}{t_3 - z_1} + \frac{N_2}{t_3 - z_2} + \frac{N_3}{t_3 - z_3}\right)\hat{Y}, \qquad (98)$$

where $\hat{Y} = {}^t(\tilde{\mathbf{y}}_1, \tilde{\mathbf{y}}_2)$ and

$$\begin{cases}
N_1 = \begin{pmatrix}
-\dfrac{P_1(s)}{640s^3(s+1)^3(3s-1)} & -\dfrac{\left(s^2+4s-1\right)\left(7s^2+4s+1\right)P_1(s)t_1{}^2}{1280s^3(s+1)^3(3s-1)\left(3s^2+1\right)^2} \\[2ex]
\dfrac{\left(s^2+4s-1\right)\left(3s^2+1\right)^2\left(7s^2+4s+1\right)}{320s^3(s+1)^3(3s-1)t_1{}^2} & \dfrac{\left(s^2+4s-1\right)^2\left(7s^2+4s+1\right)^2}{640s^3(s+1)^3(3s-1)}
\end{pmatrix}, \\[4ex]
N_2 = \begin{pmatrix}
\dfrac{P_2(s)}{640(s-1)^3s^3(3s+1)} & \dfrac{\left(s^2-4s-1\right)\left(7s^2-4s+1\right)P_2(s)t_1{}^2}{1280(s-1)^3s^3(3s+1)\left(3s^2+1\right)^2} \\[2ex]
-\dfrac{\left(s^2-4s-1\right)\left(3s^2+1\right)^2\left(7s^2-4s+1\right)}{320(s-1)^3s^3(3s+1)t_1{}^2} & -\dfrac{\left(s^2-4s-1\right)^2\left(7s^2-4s+1\right)^2}{640(s-1)^3s^3(3s+1)}
\end{pmatrix}, \\[4ex]
N_3 = \begin{pmatrix}
\dfrac{119s^8-588s^6+314s^4-108s^2+7}{40(s-1)^3(s+1)^3(3s-1)(3s+1)} & -\dfrac{\left(s^2+3\right)\left(5s^2-1\right)P_3(s)t_1{}^2}{80(s-1)^3(s+1)^3(3s-1)(3s+1)\left(3s^2+1\right)^2} \\[2ex]
-\dfrac{\left(s^2+3\right)\left(3s^2+1\right)^2\left(5s^2-1\right)}{20(s-1)^3(s+1)^3(3s-1)(3s+1)t_1{}^2} & \dfrac{\left(s^2+3\right)^2\left(5s^2-1\right)^2}{40(s-1)^3(s+1)^3(3s-1)(3s+1)}
\end{pmatrix},
\end{cases}$$

$$(99)$$

where

$$P_1(s) = 49s^8 - 320s^7 - 884s^6 - 896s^5 + 86s^4 + 192, s^3 - 20s^2 + 1$$
$$P_2(s) = 49s^8 + 320s^7 - 884s^6 + 896s^5 + 86s^4 - 192, s^3 - 20s^2 + 1$$
$$P_3(s) = 119s^8 - 588s^6 + 314s^4 - 108s^2 + 7.$$

It is easy to show that

$$N_1 + N_2 + N_3 = \begin{pmatrix} \frac{4}{5} & 0 \\ 0 & \frac{2}{5} \end{pmatrix}.$$

Denoting $Y = (t_3 - u)^{-4/5}\hat{Y}$, we obtain the differential equation

$$\frac{d}{dt_3}Y = \left(\frac{N_1}{t_3 - z_1} + \frac{N_2}{t_3 - z_2} + \frac{N_3}{t_3 - z_3} + \frac{-\frac{4}{5}}{t_3 - u} \right) Y \qquad (100)$$

for Y, where u is a function of t_1, t_2 different from all of z_1, z_2, z_3.

We define the linear fractional transformation $L(t_3)$ of t_3 by

$$\xi = L(t_3) = \frac{(z_1 + z_2 - u)t_3 - z_1 z_2}{t_3 - u}.$$

It is easy to see that

$$L(z_1) = z_1, \quad L(z_2) = z_2, \quad L(u) = \infty.$$

We let

$$z_{3a} = L(z_3), \quad z_{4a} = L(\infty).$$

Then the differential equation (100) turns out to be

$$\frac{d}{d\xi}Y = \left(\frac{N_1}{\xi - z_1} + \frac{N_2}{\xi - z_2} + \frac{N_3}{\xi - z_{3a}} + \frac{N_4}{\xi - z_{4a}} \right) Y, \qquad (101)$$

where

$$N_4 = \begin{pmatrix} 0 & 0 \\ 0 & \frac{2}{5} \end{pmatrix}. \qquad (102)$$

It is easy to show that $0, \frac{2}{5}$ are the eigenvalues of each of the matrices N_1, N_2, N_3, N_4 and

$$N_1 + N_2 + N_3 + N_4 = \frac{4}{5}I_2.$$

4.2. $H_4(4)$ *case*

Let x_1, x_2, x_3, x_4 be the variables with weights $1/5, 1/5, 1, 1$, respectively, and let z_0 be an algebraic function of x_1, x_2, x_3 defined by the equation

$$843x_1{}^5 + 1160x_1{}^4x_2 + 530x_1{}^3x_2{}^2 + 120x_1{}^2x_2{}^3 + 15x_1x_2{}^4 - x_3$$
$$-15(x_1 + x_2)^2(3x_1 + x_2)^2z_0 + 20x_1(13x_1{}^2 + 12x_1x_2 + 3x_2{}^2)z_0{}^2$$
$$+10(x_1 + x_2)(3x_1 + x_2)z_0{}^3 + z_0{}^5 = 0.$$

$$(103)$$

We introduce the algebraic function F by

$$F$$
$$= \frac{1}{198}(-45669270x_1{}^{11} - 858257224x_1{}^{10}x_2 - 1955070535x_1{}^9x_2{}^2 - 2004227280x_1{}^8x_2{}^3$$
$$-1191552120x_1{}^7x_2{}^4 - 458678880x_1{}^6x_2{}^5 - 120561210x_1{}^5x_2{}^6 - 21740400x_1{}^4x_2{}^7$$
$$-2569050x_1{}^3x_2{}^8 - 178200x_1{}^2x_2{}^9 - 4455x_1x_2{}^{10} + 198x_2x_3x_4 + 99x_1x_4{}^2)$$
$$+5x_1(x_1 + x_2)^2(3x_1 + x_2)^2(9035x_1{}^5 + 32316x_1{}^4x_2 + 30270x_1{}^3x_2{}^2 + 11760x_1{}^2x_2{}^3$$
$$+2115x_1x_2{}^4 + 180x_2{}^5)z_0 - \frac{5}{6}(994315x_1{}^9 + 4563564x_1{}^8x_2 + 7356636x_1{}^7x_2{}^2$$
$$+6239424x_1{}^6x_2{}^3 + 3321450x_1{}^5x_2{}^4 + 1252440x_1{}^4x_2{}^5 + 356940x_1{}^3x_2{}^6$$
$$+73440x_1{}^2x_2{}^7 + 9315x_1x_2{}^8 + 540x_2{}^9)z_0{}^2 + \frac{5}{3}(x_1 + x_2)(3x_1 + x_2)(15815x_1{}^6$$
$$+37788x_1{}^5x_2 + 58125x_1{}^4x_2{}^2 + 42360x_1{}^3x_2{}^3 + 13185x_1{}^2x_2{}^4 + 1260x_1x_2{}^5$$
$$-45x_2{}^6)z_0{}^3 - 100(1195x_1{}^7 + 2450x_1{}^6x_2 + 1283x_1{}^5x_2{}^2 - 650x_1{}^4x_2{}^3 - 1015x_1{}^3x_2{}^4$$
$$-450x_1{}^2x_2{}^5 - 87x_1x_2{}^6 - 6x_2{}^7)z_0{}^4 + (-74420x_1{}^6 - 181952x_1{}^5x_2 - 174765x_1{}^4x_2{}^2$$
$$-80920x_1{}^3x_2{}^3 - 17130x_1{}^2x_2{}^4 - 840x_1x_2{}^5 + 135x_2{}^6)z_0{}^5$$
$$-\frac{35}{3}(x_1 + x_2)(3x_1 + x_2)(305x_1{}^3 + 336x_1{}^2x_2 + 123x_1x_2{}^2 + 12x_2{}^3)z_0{}^6$$
$$-10(175x_1{}^4 + 344x_1{}^3x_2 + 236x_1{}^2x_2{}^2 + 64x_1x_2{}^3 + 5x_2{}^4)z_0{}^7$$
$$-5(95x_1{}^3 + 124x_1{}^2x_2 + 57x_1x_2{}^2 + 8x_2{}^3)z_0{}^8 - \frac{125}{9}(x_1 + x_2)(3x_1 + x_2)z_0{}^9$$
$$+\frac{1}{2}(-5x_1 - 4x_2)z_0{}^{10} - \frac{25}{33}z_0{}^{11}.$$

It is easy to see that F is weighted homogeneous, namely if

$$E = \frac{1}{5}(x_2\partial_{x_1} + x_2\partial_{x_2}) + x_3\partial_{x_3} + x_4\partial_{x_4},$$

then $EF = \frac{11}{5}F$. It can be shown that F ia a solution to the WDVV equation, or equivalently, F is an algebraic potential.

Using the matrix T, we define $f_0 = \det(T)$. Then f_0 is regarded as a polynomial of x_1, x_2, x_4, z_0 by eliminating x_3 by the relation (103). In the sequel, we regard f_0 as a polynomial of x_4 and write $f_0(x_4)$ when we stress the variable x_4. Let $\mathcal{Q} = \mathbf{Q}(x_1, x_2, z_0)$ be the rational function field over $\mathbf{Q}$. Then f_0 is an element of $\mathcal{Q}$. Let $\mathcal{K}$

be the field extension of Q attached by all the roots u_1, u_2, u_3, u_4 of the algebraic equation $f_0(x_4) = 0$.

We let

$$B_\infty^{(4)} = \begin{pmatrix} r + \frac{1}{5} & 0 & 0 & 0 \\ 0 & r + \frac{1}{5} & 0 & 0 \\ 0 & 0 & r + 1 & 0 \\ 0 & 0 & 0 & r + 1 \end{pmatrix}$$

and consider the ordinary differential equation

$$\frac{d}{dx_4}\tilde{Z} = -T^{-1}B_\infty^{(4)}\tilde{Z}, \tag{104}$$

where $\tilde{Z} = {}^t(\tilde{\mathbf{z}}_1, \tilde{\mathbf{z}}_2, \tilde{\mathbf{z}}_3, \tilde{\mathbf{z}}_4)$. We treat the case $r = -1$ of the equation (104). We let $\tilde{R} = (-T^{-1}B_\infty^{(4)})_{r=-1}$. Let $\tilde{R}(i,j)$ be the (i,j)-entry of $\tilde{R}$ and define the 2×2 matrix R by

$$R = \begin{pmatrix} \tilde{R}(1,1) & \tilde{R}(1,2) \\ \tilde{R}(2,1) & \tilde{R}(2,2) \end{pmatrix}.$$

Since it follows from the definition of B_∞ that $\tilde{R}(i,j) = 0$ for $i = 1, 2, 3, 4, j = 3, 4$, we conclude that

$$\frac{d}{dx_4}Z = RZ \tag{105}$$

is well-defined, where $Z = {}^t(\tilde{\mathbf{z}}_1, \tilde{\mathbf{z}}_2)$. Denoting

$$R_{ja} = \lim_{x_4 \to u_j}(x_4 - u_j)R \quad (j = 1, 2, 3, 4),$$

we obtain the differential equation

$$\frac{d}{dx_4}Z = \left(\frac{R_{1a}}{x_4 - u_1} + \frac{R_{2a}}{x_4 - u_2} + \frac{R_{3a}}{x_4 - u_3} + \frac{R_{4a}}{x_4 - u_4}\right)Z, \tag{106}$$

which is the same as (105). It is possible to show the following properties of $R_{1a}, R_{2a}, R_{3a}, R_{4a}$:

(P0) $R = \sum_{j=1}^{4} \frac{R_{ja}}{x_4 - u_j}$.

(P1) $0, \frac{2}{5}$ are the eigenvalues of each of $R_{1a}, R_{2a}, R_{3a}, R_{4a}$.

(P2) $R_{1a} + R_{2a} + R_{3a} + R_{4a} = \frac{4}{5}I_2$.

It is hard to write down the concrete forms of $R_{1a}, R_{2a}, R_{3a}, R_{4a}$ because of their length.

We now write

$$R = \frac{1}{f_0} \begin{pmatrix} a_{11} & a_{12} \\ a_{21} & a_{22} \end{pmatrix}.$$

Then, by direct computation, we find that $\det(R)$ is divisible by f_0.

We left $a_{ij}^{(k)} = a_{ij}|_{x_4=u_k}$. Then

$$R_{ka} = \frac{1}{f_0'(u_k)} \begin{pmatrix} a_{11}^{(k)} & a_{12}^{(k)} \\ a_{21}^{(k)} & a_{22}^{(k)} \end{pmatrix}.$$

It is easy to show that

$$f_0'(u_1) = (u_1 - u_2)(u_1 - u_3)(u_1 - u_4).$$

We now introduce

$$\delta_k = f_0'(u_k),$$
$$\varphi_k = a_{11}^{(k)},$$
$$\psi_k = a_{21}^{(k)}.$$

Then it follows from direct computation that

$$a_{12}^{(k)} = \varphi_k \left(-\varphi_k + \frac{2}{5}\delta_k \right) /\psi_k,$$
$$a_{22}^{(k)} = -\varphi_k + \frac{2}{5}\delta_k$$

and that

$$R_{ka} = M_k^{-1} \begin{pmatrix} 0 & 0 \\ 0 & \frac{2}{5} \end{pmatrix} M_k = \frac{1}{\delta_k} \begin{pmatrix} \varphi_k & \varphi_k(\frac{2}{5}\delta_k - \varphi_k)/\psi_k \\ \psi_k & \frac{2}{5}\delta_k - \varphi_k \end{pmatrix},$$

where

$$M_k = \begin{pmatrix} -\psi_k & \varphi_k \\ \psi_k & -\varphi_k + \frac{2}{5}\delta_k \end{pmatrix}.$$

Letting $R_{kb} = M_4 R_{ka} M_4^{-1}$ ($k = 1, 2, 3, 4$), and $\hat{Z} = M_4 Z$, we find that the equation (106) for Z turns out to be the equation for $\hat{Z}$

given by

$$\frac{d}{dx_4}\hat{Z} = \left(\frac{R_{1b}}{x_4 - u_1} + \frac{R_{2b}}{x_4 - u_2} + \frac{R_{3b}}{x_4 - u_3} + \frac{R_{4b}}{x_4 - u_4}\right)\hat{Z}. \qquad (107)$$

It is underlined here that $R_{4b} = \begin{pmatrix} 0 & 0 \\ 0 & \frac{2}{5} \end{pmatrix}$. Denoting

$$R_{kb} = \frac{5}{2\delta_k \delta_4 \psi_k \psi_4} \begin{pmatrix} r_{11}^{(k)} & r_{12}^{(k)} \\ r_{21}^{(k)} & r_{22}^{(k)} \end{pmatrix},$$

we have

$$r_{11}^{(k)} = (-\varphi_4\psi_k + \varphi_k\psi_4)\left(-\varphi_4\psi_k + \varphi_k\psi_4 + \frac{2}{5}(\delta_4\psi_k - \delta_k\psi_4)\right),$$

$$r_{12}^{(k)} = (-\varphi_4\psi_k + \varphi_k\psi_4)\left(-\varphi_4\psi_k + \varphi_k\psi_4 - \frac{2}{5}\delta_k\psi_4\right),$$

$$r_{21}^{(k)} = -\left(-\varphi_4\psi_k + \varphi_k\psi_4 + \frac{2}{5}\delta_4\psi_k\right)\left(-\varphi_4\psi_k + \varphi_k\psi_4 + \frac{2}{5}(\delta_4\psi_k - \delta_k\psi_4)\right),$$

$$r_{22}^{(k)} = -\left(-\varphi_4\psi_k + \varphi_k\psi_4 + \frac{2}{5}\delta_4\psi_k\right)\left(-\varphi_4\psi_k + \varphi_k\psi_4 - \frac{2}{5}\delta_k\psi_4\right).$$

It is hard to continue the argument, because the computation is beyond computer capacity. For this reason, we restrict our attention to the case

$$x_2 = -2x_1 \qquad (108)$$

in the rest of this section. For simplicity, we let

$$A = -613188{x_1}^{10} - 70110{x_1}^9 z_0 + 110625{x_1}^8 {z_0}^2 - 75000{x_1}^7 {z_0}^3$$
$$+ 12480{x_1}^6 {z_0}^4 + 7980{x_1}^5 {z_0}^5 - 2550{x_1}^4 {z_0}^6 - 120{x_1}^3 {z_0}^7$$
$$+ 60{x_1}^2 {z_0}^8 - 30 x_1 {z_0}^9 - 3{z_0}^{10},$$

$$B = \left(11{x_1}^2 - x_1 z_0 - {z_0}^2\right)^2 \left(9{x_1}^2 - 4x_1 z_0 + {z_0}^2\right)^3$$
$$\left(13{x_1}^2 + 4x_1 z_0 + {z_0}^2\right)^5.$$

Then it is easy to show that

$$f_0(= \det(T)) = y_4^4 - 2A y_4^2 + A^2 - 12B,$$

where

$$y_4 = 2\left(11{x_1}^2 - x_1 z_0 - {z_0}^2\right)\left(7{x_1}^3 + 2{x_1}^2 z_0 - x_1 {z_0}^2 + {z_0}^3\right) + x_4.$$
$$\qquad (109)$$

We regard f_0 as a polynomial of y_4. Then we may take the solutions u_1, u_2, u_3, u_4 of $f_0 = 0$ by

$$u_1 = \sqrt{A + 2\sqrt{3}B},$$

$$u_2 = -\sqrt{A + 2\sqrt{3}B},$$

$$u_3 = \sqrt{A - 2\sqrt{3}B},$$

$$u_4 = -\sqrt{A - 2\sqrt{3}B}.$$

Since

$$A^2 - 12B = -3(x_1 - z_0)^5(7x_1 - z_0)^5(3x_1 + z_0)^3(5x_1 + z_0)^3$$
$$\times \left(10x_1{}^2 - 2x_1 z_0 + z_0{}^2\right)^2,$$

it follows that

$$u_1 u_3 = \sqrt{\frac{-3(x_1 - z_0)^5(7x_1 - z_0)^5(3x_1 + z_0)^3(5x_1 + z_0)^3}{\left(10x_1{}^2 - 2x_1 z_0 + z_0{}^2\right)^2}}.$$

We will now compare (107) with (101) under the condition (108). We let

$$\eta = \frac{z_0}{x_1}.$$

In spite of the fact that the concrete form of the matrix R_{3b} is not given because of its length, letting $P = \begin{pmatrix} 1 & 0 \\ 0 & g_0 \end{pmatrix}$, where

$$g_0 = \frac{15\left(\eta^2 - 4\eta + 9\right)^2\left(\eta^2 + \eta - 11\right)\left(\eta^2 + 4\eta + 13\right)^4}{(\eta + 2)(\eta + 8)\left(\eta^2 - 2\eta + 10\right)\begin{pmatrix} 5\eta^{10} + 48\eta^9 - 42\eta^8 + 168\eta^7 + 2856\eta^6 \\ -7314\eta^5 - 12906\eta^4 + 59484\eta^3 - 75021\eta^2 \\ -11162\eta + v_1{}^2 + 201348 \end{pmatrix}}$$

and $R_{3c} = PR_{3b}P^{-1}$, we find by direct computation that

$$R_{3c} = \begin{pmatrix} \dfrac{2P_3}{15(\eta^2 - 4\eta + 9)(\eta^2 + 4\eta + 13)^3} & \dfrac{4(\eta+2)^2(\eta+8)^2\left(\eta^2 - 2\eta + 10\right)^2 P_3}{225(\eta^2 - 4\eta + 9)^2(\eta^2 + 4\eta + 13)^6} \\ 1 & \dfrac{2(\eta+2)^2(\eta+8)^2\left(\eta^2 - 2\eta + 10\right)^2}{15(\eta^2 - 4\eta + 9)(\eta^2 + 4\eta + 13)^3} \end{pmatrix},$$

$$(110)$$

where

$$P_3 = 2\eta^8 + 8\eta^7 + 68\eta^6 + 176\eta^5 - 214\eta^4 - 712\eta^3 + 6248\eta^2 + 6632\eta + 33719.$$

We recall the definition of matrices N_1, N_2, N_3, N_4 of the case of the reflection group of type H_3 (cf. ((99)), (102)). Denoting $Q = \begin{pmatrix} 1 & 0 \\ 0 & h_0 \end{pmatrix}$, where $h_0 = -\dfrac{20(s-1)^3(s+1)^3(3s-1)(3s+1)y_1{}^2}{(s^2+3)(3s^2+1)^2(5s^2-1)}$ and $N_3{}' = QN_3Q^{-1}$, we find that

$$N_3{}' = \begin{pmatrix} \frac{119s^8-588s^6+314s^4-108s^2+7}{40(s-1)^3(s+1)^3(3s-1)(3s+1)} & \frac{\left(s^2+3\right)^2\left(5s^2-1\right)^2\left(119s^8-588s^6+314s^4-108s^2+7\right)}{1600(s-1)^6(s+1)^6(3s-1)^2(3s+1)^2} \\ 1 & \frac{\left(s^2+3\right)^2\left(5s^2-1\right)^2}{40(s-1)^3(s+1)^3(3s-1)(3s+1)} \end{pmatrix}.$$

Then if

$$s^2 = -\frac{(-7+\eta)(-1+\eta)}{3(3+\eta)(5+\eta)}, \tag{111}$$

we find that $R_{3c} = N_3{}'$. Under the condition (111), it follows that

$$A^2 - 12B = 9(\eta - 7)^4(\eta - 1)^4(\eta + 3)^4(\eta + 5)^4 \left(\eta^2 - 2\eta + 10\right)^2 s^2 x_1{}^{20}. \tag{112}$$

Noting that $(u_1 u_3)^2 = A^2 - 12B$, we take

$$u_1 u_3 = 3(\eta - 7)^2(\eta - 1)^2(\eta + 3)^2(\eta + 5)^2 \left(\eta^2 - 2\eta + 10\right) s x_1{}^{10}. \tag{113}$$

Then, after a little lengthy computation with the help of Mathematica, we conclude the following:

Theorem 7. *We let $\eta = \frac{z_0}{x_1}$. Then if $s^2 = -\dfrac{(-7+\eta)(-1+\eta)}{3(3+\eta)(5+\eta)}$ holds, it follows that $R_{3c} = N_3{}'$. Moreover, if*

$$u_1 u_3 = 3(\eta - 7)^2(\eta - 1)^2(\eta + 3)^2(\eta + 5)^2 \left(\eta^2 - 2\eta + 10\right) s x_1{}^{10},$$

it follows that

$$PR_{jb}P^{-1} = QN_jQ^{-1} \quad (j = 1, 2).$$

We omit the proof.

As in the previous cases, we define the linear transformation

$$S(y_4) = \frac{(z_1 - z_2)y_4 + u_1 z_2 - u_2 z_1}{u_1 - u_2}.$$

Then it is clear that $S(u_1) = z_1$, $S(u_2) = z_2$.

Theorem 8. *If u is defined by the equation*

$$u = \frac{t_1{}^5 \left(\begin{array}{c} \left(s^2 - 4s - 1\right)\left(s^2 + 4s - 1\right)\left(357s^6 + 301s^4 - 89s^2 + 7\right)u_1 \\ -4096s^5\left(5s^2 - 1\right)u_3 \end{array} \right)}{10\left(3s^2 + 1\right)^5 u_1},$$

then

$$S(u_3) = L(z_3), \quad S(u_4) = L(\infty).$$

5.   Miscellaneous Remarks

Remark 10. In the Introduction, the author formulated the definition of algebraic potentials of types $D_4(1)$, $F_4(1)$, $H_4(4)$. It is not clear to him whether any algebraic potential of the type $D_4(1)$, $F_4(1)$, $H_4(4)$ is transformed to that in this definition or not. This formulation is available in treating other cases of algebraic potentials, but there are exceptional cases for which the definitions are not available (cf. [14]). For this reason, it is worthwhile to formulate the correct definition of the "algebraic potential" of a Frobenius manifold.

Remark 11. Applying the procedure explained in [5] to the equation (16), we obtain an algebraic solution to the Painlevé VI equation which is given by

$$t = \frac{\nu^3(\nu + 2)}{2\nu + 1}, \quad y = \frac{\nu^2(\nu^2 + 7\nu + 1)}{5(2\nu + 1)(\nu^2 + \nu + 1)},$$

where t is the variable and y is the solution to the Painlevé VI equation.

Similarly, in the case (58), we obtain an algebraic solution

$$t = \frac{(s-1)^2(s+2)}{(s-2)(s+1)^2}, \quad y = \frac{(s-1)s\left(3s^2-4\right)}{(s-2)(s+1)\left(3s^2+4\right)}.$$

Also in the case (98), we obtain an algebraic solution

$$t = \frac{(s-1)^5(3s+1)^3\left(s^2+4s-1\right)}{(s+1)^5(3s-1)^3\left(s^2-4s-1\right)},$$

$$y = \frac{(s-1)^2(3s+1)^2\left(s^2+3\right)\left(119s^8-588s^6+314s^4-108s^2+7\right)}{7(s+1)^3(3s-1)\left(s^2-4s-1\right)\left(3s^2+1\right)\left(17s^6+209s^4-37s^2+3\right)}.$$

Remark 12. It is known that there are three kinds of potentials related with the reflection group of type H_3. In this chapter, we treated one of them. There are two other potentials constructed by Dubrovin and Mazzocco [4] denoted by $(H_3)'$, $(H_3)''$. They are algebraic and each of them is not reduced to a polynomial. By the result of Romano [12], there exist two algebraic potentials having tri-Hamiltonian structure in dimensions four corresponding to $(H_3)'$, $(H_3)''$.

The author was informed by Douvropoulos of the data on the primitive conjugacy classes of real reflection groups of exceptional type including the group of type H_4. (See [14] for the details of his data.) His data suggest the existence of ten kinds of potentials corresponding to primitive conjugacy classes of the reflection group of type H_4. They are denoted by $H_4(j)$ $(0 \leq j \leq 10, j \neq 5)$. The class $H_4(0)$ corresponds to the Coxeter transformation. Among these primitive conjugacy classes, the author constructed seven algebraic potentials corresponding to $H_4(j)$ $(1 \leq j \leq 9, j \neq 5,8)$. Comparing his data with [12], the author recognized that there are three classes denoted by $H_4(4)$, $H_4(8)$, $H_4(10)$ which satisfy the tri-Hamiltonian condition of Romano and conjectured that the correspondences

$$H_3 \quad \Longleftrightarrow \quad H_4(4)$$
$$(H_3)' \quad \Longleftrightarrow \quad H_4(10)$$
$$(H_3)'' \quad \Longleftrightarrow \quad H_4(8)$$

are interpreted by three-dimensional Frobenius manifolds/PVμ to four-dimensional tri-Hamiltonian Frobenius manifolds. It is still an open problem to construct algebraic potentials of types $H_4(8)$, $H_4(10)$.

Remark 13. It is already established by Kawakami and Mano that there is a correspondence between generic solutions to the Painlevé VI equation and generic solutions to the extended WDVV equations with additional conditions including the condition similar to the tri-Hamiltonian structure. (Refer §5 of [10] for the precise statement and related results.) One of the interesting problems is to construct algebraic potential vector fields having "tri-Hamiltonian-like structures. (Refer to [6–9] for the definition of the generalized WDVV equation, potential vector fields.)

Acknowledgment

The author thanks Y. Dinar, T. Suzuki and T. Mano for useful comments. He also thanks the referee for his valuable suggestion and among others he includes Section 3.4 from the first draft to answer the question by the referee.

This work was partially supported by JSPS KAKENHI Grant Number 17K05269.

References

1. Y. Dinar, On classification and construction of algebraic Frobenius manifolds. *J. Geometry Phys.*, **58**(2008), 1171–1185.
2. Y. Dinar, Frobenius manifolds from subregular classical W-algebras. *Int. Math. Res. Not.*, **17**(2013), 2822–2861.
3. B. Dubrovin, Geometry of 2D topological field theories. In: *Integrable Systems and Quantum Groups (Montecatini, Terme 1993)*, M. Francoviglia, S. Greco (eds.), Lecture Notes in Math. 1620. Springer-Verlag, Berlin Heidelberg, 1996, pp. 120–348.
4. B. Dubrovin and M. Mazzocco, Monodromy of certain Painlevé VI transcendents and reflection groups. *Invent. Math.*, **141**(2000), 55–147.
5. M. Jimbo and T. Miwa, Monodromy preserving deformation of linear ordinary differential equations with rational coefficients. II. *Physica D*, **2**(1981), 407–448.
6. M. Kato, T. Mano and J. Sekiguchi, Flat structures without potentials. *Rev. Roumaine Math. Pures Appl.*, **60**(2015), 481–505.

7. M. Kato, T. Mano and J. Sekiguchi, Flat structure and potential vector fields related with algebraic solutions to Painlevé VI equation. *Opuscula Math.*, **38**(2018), 201–252.

8. M. Kato, T. Mano and J. Sekiguchi, Solutions to the extended WDVV equations and Painlevé VI equation. In: *Complex Differential and Difference Equations*, De Gruyter Proceedings in Math. 2019, pp. 343–364. https://www.degruyter.com/document/doi/10.1515/978 3110611427/html.

9. M. Kato, T. Mano and J. Sekiguchi, Flat structure on the space of isomonodromic deformations. *SIGMA*, **16**(2020), 110, 36 pages, arXiv:1511.01608.

10. H. Kawakami and T. Mano, Regular flat structure and generalized Okubo system. *Commun. Math. Phys.*, **369**(2019), 403–431.

11. O. Pavlyk, Solutions to WDVV from generalized Drinfeld-Sokolov hierarchies, arXiv:math-ph/0003020v1.

12. S. Romano, 4 dimensional Frobenius manifolds and Painleve' VI. *Math. Ann.*, **360**(2014), 715–751.

13. C. Sabbah, *Isomonodromic Deformations and Frobenius Manifolds. An Introduction.* Universitext. Springer-Verlag London, Ltd., London, 2007.

14. J. Sekiguchi, The construction problem of algebraic potentials and reflection groups. Preprint.

© 2022 World Scientific Publishing Europe Ltd.
https://doi.org/10.1142/9781800611368_0003

Chapter 3

Invariant Coordinate Subspaces of Normal Form of Hamiltonian System

Alexander B. Batkhin

Keldysh Institute of Applied Mathematics of RAS
Miusskaya sq. 4, Moscow 125047, Russia
batkhin@gmail.com

We consider an autonomous system of ordinary differential equations (ODEs) with non-degenerate linear part near its stationary point in two cases: in general case and in Hamiltonian case. For these two cases, the problem of existence of an invariant coordinate subspace in the coordinates of normal form is considered. The theorems of existence of invariant coordinate subspaces with explicit conditions are proven. Some examples with different cases of resonances between eigenvalues of the linear part of the system of ODE are considered. A technique for determination of resonance relations with the help of q-subdiscriminants is presented. An example of determination of resonance relations is given for a certain model system with six degrees of freedom.

1. Introduction

An approach of Poincaré for investigation of systems of nonlinear ordinary differential equations (ODEs) was based on the maximal simplification of the right-hand sides of these equations by invertible transformations. This approach led to the theory of normal forms (NFs) of the general system, developed in works of H. Dulac and A. D. Bruno, and in particular of the Hamiltonian ones, developed in

87

works of G. D. Birkhoff, T. M. Cherry, F. G. Gustavson, C. L. Siegel, J. Moser, A. D. Bruno and others (see [1], Chs. I, II for more details).

Even though the NF of ODE system in the vicinity of an invariant manifold (stationary point, periodic solution, k-dimensional invariant torus) is a formal object, it can be effectively used to study the stability of the corresponding invariant manifold (see [2,3]), Part II local integration of the system (see [4]), searching for periodic solutions, for first integrals (see [5,6]), and asymptotic integration of the ODE system (see [7]).

The goal of the presented work is to investigate invariant coordinate subspaces of NF of a real Hamiltonian system with non-degenerated linear part. The existence of invariant subspace can reduce the phase flow in the whole all space into the subspace of less dimension and in some cases can give information about periodic solutions of the whole system.

The work consists of an introduction, four sections and references. Section 2 defines the normal form of the general ODE system and formulates the condition of an invariant coordinate subspace. Section 3 discusses the normal form of the Hamiltonian system and formulates the condition for the existence of an invariant coordinate subspace in this case. Section 4 contains four model examples demonstrating typical situations with eigenvalues of the linear part of the system. Finally, Section 5 contains a methodology for applying q-analogs of such classical objects as the polynomial subdiscriminants and Jackson derivative to analyze the resonance relations between eigenvalues of the linear part of the ODE system in a situation where these eigenvalues form pairs of rationally commensurable values. The described algorithms can be implemented in various computer algebra systems, such as `Wolfram Mathematica`, `Maplesoft Maple`, `SymPy`. A model example of the determination of resonance relations for an oscillating system with six degrees of freedom is given.

Remark on Notations

- Boldface symbols like $\mathbf{x}, \mathbf{y}, \mathbf{u}, \mathbf{v}$ denote column-vectors in n-dimensional real $\mathbb{R}^n$ or complex $\mathbb{C}^n$ spaces.
- Boldface symbols like $\mathbf{p}, \mathbf{q}$ denote vectors in n-dimensional integer lattice $\mathbb{Z}^n$.
- $|\mathbf{p}| = \sum_{j=1}^{n} |p_j|$ denotes vector norm.

- For $\mathbf{x} = (x_1, \ldots, x_n)^{\mathrm{T}}$ and $\mathbf{p} = (p_1, \ldots, p_n)^{\mathrm{T}}$, we denote by $\mathbf{x}^{\mathbf{P}} \equiv \prod_{j=1}^{n} x_j^{p_j}$ and by $\langle \mathbf{p}, \mathbf{x} \rangle \equiv \sum_{j=1}^{n} p_j x_j$.

2. Invariant Subspaces of a Normal Form for General System

Consider an analytic system of ODE

$$\dot{\mathbf{x}} = \mathbf{f}(\mathbf{x}) \tag{1}$$

near its stationary point

$$\mathbf{x} = 0,$$

coinciding with the coordinates origin.

Let the linear part

$$\dot{\mathbf{x}} = A\mathbf{x}, \quad A = \left. \frac{\partial \mathbf{f}}{\partial \mathbf{x}} \right|_{\mathbf{x}=0} \tag{2}$$

of the system (1) be non-degenerated. Then the matrix A has n eigenvalues at least, one of which is non-zero

$$\boldsymbol{\lambda} = (\lambda_1, \ldots, \lambda_n).$$

According to [8] Theorem I there exists a formal invertible transformation $\mathbf{g} : \mathbf{x} \to \mathbf{y}$

$$\mathbf{x} = \mathbf{g}(\mathbf{y}),$$

represented in the form of power series, which reduces the initial system (1) into its *normal form*:

Definition 1 ([8]). *Normal form of the initial system* (1) *is a system of the form*

$$\dot{y}_j = y_j h_j(\mathbf{y}), \quad j = 1, \ldots, n, \tag{3}$$

right-hand sides $y_j h_j(\mathbf{y})$ *of which are power series*

$$y_j h_j(\mathbf{y}) = y_j \sum_{\mathbf{q}} h_{j\mathbf{q}} \mathbf{y}^{\mathbf{q}}, \quad h_{j0} = \lambda_j, \ j = 1, \ldots, n, \tag{4}$$

containing only resonant terms with

$$\langle \mathbf{q}, \boldsymbol{\lambda} \rangle = 0. \tag{5}$$

Here $h_{j\mathbf{q}}$ *are constant coefficients and in* $y_j h_j(\mathbf{y})$ *integral power exponents of coordinates* $q_j \geqslant -1$, *but others* $q_k \geqslant 0$.

Let $I = \{i_1, \ldots, i_k\}$ be a set of increasing indices $1 \leqslant i_1,\, i_k \leqslant n$, $k \leqslant n$. By K_I we denote the *coordinate subspace*

$$K_I = \{\mathbf{y} : y_j = 0 \text{ for all } j \notin I\}.$$

All non-zero coordinates y_j, $j \in I$, of the subspace K_I we call *internal coordinates* and denote them shortly by $\mathbf{y}_I$, others we call *external coordinates*. The eigenvalues λ_j, $j \in I$, corresponding to the internal coordinates $\mathbf{y}_I$ we call *internal eigenvalues* and denote them by $\boldsymbol{\lambda}_I$. Others λ_j, $j \notin I$, are called *external eigenvalues*.

Problem 1. *Which subspaces K_I are invariant in the normal form* (3), (4), (5)?

The solution of the problem 1 gives the following theorem.

Theorem 1. *The coordinate subspace K_I of dimension k is invariant in the normal form* (3)–(5) *if each external eigenvalue $\lambda_j \notin \boldsymbol{\lambda}_I$ satisfies the following condition*

$$\lambda_j \neq \langle \mathbf{p}, \boldsymbol{\lambda}_I \rangle, \tag{6}$$

for all nonnegative integer vectors $\mathbf{p} \geqslant 0$, $\mathbf{p} \in \mathbb{Z}^k$.

Proof. Under condition (6) it follows that each series $h_j(\mathbf{y})$ for $j \notin I$ does not contain any term $h_{j\mathbf{q}}\mathbf{y}^{\mathbf{q}}$ with indices $q_j = -1$, $q_i \geqslant 0$, $i \neq j \notin I$.

Since the external variables y_j, $j \notin I$, are equal to zero in the subspace K_I, it follows that for $\mathbf{y} \in K_I$

$$y_j h_j(\mathbf{y}) = 0, \text{ for all } j \notin I.$$

So the subspace K_I is invariant in the normal form (3)–(5). $\square$

3. Invariant Subspaces of a Normal Form for Hamiltonian System

We consider an analytic Hamiltonian system

$$\dot{\mathbf{x}} = \frac{\partial H}{\partial \mathbf{y}}, \quad \dot{\mathbf{y}} = -\frac{\partial H}{\partial \mathbf{x}} \tag{7}$$

with n degrees of freedom near its stationary point

$$\mathbf{x} = \mathbf{y} = 0.$$

The Hamiltonian function $H(\mathbf{x}, \mathbf{y})$ is expanded into convergent power series

$$H(\mathbf{x}, \mathbf{y}) = \sum H_{\mathbf{pq}} \mathbf{x}^{\mathbf{p}} \mathbf{y}^{\mathbf{q}} \tag{8}$$

with constant coefficients $H_{\mathbf{pq}}$, $\mathbf{p}, \mathbf{q} \geqslant 0$, $|\mathbf{p}| + |\mathbf{q}| \geqslant 2$.

Canonical transformations of coordinates $\mathbf{x}, \mathbf{y}$

$$\mathbf{x} = \mathbf{f}(\mathbf{u}, \mathbf{v}), \quad \mathbf{y} = \mathbf{g}(\mathbf{u}, \mathbf{v}), \tag{9}$$

preserve the Hamiltonian character of the initial system (7).

Let us denote by $\mathbf{z} = (\mathbf{x}, \mathbf{y}) \in \mathbb{R}^{2n}(\mathbb{C}^{2n})$ the phase vector. Then the linear part of the system (7) can be written in the form

$$\dot{\mathbf{z}} = B\mathbf{z}, \quad B = \frac{1}{2} \left. \begin{pmatrix} \dfrac{\partial^2 H}{\partial \mathbf{y} \partial \mathbf{x}} & \dfrac{\partial^2 H}{\partial \mathbf{y} \partial \mathbf{y}} \\ -\dfrac{\partial^2 H}{\partial \mathbf{x} \partial \mathbf{x}} & -\dfrac{\partial^2 H}{\partial \mathbf{x} \partial \mathbf{y}} \end{pmatrix} \right|_{\mathbf{x} = \mathbf{y} = 0}. \tag{10}$$

Let $\lambda_1, \ldots, \lambda_{2n}$ be eigenvalues of the matrix B, which can be reordered in such a way that $\lambda_{j+n} = -\lambda_j$, $j = 1, \ldots, n$. Denote by $\boldsymbol{\lambda} = (\lambda_1, \ldots, \lambda_n)$.

According to [9], Theorem 12 in §12 there exists a canonical formal transformation (9), where all $\mathbf{f}$ and $\mathbf{g}$ are power series, which reduces the initial system (7) into its *normal form*

$$\dot{\mathbf{u}} = \partial h / \partial \mathbf{v}, \qquad \dot{\mathbf{v}} = -\partial h / \partial \mathbf{u} \tag{11}$$

defined by the normalized Hamiltonian $h(\mathbf{u}, \mathbf{v})$

$$h(\mathbf{u}, \mathbf{v}) = \sum_{j=1}^{n} \lambda_j u_j v_j + \sum h_{\mathbf{pq}} \mathbf{u}^{\mathbf{p}} \mathbf{v}^{\mathbf{q}} \tag{12}$$

containing only resonant terms $h_{\mathbf{pq}} \mathbf{u}^{\mathbf{p}} \mathbf{v}^{\mathbf{q}}$ with

$$\langle \mathbf{p} - \mathbf{q}, \boldsymbol{\lambda} \rangle = 0. \tag{13}$$

Here $0 \leqslant \mathbf{p}, \mathbf{q} \in \mathbb{Z}^n$, $|\mathbf{p}| + |\mathbf{q}| \geqslant 2$ and $h_{\mathbf{pq}}$ are constant coefficients.

Resonance equation (13) has two types of solutions, which correspond to two types of resonant terms in NF (12):

(1) *secular terms* of the form $h_{\mathbf{pp}}\mathbf{u}^{\mathbf{p}}\mathbf{v}^{\mathbf{p}}$, which are always present in the NF due to the special structure of the matrix B of the system (9);
(2) *pure resonant terms* corresponding to nontrivial integral solutions to equation

$$\langle \mathbf{p}, \boldsymbol{\lambda} \rangle = 0. \tag{14}$$

According to [1], Ch. I, §3 let's denote by $\mathfrak{k}$ the *multiplicity of resonance* as the number of linearly independent solutions $\mathbf{p} \in \mathbb{Z}^n$ to equation (14) and by $\mathfrak{q}$ the *order of resonance*, where $\mathfrak{q} = \min |\mathbf{p}|$ over $\mathbf{p} \in \mathbb{Z}^n$, $\mathbf{p} \neq 0, \langle \mathbf{p}, \boldsymbol{\lambda} \rangle = 0$.

The main difference between the NF of the Hamilton system (7) and the NF of the common system of ODE (1) is that the first one always contains secular terms even in the absence of nontrivial solutions to equation (14). In this case, we have the so-called *Birkhoff normal form* [10].

Let $I = \{i_1, \ldots, i_k\}$ be a set of increasing indices $1 \leqslant i_1$, $i_k \leqslant n$, $k \leqslant n$. By L_I we denote the *coordinate subspace*

$$L_I = \{\mathbf{u}, \mathbf{v} : u_j = v_j = 0 \text{ for all } j \notin I\}.$$

All non-zero coordinates $w_j = (u_j, v_j)$, $j \in I$, of a subspace L_I we call *internal coordinates* and denote them shortly by $\mathbf{w}_I$, others we call *external coordinates*.

The eigenvalues λ_j, $j \in I$, corresponding to the internal coordinates $\mathbf{w}_I$ we call *internal eigenvalues* and denote by $\boldsymbol{\lambda}_I$. Others λ_j, $j \notin I$, are called *external eigenvalues*.

Problem 2. *Which subspaces L_I are invariant in the normal form* (11), (12), (13)?

Theorem 2. *The coordinate subspace L_I of dimension $2k$ is invariant in the normal form* (11)–(13) *if each external eigenvalue $\lambda_j \notin \boldsymbol{\lambda}_I$ satisfies the following condition:*

$$\lambda_j \neq \langle \mathbf{p}, \boldsymbol{\lambda}_I \rangle, \tag{15}$$

for any integer vector $\mathbf{p} \neq 0$, $\mathbf{p} \in \mathbb{Z}^k$.

Proof. It is an evident consequence of Theorem 1. $\qquad\square$

Remark 1. The principal difference between condition (6) in Theorem 1 and condition (15) in Theorem 2 is that any non-zero vector $\mathbf{p}$ is taken from the lattice $\mathbb{Z}^k$ in the Hamiltonian case but it is taken for $0 \leqslant \mathbf{p} \in \mathbb{Z}^k$ in the general case.

4. Examples

Example 1. Let the eigenvalues of the linear part of the common ODE system be $\lambda_1 = 1$, $\lambda_2 = \sqrt{2}$, $\lambda_3 = 1 + \sqrt{2}$.

There are three one-dimensional invariant subspaces K_j, $j = 1, 2, 3$, corresponding to each of the eigenvalues λ_j, because none of the ratios λ_i/λ_j with $i \neq j$ is a natural number.

Among three two-dimensional subspaces, the subspaces K_{13} and K_{23} are invariant, but K_{12} is not invariant, since there is a ratio of $\lambda_3 = \lambda_1 + \lambda_2$ between eigenvalues, while for no vectors $\mathbf{p} \geqslant 0$, $\mathbf{p} \in \mathbb{Z}^2$ do not perform the ratio $\lambda_1 = p_1\lambda_2 + p_2\lambda_3$ and $\lambda_2 = p_1\lambda_1 + p_2\lambda_3$.

Example 2. Let all $\lambda_j/\lambda_1 \notin \mathbb{Z}$, $j = 2, \ldots, n$. Then the Hamiltonian normal form has two-dimensional invariant subspace $L_1 = \{u_j = v_j = 0\, j \notin I_1\}$, where $I_1 = \{1\}$. On the subspace L_1, normal form (12) induces a Hamiltonian normal form with one degree of freedom and the normalizing transformation converges.

If $\lambda_1 \neq 0$ and is purely imagined, then for real Hamiltonian system (7) the real subspace L_1 is a family of periodic solutions. This fact was found by Lyapunov [11] and was described with the help of Hamiltonian formalism by Siegel [12], §§16, 17.

Example 3. Let in Hamiltonian case there exist the only pair of eigenvalues λ_1, λ_2 with property

$$\lambda_2/\lambda_1 = r/s,$$

where $r, s \in \mathbb{N}$, $s \neq 0$, $\mathrm{GCD}(r, s) = 1$, i.e. the multiplicity $\mathfrak{k}$ of the resonance is equal to 1 and the order $\mathfrak{q}$ of resonance is equal to $|r| + |s|$. Then equation (14) has a one-parameter family of solutions $\mathbf{p} = (lr, -ls, 0, \ldots)$, $l \in \mathbb{Z}\backslash\{0\}$, and the normal form, besides the secular terms, contains purely resonant terms. These lowest order terms with $l = 1$ have the form $h_{12}u_1^r v_2^s$ and $h_{21}v_1^r u_2^s$.

Let, for instance, the number $s = 1$. Then the normal form (12) contains the resonant terms $u_1^r v_2$ and $v_1^r u_2$. It means that the subspace L_1 can be not invariant, because the right-hand sides of equations for variables u_2, v_2 have terms depending on variables u_1, v_1 and these right-hand sides cannot be always equal to zero for the case $u_2 = v_2 = 0$.

Example 4. Let the matrix B of the linear part of the Hamilton system with four degrees of freedom have four pairs of eigenvalues: one pair of real ± 1, second one of purely imagined $\pm i$ and the third and fourth ones of complex $\pm 1 \pm i$. Let's reorder these eigenvalues in such a manner:

$$\lambda_1 = 1, \quad \lambda_2 = i, \quad \lambda_3 = 1 + i, \quad \lambda_4 = 1 - i.$$

It is evident that there exist four two-dimensional invariant subspaces L_1, L_2, L_3, L_4 due to the fact that all the relations $\lambda_i / \lambda_j \notin \mathbb{Z}$, $i \neq j$.

There are six subspaces of dimension 4:

L_{12} is not invariant, because the external eigenvalue $\lambda_3 = \lambda_1 + \lambda_2$.
L_{13} is not invariant, because the external eigenvalue $\lambda_2 = \lambda_3 - \lambda_1$.
L_{14} is not invariant, because the external eigenvalue $\lambda_2 = \lambda_1 - \lambda_4$.
L_{23} is not invariant, because the external eigenvalue $\lambda_1 = \lambda_3 - \lambda_2$.
L_{24} is not invariant, because the external eigenvalue $\lambda_1 = \lambda_2 + \lambda_4$.
L_{34} is *invariant*, because neither λ_1 nor λ_2 can be obtained as a linear combination of λ_3 and λ_4 with integer coefficients. Thus, the condition (15) of Theorem 2 is satisfied.

Finally, there are no invariant subspaces of dimension 6.

5.   Applying q-Subdiscriminants to Determine Resonances of the Linear Part of the ODE System

As it follows from the reasoning of Sections 2 and 3, the type of resonance terms of the normal form of general (1) or Hamiltonian (7) systems is determined exclusively by the eigenvalues $\boldsymbol{\lambda}$ of matrix A or B, respectively. This means that to determine the existence of invariant subspaces in the normal form (3)–(5) or (11)–(13) it is sufficient to find out the structure of eigenvalues of matrix A or B, namely whether the corresponding conditions (6) or (15) are fulfilled. A partial answer to this question can be obtained using the technique q-subdiscriminats of the characteristic polynomial $f(\lambda)$ matrix A or B.

Let us limit ourselves to the case when the vector of eigenvalues $\boldsymbol{\lambda}$ of the matrices A or B contains only real and/or purely imaginary values. In addition, suppose that if eigenvectors $\boldsymbol{\lambda}$ satisfy the resonant equation (5) or (14), all λ_j included in these equations form pairs of rationally commensurable values. In other words, there is no situation with eigenvalues like in examples 1 and 4.

5.1. *q-subdiscriminants and their properties*

Let us first briefly recall the definition of a kth order subdiscriminant of an arbitrary monic polynomial of n degree (for more details, see [13,14]).

Definition 2. Let

$$f_n(x) = x^n + a_1 x^{n-1} + \cdots + a_{n-1}x + a_n \tag{16}$$

be some monic polynomial from the x variable. Then its k-th subdiscriminant $D^{(k)}(f_n)$,

$$D^{(k)}(f_n) = \sum_{\substack{I \subset \{1,\dots,n\} \\ \#(I)=n-k}} \prod_{\substack{(j,l) \in I \\ l>j}} (x_j - x_l)^2 ,$$

where x_j is the roots of the polynomial (16), $\#(I)$ is the power of the I set. For $k = n - 1$, we put $D^{(n-1)}(f_n) = n$ and for $k = n$, we put $D^{(n)}(f_n) = 1$. For $k = 0$, we get $D^{(0)}(f_n) = D(f_n)$.

The application of q-subdiscriminants $D_q^{(k)}(f_n)$ of the characteristic polynomial f_n allows not only to find out if it has commensurable roots, but also, under certain conditions, to find these roots without having to calculate all its eigenvalues. If the right parts of the systems (1) or (7) depend on parameters, both q-subdiscriminants are functions of these parameters, which makes it possible to determine at what values of q resonance takes place and to find its multiplicity and order.

To determine the rational comparability of the roots of the polynomial, we will use q-analogues of the classic derivative and subdiscriminant.

Recall the basic q-objects (see, for example, [15–17]) used as follows.

Definition 3. Let's define the following objects:

- q-**number** a: $[a]_q = \dfrac{q^a - 1}{q - 1}$, $a \in \mathbb{R}\backslash\{0\}$,
- **shifted q-factorial (q-Pochhammer symbol):**

$$(a; q)_n = \prod_{k=0}^{n-1} \left(1 - aq^k\right), \, (a; q)_0 = 1,$$

- q-**factorial** $[n]_q! = \prod_{k=1}^{n} [k]_q = \dfrac{(q; q)_n}{(1 - q)^n}, \, q \neq 1$,
- q-**binomial (Gaussian) coefficients:**

$$\binom{n}{k}_q = \frac{[n]_q!}{[n - k]_q! \, [k]_q!} = \prod_{i=1}^{k} \frac{q^{n-i+1} - 1}{q^i - 1},$$

- q-**binomial:**

$$\{x; a\}_{n;q} \stackrel{\text{def}}{=} \prod_{i=0}^{n-1} \left(x - aq^i\right), \quad \{x; t\}_{0;q} = 1,$$

- **Jackson derivative** (q-derivative, q-differential Jackson operator):

$$(\mathcal{A}_q f)(x) \stackrel{\text{def}}{=} \begin{cases} \dfrac{f(qx) - f(x)}{(q - 1)x}, & x \neq 0, \\ f'(0), & x = 0, \end{cases} \quad q \notin \{0, 1\}.$$

As $q \to 1$, all objects defined above become classic.

Jackson derivative has all the properties of an ordinary derivative. An exception is the rule of differentiation of a complex function, which is absent in the case of a q-derivative. In addition to the above properties, we note that q-derivative function x^n is equal to $[n]_q x^{n-1}$, and applying q-derivative to q-binomial of degree n again gives q-binomial of degree $n - 1$ multiplied by $[n]_q$.

With the q-derivative, the q-analog of the classic polynomial discriminant is now determined.

Definition 4. Define q-discriminant $D_q(f_n)$ of polynomial $f_n(x)$ as the resultant of a polynomial pair $f_n(x)$ and $(\mathcal{A}_q f_n)(x)$:

$$D_q(f_n) = (-1)^{n(n-1)/2} \operatorname{Res}_x(f_n(x), (\mathcal{A}_q f_n)(x)).$$

The equality to zero of the q-discriminant of polynomial $f_n(x)$ with fixed q is a signature of the existence of at least one pair of q-commensurable roots, however, the detailed structure of all commensurable roots can be obtained using the sequence of q-subdiscriminants of various orders of the polynomial $f_n(x)$.

$$\mathsf{S}_q(f_n) \overset{\text{def}}{=} \left(D_q^{(0)}(f_n), D_q^{(1)}(f_n), \ldots, D_q^{(n-1)}(f_n) \right). \tag{17}$$

A special case of the result proved for the Hahn operator generalizing q-derivative, in [16], is the following theorem.

Theorem 3. *The $f_n(x)$ polynomial has exactly $n - d$ of different sequences q-commensurable roots if and only if the first non-zero q-subdiscriminant in sequence (17) of k-th q-subdiscriminants $D_q^{(k)}(f_n)$, $k = 0, \ldots, n - 2$, has the index d.*

All commensurable roots of the polynomial $f_n(x)$ are the roots of the largest common polynomial divisor $f_n(x)$ and its q-derivative $(\mathcal{A}_q f_n)(x)$:

$$\tilde{f}_q(x) = \operatorname{GCD}(f_n(x), (\mathcal{A}_q f_n)(x)).$$

Theorem 3 states that the degree of the polynomial $\tilde{f}_q(x)$ is equal to the number d of the first non-zero q-subdiscriminant in the sequence $\mathsf{S}_q(f_n)$.

Note that q-subdiscriminants are calculated using any of the matrix methods to calculate the classical subresultants of the polynomial pair $f_n(x)$ and $(\mathcal{A}_q f_n)(x)$ [18]. For example, if we compose a Sylvester matrix of $\mathbf{Sylv}_q(f_n)$ of size $(2n-1) \times (2n-1)$ in the form of Sylvester–Habicht, then the kth q-subdiscriminant $D_q^{(k)}(f_n)$ is equal to the determinant of the k-inner (see [19], Chapter I) of this matrix.

Let in the conditions of Theorem 3 the first non-zero q-discriminant have the index d, $0 < d < n-1$. Let's denote with $\mathbf{M}_d^{(i)}$,

$i = 1, \ldots, d$, d the modified q-Sylvester matrix $\mathbf{Sylv}_q(f_n)$ where the column with the number $2n - 1 - d$ is replaced by its column with the number $2n - 1 - d + i$ and with $M_d^{(i)}$ which is the determinant of this index. Then, as shown in [16], the following Proposition takes place.

Proposition 1. *If in the sequence* (17), *the first one different from zero q-subdiscriminant $D_q^{(d)}(f_n)$ has the index d, then*

$$\tilde{f}_q(x) \equiv D_q^{(d)} x^d + M_d^{(1)} x^{d-1} + \cdots + M_d^{(i)}. \tag{18}$$

Let's specify here some important properties of the q-subdiscriminants of the monic polynomial (16).

(1) The k-th q-subdiscriminant $D_q^{(k)}(f_n)$ is a quasi-homogeneous polynomial of the variables $a_1, \ldots, a_n$, such that

$$\sum_{j=1}^{m} j a_j \frac{\partial D_q^{(k)}(f_n)}{\partial a_j} = (m - k)(m - k - 1) D_q^{(k)}(f_n).$$

In this case, the k-th subdiscriminant depends on no more than m first polynomial coefficients (16), where $m = \min(2(n-k-1), n)$.

(2) The kth q-subdiscriminant $D_q^{(k)}(f_n)$ is an even-order reciprocal polynomial in the variable q. The reciprocity follows from the fact that root commensurability of $\lambda_i/\lambda_j = q$, $i \neq j$, implies the commensurability of $\lambda_j/\lambda_i = 1/q$. This property allows, on the one hand, to halve the degree of the q-subdiscriminant, as a polynomial in q using a substitution

$$q + \frac{1}{q} = Q, \tag{19}$$

on the other hand, to avoid double-checking the commensurability of the root pair $\lambda_i/\lambda_j = q$ and $\lambda_j/\lambda_i = 1/q$. Due to the above degree of $D_q(f_n)$ as a polynomial in q being equal to $n(n - 1)$.

5.2. *Algorithm for resonance search*

Here the algorithm of investigation of the linear part (2) of the general system of ODE to find resonances is presented. The *Algorithm* is described as a sequence of steps.

Step 1. For the A matrix of the system (7) we calculate the characteristic polynomial $f_n(\lambda)$ and optionally check if it has only real and/or purely imaginary roots.

Step 2. By the polynomial $f_n(\lambda)$ we calculate the sequence $\mathsf{S}_q(f)$ (17) of q-subdiscriminants of orders from 0 to $n-2$. At this stage, their degree as polynomials of the variable q can be halved as mentioned in Proposition 2.

The next steps depend on how easy it is to find the rational roots of the q-discriminant $D_q(f_n)$ as a polynomial in q.

Step 3a. Consider $D_q(f_n)$ as a polynomial from variable q in the ring $\mathbb{Z}[q]$. Then one should try to factorize it using different algorithms (e.g. Cantor–Zassenhaus or Berlekamp, see [20], Ch. 6 or [21]), Ch. III. If successful, we get all (or part of) the q values for which the roots of the polynomial (16) are commensurable.

Step 3b. If it is not possible to apply the previous step, we will apply some sort of brute-force searching. We limit ourselves to a certain maximum order of m resonance. Let's make an ordered tuple $\mathcal{P}_m$ of all kinds of natural pairs (r, s), such as $r, s \in \mathbb{N}$, $\mathrm{GCD}(r, s) = 1, r \geqslant s, r+s \leqslant m$. Now, sequentially going through the elements of the tuple $\mathcal{P}_m$, for each pair (r, s) check the zero value of the q-discriminant $D_q(f)$ for certain values $q = r/s$.

Step 4. Let for some rational $q_1^* \in \mathbb{Q}$ condition $D_q(f_n) = 0$ be satisfied. Then we calculate by the sequence (17) the degree of the polynomial $\tilde{f}_{q_1^*}(x)$, by the formula (18) this polynomial itself and define the structure of its roots. If the degree of this polynomial is small, then commensurable roots can be found according to the corresponding formulas, and the original polynomial $f_n(\lambda)$ can be represented as $f_n(\lambda) = u(\lambda)v(\lambda)$, where $u(\lambda)$ is a factor with already known roots and $v(\lambda)$ is a factor with still unknown roots, but among which there are no q_1^*-commensurable ones.

Step 5. The next steps depend on whether the factors $u(x)$ and $v(x)$ have q-commensurable roots or not and whether the factor $v(x)$ has q-commensurable roots as $q \neq q_1^*$.

Remark 2. When investigating the linear part (10) of the Hamilton system, one should take into account that the characteristic polynomial $f_{2n}(\lambda)$ of the matrix B is a polynomial consisting of only even powers of variable λ. It is therefore possible to halve its order and investigate the polynomial named in [22] as *semi-characteristic*,

$\hat{f}_n(\mu) = f_{2n}(\lambda)$, where $\mu = \lambda^2$. In this case, the tuple $\mathcal{P}_m$ is made up of pairs of (r^2, s^2) natural mutually prime numbers r and s.

Note that when it is possible to explicitly express eigenvalues of the matrix B, the square part of the Hamiltonian (8) can be brought to normal form by one of the methods described in the books [7,23].

Remark 3. The method described above has two drawbacks. First, if there is a multiplicity resonance of $\mathfrak{k} > 1$, there may be a situation where the pairwise commensurability of eigenvalues is set to large values of q, but the resonance itself may have a small order. For example, if three eigenvalues $\lambda_1, \lambda_2, \lambda_3$ are considered as natural numbers $2k - 1, 2k, 2k + 1$, $k \in \mathbb{N}$, that is $\lambda_1 : \lambda_2 : \lambda_3 = (2k - 1) : 2k : (2k + 1)$, the smallest order of pairwise commensurability is equal to $4k - 1$. However, at the same time, there is a resonance ratio of $\lambda_3 = 2\lambda_2 - \lambda_1$ of order 4. Secondly, the method is unable to find resonances that involve three or more of its eigenvalues, but these numbers are not pairwise commensurable.

Remark 4. The methods described in this section to study the characteristic polynomial $f(\lambda)$ of the linear part of ODE are easily algorithmized and can be implemented in many computer algebra systems.The author has written procedure libraries in computer algebra systems `Maple` and `SymPy` designed to calculate q-subdiscriminants of a polynomial and to investigate resonance relations between the roots of the latter without their explicit computations. There is an example of such calculations in what follows.

5.3. *Model example*

Let us consider six sympathetic mathematical pendulums (i.e. the same mass m and length l), the suspension points of which are located at equal distances d on a horizontal line. Let the pendulums be connected with each other by weightless linearly elastic springs of rigidity k of length d in an undistorted state. The spring fixing points are located at a distance of $b \leqslant d$ from the pendulum suspension points (see Fig. 1).

Following [24], which considers a pair of such pendulums, it is not difficult to obtain the square part of Π_2 of potential energy in the vicinity of the equilibrium position, when all angles φ_i, $i = 1, \ldots, 6$,

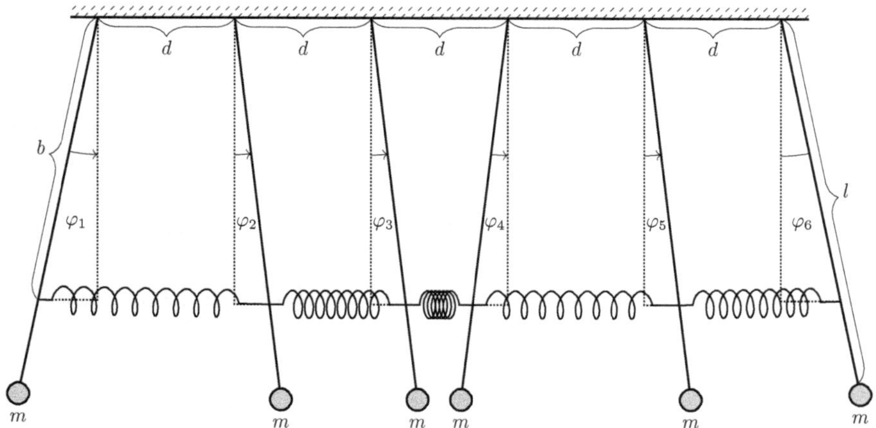

Fig. 1. Six sympathetic pendulums.

the deviations of pendulums from the vertical are equal to zero. By passing to the canonical variables and scaling the time $\tau = t\sqrt{g/l}$, we can obtain a semi-characteristic polynomial (secular equation) in the form of

$$
\begin{aligned}
\hat{f}_6(\mu) = {}& \mu^6 - 2\left(5\beta + 3\right)\mu^5 + \left(36\beta^2 + 50\beta + 15\right)\mu^4 \\
& - 2\left(28\beta^3 + 72\beta^2 + 50\beta + 10\right)\mu^3 \\
& + \left(35\beta^4 + 168\beta^3 + 216\beta^2 + 100\beta + 15\right)\mu^2 \\
& - 2\left(3\beta^5 + 35\beta^4 + 84\beta^3 + 72\beta^2 + 25\beta + 3\right)\mu \\
& + (2\beta + 1)(3\beta + 1)(\beta + 1)(\beta^2 + 4\beta + 1),
\end{aligned}
\tag{20}
$$

where $\mu = \lambda^2$ and the only parameter $\beta = \frac{kb^2}{mgl} > 0$. Due to the stability of the equilibrium position, the polynomial (20) has six negative real roots according to the number of degrees of freedom of the original mechanical system.

Following step 2 of the *Algorithm*, let's calculate the first q-subdiscriminant $D_q(\hat{f}_6)$ and, using the (19) substitution, let's simplify it. As a result, we get a polynomial in the variable Q of degree 15. Here, its explicit expression is not given due to its cumbersomeness. In the system `Maplesoft Maple`, this polynomial has been factorized into linear and square multipliers which made it easy to find all commensurability between the roots of the polynomial (20). In this way, step 3a of the *Algorithm* is done and it is possible to find all the roots of the polynomial (20) and therefore to check the fulfillment of the resonance relations between them.

In the system SymPy, factorization of q-discriminant $D_q(\hat{f}_6)$ into multipliers failed, so we will apply steps 3b–5 of the *Algorithm* for some fixed value of the parameter $\beta = 48/25$. Let's choose the maximum resonance order of $m = 20$ and according to the remark 2, make up a tuple of $\mathcal{P}_{20}$ of all kinds of pairs of (r^2, s^2), such as $\mathrm{GCD}(r, s) = 1$ and $r + s \leqslant 20$, $r, s \in \mathbb{N}$. Looking through all such pairs from $\mathcal{P}_{20}$, we find that with $q_1^* = 121/25$ and $q_2^* = 169/25$ q-discriminant is zero and the rest of the q-subdiscriminants are not. The polynomials $\tilde{f}_{q_1}(\mu)$ and $\tilde{f}_{q_2}(\mu)$ calculated by the formula (18) have the same root $\mu_1 = 1$. Therefore, one more value $q_3^* = q_2^*/q_1^* = 169/121$ must exist at which the q-discriminant vanishes. Thus, there are three roots of the polynomial (20): $\mu_1 = 1$, $\mu_2 = q_1^*$, $\mu_3 = q_2^*$, and therefore, $f_6(\mu) = (\mu - 1)(\mu - q_1^*)\,(\mu - q_2^*)\,v_3(\mu)$. The other irrational roots μ_k, $k = 4, 5, 6$, of the cubic polynomial $v_3(\mu)$ can be easily found using the Vieta trigonometric formula.

So, there is a resonant relation (14)

$$5p_1 + 11p_2 + 13p_3 = 0 \tag{21}$$

between frequencies $\lambda_1 = 1, \lambda_2 = 11/5, \lambda_3 = 13/5$. The relation (21) has three integer solutions $\mathbf{p}_1 = (1, -4, 3)$, $\mathbf{p}_2 = (3, 1, -2)$ and $\mathbf{p}_3 = (4, -3, 1)$, which means that the condition (15) of Theorem 2 is not fulfilled. Thus, there is an invariant coordinate subspace of L_I, $I = \{1, 2, 3\}$, of dimensions six in the normal form corresponding to the three eigenvalues $\lambda_1, \lambda_2, \lambda_3$. For each of the remaining eigenvalues of λ_i, $i = 4, 5, 6$, there corresponds a two-dimensional invariant coordinate subspace of L_i. Since the orders of the integer vectors $\mathbf{p}_1, \mathbf{p}_2, \mathbf{p}_3$ are equal to 8, 6, 8, respectively, the investigation of the initial system dynamics in the subspace L_I is possible only when the nonlinear normalization of the initial system is performed at least up to the sixth order.

Acknowledgment

I am grateful to Professor A.D. Bruno for his support and fruitful discussion of this work.

References

1. A. D. Bruno, *The Restricted 3–body Problem: Plane Periodic Orbits*, Walter de Gruyter, Berlin, 1994; Nauka, Moscow, 1990, 296 p. (in Russian).
2. A. P. Markeev, *Libration Points in Celestial Mechanics and Cosmodynamics*. Nauka, Moscow, 1978 (in Russian).
3. A. D. Bruno, *Local Methods in Nonlinear Differential Equations*, Springer-Verlag, Berlin, Heidelberg, New York, London, Paris, Tokyo, 1989.
4. A. D. Bruno, V. F. Enderal and V. G. Romanovski, On new integrals of the Algaba-Gamero-Garcia system. *Comput. Algebra in Scientific Computing*. Springer, Berlin Heidelberg, 2017, doi:10.1007/978-3-642-32973-9.
5. A. D. Bruno, Normal form of a Hamiltonian system with a periodic perturbation. *Computational Mathematics and Mathematical Physics* **60**(1) (2020a), 36–52, doi:10.1134/S09655425200100066.
6. A. D. Bruno, Normalization of the periodic Hamiltonian system. *Program. Comput. Softw.*, **46**(2) (2020b), 76–83, doi:10.31857/S0132347420020053.
7. V. F. Zhuravlev, A. G. Petrov and M. M. Shunderyuk, *Selected Problems of Hamiltonian Mechanics*. LENAND, Moscow, 2015 (in Russian).
8. A. D. Bruno, Analytical form of differential equations (I). *Trans. Moscow Math. Soc.*, **25**(1971), 131–288.
9. A. D. Bruno, Analytical form of differential equations (II). *Trans. Moscow Math. Soc.*, **26**(1972), 199–239.
10. G. D. Birkhoff, *Dynamical Systems, Colloquim Publications*, Vol. 9, revised edn. AMS, Providence, Rhode Island, 1966.
11. A. M. Lyapunov, Problème général de la stabilité du mouvement. *Ann. Fac. Sci. Toulouse Math.*, **9**(2) (1892), 204–474; *Stability of Motion*. Academic Press, New York, London. 1966 [English].
12. C. L. Siegel and J. K. Moser, *Lectures on Celestial Mechanics*, Springer-Verlag, Berlin, Heidelberg, New York, 1971.
13. S. Basu, R. Pollack and M.-F. Roy, *Algorithms in Real Algebraic Geometry*, Algorithms and Computations in Mathematics 10. Springer-Verlag, Berlin, Heidelberg, New York, 2006.
14. A. B. Batkhin, Parameterization of the discriminant set of a polynomial. *Program. Comput. Softw.*, **42**(2) (2016), 65–76, doi:10.1134/S0361768816020031.
15. V. Kac and P. Cheung, *Quantum Calculus*. Springer-Verlag, New York, Heidelberg, Berlin, 2002.

16. A. B. Batkhin, Parameterization of a set determined by the generalized discriminant of a polynomial. *Program. Comput. Softw.*, **44**(2) (2018), 75–85, doi:https://doi.org/10.1134/S0361768818020032.

17. A. B. Batkhin, Computation of the resonance set of a polynomial under constraints on its coefficients, *Program. Comput. Softw.*, **45**(2) (2019), 27–36, doi:https://doi.org/10.1134/S0361768819020038.

18. J. von zur Gathen and T. Lücking, Subresultants revisited, *Theoret. Comput. Science*, **297**(2003), 199–239, doi:10.1016/S0304-3975(02)00639-4.

19. E. I. Jury, *Inners and Stability of Dynamic Systems*. John Wiley & Sons, New York, 1974.

20. A. G. Akritas. *Elements of Computer Algebra with Applications*. John Wiley & Sons, Inc., New York, 1989.

21. J. von zur Gathen and J. Gerhard, *Modern Computer Algebra*, 3rd edition edn. Cambridge University Press, Cambridge, New York, Melbourne, Madrid, Cape Town, Singapore, Sao Paulo, Delhi, Mexico City, 2013.

22. A. B. Batkhin, A. D. Bruno and V. P. Varin, Stability sets of multiparameter Hamiltonian systems. *J. App. Math. Mech.*, **76**(1) (2012), 56–92, doi:10.1016/j.jappmathmech.2012.03.006.

23. A. P. Markeev, *Linear Hamiltonian Systems and Some Problems of Stability of the Satellite Center of Mass*. Institute of Computer Science, Izhevsk, 2009 (in Russian).

24. A. P. Markeev, Nonlinear oscillations of sympathetic pendulums, *Rus. J. Nonlin. Dyn.*, **6**(3) (2010), 605–621, doi:10.20537/nd1003009 (in Russian).

© 2022 World Scientific Publishing Europe Ltd.
https://doi.org/10.1142/9781800611368_0004

Chapter 4

Non-Standard Discretizations of a Hamiltonian System Related to the Fourth Painlevé Equation

Galina Filipuk[*,‡] and Thomas Kecker[†,§]

*Faculty of Mathematics, Informatics and Mechanics
University of Warsaw
Banacha 2, 02-098, Warsaw, Poland
†School of Mathematics and Physics
University of Portsmouth
Lion Gate Building, Lion Terrace
Portsmouth, PO1 3HF, UK
‡filipuk@mimuw.edu.pl
§thomas.kecker@port.ac.uk

We consider non-standard discretizations of a cubic, non-autonomous Hamiltonian system, studied by the authors in an earlier article, related to the fourth Painlevé equation (a class of integrable, second-order non-linear, non-autonomous differential equations). Discretizing a differential system is a non-unique process and we employ different methods such as the (symmetric) Kahan method and a more general class of discretizations by introducing further parameters to the equations, with the aim of reducing the algebraic entropy of the system, which is a measure of complexity of the solutions of the system. Integrability of discrete systems is usually associated with zero algebraic entropy. Using the concept of singularity confinement, we can identify the cases which have reduced, though non-zero, algebraic entropy as compared with generic discretization.

1. Introduction

When studying differential equations, an important question concerns the long-term behavior of the solutions, that is to decide whether an equation exhibits chaotic behavior or if the solutions can be given, in some sense, exactly. In the latter case, the system described by these equations is said to be integrable. There are different notions of integrability, e.g. integrability in classical mechanics in the sense of Liouville, where a Hamiltonian system can be completely integrated if there exists a sufficient number of independent conserved quantities. For complex differential equations, i.e. ODEs for which we are interested in analytic solutions in the complex plane, the integrability is closely related to their singularity structure in the complex plane, and the usual integrability criterion is for an equation to have the Painlevé property. An equation is said to possess the Painlevé property if all movable singularities (i.e. those for which the location changes with the initial conditions) of all its solutions are poles. In the class of equations $w'' = R(z, w, w')$, where R is a rational function in its arguments w, w', with coefficients that are analytic in z, all equations with this property can either be linearized, solved in terms of elliptic functions, or can be reduced to a set of six (nonlinear) second-order equations known as the Painlevé equations. Furthermore, the solutions, known as Painlevé transcendents, can be obtained exactly by the inverse scattering transform (IST), a nonlinear analogue of the Fourier transform. In this sense, the six Painlevé equations deserve to be called integrable. Prior to the classification that led to the discovery of the six nonlinear ODEs by Painlevé [15] and Gambier [4], already in the 1880s. Kovalevskaya [14] identified all integrable cases of the equations of motion of a rigid body around a fixed point with methods that are very much related to the Painlevé classification, known as the Painlevé test. That is, Kovalevskaya showed that four conserved quantities can be found for the dynamical system exactly in the cases of certain ratios of the principle moments of inertia of the body when the solutions, considered in the complex plane, are analytic with the exception of movable poles, i.e. there are no movable critical points (such as branch points or essential singularities). Apart from the previously known cases of the Euler top and Lagrange top, Kovalevskaya was able to identify one additional case, now known as the Kovalevskaya top. For the

Painlevé test in general, one is seeking a sufficiently large family of formal Laurent series solutions of a given class of equations which allows to fix certain parameters in this class. Although the Painlevé test is not a sufficient criterion for an equation to have the Painlevé property, it is a very useful detector for integrability, with any equation found in this way to be analyzed by more involved methods to decide whether or not it has the Painlevé property.

Discretizing a dynamical system, e.g. for numerical calculations, is a non-unique process, i.e. there are in general infinitely many discrete systems that correspond to the same continuous system which can be recovered by taking an appropriate continuous limit. It is of importance to perform the discretization in such a way that the characteristic properties of the system, such as its integrability, and therefore the stability of its solutions, are preserved. In the case of the equations of motion for the spinning top, this was achieved by Hirota and Kimura for the Euler top [10] and the Lagrange top [11] using a non-standard discretization method leading to systems that are symmetric under time reversal and possess a sufficient number of first integrals which are given explicitly, thus proving the integrability of these discrete systems. Integrability preserving discretizations are also important for numerical calculations to guarantee the stability of the approximate solutions. Kahan and Li [12] devised a non-standard numerical scheme which serves this purpose and which they have applied to various systems. Incidentally, this scheme is equivalent to the Hirota–Kimura method, and for quadratic vector fields, also identical to a well-known Runge–Kutta method. [1] The method was applied also to various non-autonomous systems [2] for which it is shown that integrability is preserved.

Discrete versions of the Painlevé equations are known for all six types of equations, obtained by de-autonomization of certain discrete systems in the family of QRT maps [16] using the method of singularity confinement [17]. Similar as for differential equations, the singularity confinement test can be seen as an analogue of the Painlevé test for discrete systems, which, though not a sufficient criterion for integrability, serves as a useful detector.

In this chapter, we perform a class of non-standard discretizations on a certain non-autonomous Hamiltonian system related to the fourth Painlevé equation, introduced by one of the authors [13]. The class of discretizations considered, with various parameters, is a generalization of the Kahan method when applied to quadratic vector fields. We will employ the method of singularity confinement to fix the parameters of the discretized equations. Roughly speaking, the singularity confinement property expresses the fact that any movable (i.e. spontaneously arising) singularity disappears after a finite number of iterations of the discrete equations. This will be explained in Section 2.1.

Similar to the Painlevé test, the singularity confinement property is not sufficient for an equation to be integrable (to have the Painlevé property in the continuous case). A more refined measure of integrability is obtained by computing the algebraic entropy of the system. This is a measure of complexity of rational expressions obtained by iterating the discrete system on an initial set of rational functions of some low degree, see Section 2.2.

Kahan's method was applied to some of the Painlevé equations [2], namely

$$P_I : y'' = 6y^2 + z, \tag{1}$$

$$P_{II} : y'' = 2y^3 + zy + \alpha, \tag{2}$$

$$P_{IV} : yy'' = \frac{(y')^2}{2} + \frac{3}{2}y^4 + 4zy^3 + 2(z^2 - \alpha)y^2 + \beta, \tag{3}$$

which are the instances of the Painlevé equations which have meromorphic solutions in the whole complex plane (the equations P_{III}, P_V and P_{VI} also have fixed singularities). The conclusion, however, is that apart from a few exceptional cases with special values of the parameters in the equations, the resulting discrete equations do not exhibit singularity confinement.

We applied Kahan's method to a Hamiltonian system with cubic Hamiltonian [3],

$$H(z, q, p) = \frac{1}{3} \left(p^3 + q^3 \right) + zpq + \alpha p + \beta q, \tag{4}$$

leading to the system of equations

$$q' = p^2 + zq + \alpha, \quad p' = -q^2 - zp - \beta, \tag{5}$$

where $q' = \frac{dq}{dz}$, $p' = \frac{dp}{dz}$, z is the (complex) independent variable and α, β are arbitrary complex parameters. This system of equations is closely related to the Painlevé IV equation (3), though with different parameters α and β. Namely, if $w = \rho p + \bar{\rho} q - z$, where $\rho \in \{1, \omega, \bar{\omega}\}$, $\omega = \frac{-1+i\sqrt{3}}{2}$ being the third root of unity, then w satisfies the equation

$$ww'' = (w')^2 - w^4 - 4zw^3 - (2\rho\alpha + 2\bar{\rho}\beta + 3z^2)w^2 - (\rho\alpha - \bar{\rho}\beta + 1)^2. \quad (6)$$

Equation (6) can be transformed into equation (3) by a re-scaling of the dependent and independent variables. The solutions of the system (5) can thus be expressed by a linear combination of Painlevé IV transcendents with different sets of parameters. In particular, the Hamiltonian system (5) has the Painlevé property, i.e. all movable singularities of all its solutions are (simple) poles. Since it does not have any fixed singularities, this means that any local solution can be analytically continued to a meromorphic function of the whole complex plane. We remark that the system (5) retains the Painlevé property if z is replaced by a linear function $az + b$, which can also be achieved by a re-scaling and shifting of variables.

2. Indicators of Discrete Integrability

Integrability, in the case of (continuous) differential equations as well as discrete equations, is related to low complexity of the solutions, measured appropriately. For example, all the solutions of Painlevé's equations I, II and IV (equations (1)–(3)) have a finite order of growth, in the sense of Nevanlinna theory (see Section 2.3). In Sections 2.1 and 2.2, we will briefly discuss some standard indicators of discrete integrability, namely, singularity confinement and zero algebraic entropy. Other measures of complexity, namely the growth of meromorphic solutions of the discrete system, can also be used to detect discrete integrability [6,7]. Usually, several of these methods are used in conjunction to identify integrable discrete equations.

2.1. *Singularity confinement*

We briefly describe here the idea of singularity confinement, which serves as a detector of discrete integrable equations. It was used to de-autonomize certain discrete equations with elliptic solutions to obtain discrete versions of all the Painlevé equations [17], that is, integrable second-order difference equations in the class

$$y_{n+1} + y_{n-1} = R(n, y_n), \tag{7}$$

where $R(n, y_n)$ is a rational expression in its arguments. As an example, we consider here the class of equations leading to the discrete Painlevé I equation,

$$y_{n+1} + y_{n-1} = \frac{a_n y_n + b_n}{y_n^2}. \tag{8}$$

The idea of singularity confinement is to study the solutions of the equation at their singularities, in analogy with the Painlevé test for differential equations which studies the local behavior at the movable singularities. The only way the dependent variable y_{n+1} in (8) can develop a singularity is when $y_n = 0$. We therefore analyze this point by considering a small perturbation of ϵ around this value, i.e. we let $y_n = \epsilon$, whereas $y_{n-1} = \kappa$ takes on an arbitrary finite value. We then study how this singular behavior propagates under the iteration given by equation (8). One obtains the following expansions around $\epsilon = 0$:

$$y_{n+1} = \frac{b_n}{\epsilon^2} + \frac{a_n}{\epsilon} - \kappa,$$

$$y_{n+2} = -\epsilon + \frac{a_{n+1}}{b_n}\epsilon^2 + O(\epsilon^3), \tag{9}$$

$$y_{n+3} = \frac{b_{n+2} - b_n}{\epsilon^2} - \frac{a_{n+2} - 2a_{n+1} + a_n}{\epsilon} + O(1).$$

The expression for y_{n+3} is infinite unless both the conditions $b_{n+2} - b_n = 0$ and $a_{n+2} - 2a_{n+1} + a_n = 0$ hold. In this case, the singularity is said to be confined. For all such singularities to be confined in this way we need the two above conditions to be satisfied identically (for all n), that is, we require

$$a_n = \alpha n + \beta, \quad b_n = \gamma + (-1)^n \delta. \tag{10}$$

Thus, we obtain the equation

$$y_{n+1} + y_{n-1} = \frac{(\alpha n + \beta)y_n + \gamma + (-1)^n \delta}{y_n^2}, \tag{11}$$

which, for $\delta = 0$, is a discrete analogue of the Painlevé I equation (1), to which it has an appropriate continuous limit. For $\delta \neq 0$, the term $(-1)^n \delta$ is not a rational expression in n and so lies outside the class of equations (7).

2.2. *Algebraic entropy*

Algebraic entropy, introduced by Hietarinta and Viallet [8], is a measure for the degree growth of rational functions under the iteration of a discrete mapping. It is considered a useful concept to detect integrability. Given an initial rational function ϕ_0, we denote the iterates by $\phi_1, \phi_2, \dots$ and their respective degrees by d_n, $n = 0, 1, 2, \dots$. The algebraic entropy of the discrete map is then defined by

$$e_{\text{alg}} = \lim_{n \to \infty} \frac{\log d_n}{n}.$$

In the generic case, the degree of the iterates will grow by a constant factor with each iteration, given by the degree of the rational map, i.e. for a quadratic map, the sequence $\langle d_n \rangle$ will grow like $1, 2, 4, 8, 16, \dots$, hence, the algebraic entropy is $\log 2$ in this case. Integrability of an equation is associated with zero algebraic entropy. For this, strong cancellations have to occur in the iteration of the discrete map, such that the sequence $\langle d_n \rangle$ grows only polynomially. This is the case, for example, for the discrete Painlevé equations. In the case of equation (11), the first discrete Painlevé equation, the sequence of degrees starts with $1, 3, 9, 19, 33, 51, 73, 99, 129, 163, 201, \dots$, and the general rule is $d_n = 2n^2 + 1$, i.e. we have polynomial growth [8]. Thus, although (11) is a degree 3 map, we have much slower growth of the iterates than in the generic case of equation (8). This is also the case for the other known discrete Painlevé equations. Zero algebraic entropy is a stronger criterion for integrability than singularity confinement. For example, the following map considered by Hietarinta and Viallet [8]

$$y_{n+1} + y_{n-1} = y_n + \frac{a}{y_n^2}, \tag{12}$$

where a is a constant, passes the singularity confinement test. Letting $y_n = \epsilon$ and $y_{n-1} = \kappa \neq 0$, we have the iterates

$$
\begin{aligned}
y_{n+1} &= \frac{a}{\epsilon^2} - \kappa + O(\epsilon), \\
y_{n+2} &= \frac{a}{\epsilon^2} - \kappa + O(\epsilon^4), \\
y_{n+3} &= -\epsilon + O(\epsilon^4), \\
y_{n+4} &= \kappa - \epsilon + O(\epsilon^2),
\end{aligned}
\tag{13}
$$

meaning that the (movable) singularity occurring in y_{n+1} disappears after a finite number of steps, i.e. this equation has the singularity confinement property. However, the map is known to exhibit chaotic behavior and has non-zero algebraic entropy. For other discrete systems, with only mild cancellations occurring in the iteration of rational solutions, the algebraic entropy can be less than in the generic case for a map of degree d, but still positive. The discrete system considered in this chapter falls within this class of equations.

2.3. *Growth of solutions in the complex plane*

A third useful criterion to identify integrability of discrete equations is the growth, in the sense of Nevanlinna theory, of meromorphic solutions of the system in the complex plane. Value-distribution theory defines the Nevanlinna characteristic function $T(r, f)$, a convex function of $\log r$, which is an appropriate measure for the growth of the function f, or rather the quantity

$$\rho(f) = \limsup_{r \to \infty} \frac{\log T(r, f)}{\log r},$$

called the order of growth of f, which may be finite or infinite. In the latter case, one can define the hyper-order

$$\varsigma(f) = \limsup_{r \to \infty} \frac{\log \log T(r, f)}{\log r}.$$

Integrability is associated with the existence of meromorphic solutions of hyper-order less than 1. Using this criterion, Halburd and Korhonen [6,7] were able to single out, within the class of discrete equations

$$y(z + 1) + y(z - 1) = R(z, y(z)),$$

where $R(z, y)$ is rational in y with analytic coefficients, all the discrete Painlevé equations contained in this class.

Halburd and Korhonen [5] discussed the different approaches to detect the integrable equations within a given class. Another criterion mentioned in this chapter is that of Diophantine integrability, which measures the growth of the logarithmic height of rational numbers, when taken as initial values, under the iteration of a discrete mapping. All three methods, algebraic entropy, the growth in the sense of

Nevanlinna theory and the logarithmic height growth, each provide some kind of measure of complexity for the solutions of the discrete system, where low complexity is associated with the integrability of a system.

3. Kahan Discretization

The discrete Painlevé equations were obtained [17] by de-autonomization of certain (autonomous) systems known as QRT maps, which are known to have elliptic functions, i.e. doubly periodic meromorphic functions, as their solutions. From the de-autonomized versions of these equations it is more or less straightforward to find the appropriate continuous limits to some corresponding Painlevé equation. In the other direction, starting from a differential equation or a system of equations, discretizing the system is a non-unique process which is not guaranteed to lead to an integrable discrete version of the equation.

Kahan and Li [12] introduced a discretization scheme for the numerical solution of a first-order system of differential equations,

$$\dot{\mathbf{x}} = \mathbf{f}(\mathbf{x}), \quad \mathbf{x}(t_0) = \mathbf{x}_0, \quad \mathbf{x} \in \mathbb{R}^N$$

by replacing $\dot{\mathbf{x}}$ by the standard forward difference, while $\mathbf{f}(\mathbf{x})$ is replaced using a symmetric rule,

$$\dot{x}_i \to \frac{\tilde{x}_i - x_i}{h}, \quad x_i \to \frac{\tilde{x}_i + x_i}{2}, \quad x_i x_j \to \frac{x_i \tilde{x}_j + \tilde{x}_i x_j}{2}.$$

Here, $\tilde{x}_i = x_i(t+h)$ denotes the forward shift of x_i. Although at first sight this looks like an implicit scheme, it can be shown that the map can be written explicitly as

$$\tilde{\mathbf{x}} = \mathbf{x} + h \left(\mathbf{I} - \frac{h}{2} D\mathbf{f}(\mathbf{x}) \right)^{-1} \mathbf{f}(\mathbf{x}),$$

where $D\mathbf{f}$ is the Jacobian of $\mathbf{f}$ and $\mathbf{I}$ denotes the $n \times n$ identity matrix. This is a birational relation, the inverse being

$$\mathbf{x} = \tilde{\mathbf{x}} - h \left(\mathbf{I} + \frac{h}{2} D\mathbf{f}(\tilde{\mathbf{x}}) \right)^{-1} \mathbf{f}(\tilde{\mathbf{x}}).$$

It was found that this scheme leads to numerically stable calculations for solutions of ODEs. Independent from this, the method was used by Hirota and Kimura to discretize the equations of motion of a rotating body around a fixed point, in particular the equations of the Euler top [10] and of the Lagrange top [11]. They have shown that integrability is preserved in this case by finding a sufficient number of conserved quantities for the discretized system. The method is thus also referred to as the Hirota–Kimura method.

For non-autonomous systems which are known to be integrable by the inverse scattering transform, such as the Painlevé equations and their Hamiltonian forms, this method was applied [2] to the Painlevé equations I, II and IV (i.e. those which have meromorphic solutions in the whole complex plane). Although for Painlevé I an integrable discretization was found, in the case of Painlevé II this has led to no success and in the case of Painlevé IV, only for special values of the parameters in the equation. In the following, we will apply Kahan's method to a more general form of the system (5),

$$\begin{aligned} x' &= y^2 + f(z)x + \alpha \\ y' &= -x^2 - f(z)y - \beta, \end{aligned} \tag{14}$$

where we have replaced z with an arbitrary function of z. Kahan's prescription gives

$$\begin{aligned} &\left(-4 + f(n)^2 - 4x(n)y(n)\right) x(n+1) \\ &\quad = -\left(4\alpha + 2(2 + f(n+1))x(n) + f(n)(2\alpha + (2 + f(n+1))x(n))\right. \\ &\qquad \left. -4\beta y(n) + 4y(n)^2 - 2f(n+1)y(n)^2\right), \\ &\left(-4 + f(n)^2 - 4x(n)y(n)\right) y(n+1) \\ &\quad = -f(n)(2\beta + (-2 + f(n+1))y(n)) + 2\left(2\beta + 2\alpha x(n)\right. \\ &\qquad \left. +(2 + f(n+1))x(n)^2 + (-2 + f(n+1))y(n)\right). \end{aligned} \tag{15}$$

Note that, in contrast to the discrete Painlevé equations, the common denominator (when solved for $x(n+1)$ and $y(n+1)$) in these equations is also n dependent. To apply the singularity confinement test, we choose a pair of values $(x(n), y(n))$ so that the denominator becomes zero and perturbs these values by some small parameter ϵ,

$$x(n) = \frac{f(n)}{2} - \sigma + \epsilon, \quad y(n) = \frac{f(n)}{2} + \sigma + \epsilon, \quad \sigma \in \{1, -1\}. \tag{16}$$

In principle, the condition $4x(n)y(n) = f(n)^2 - 4$ defines a 1-parameter curve of possible choices for $x(n), y(n)$. The choices (16), however, turn out to lead to some meaningful conditions, which we consider in the following:

The case $\sigma = 1$. Inserting the expressions (16) into the system (15), one obtains the expansions

$$x(n + 1) = \frac{(f(n) + 2)(\alpha - \beta + f(n) - f(n + 1))}{2\epsilon f(n)} + O(1),$$

$$y(n + 1) = \frac{(f(n) - 2)(\alpha - \beta + f(n) - f(1 + n))}{2\epsilon f(n)} + O(1).$$

In the next step,

$$x(n + 2) = \frac{(f(n) - 2)(f(n + 2) - 2)}{2(f(n) + 2)} + O(\epsilon),$$

$$y(n + 2) = \frac{(f(n) + 2)(f(n + 2) + 2)}{2(f(n) - 2)} + O(\epsilon).$$

Furthermore,

$$x(n+3) = \frac{(2+f(n))(\alpha - \beta + f(n) - f(1 + n))(2 + f(n + 2))}{\epsilon}$$
$$R_1(n, f(n)) + O(1),$$

$$y(n+3) = \frac{(f(n) - 2)(\alpha - \beta + f(n) - f(1 + n))(f(n + 2) - 2)}{\epsilon}$$
$$R_2(n, f(n)) + O(1),$$

where $R_1(n, f(n))$ and $R_2(n, f(n))$ are some rational expressions in their arguments which cannot vanish identically. Hence, from the expressions $x(n + 3), y(n + 3)$ we see that the singularity propagates unless the condition

$$\alpha - \beta + f(n) - f(n + 1) = 0$$

is satisfied. For all singularities to be confined in this way, we require this relation to be satisfied identically, i.e. $f(n)$ is linear in n,

$$f(n) = (\alpha - \beta)n + \gamma,$$

where γ is an arbitrary constant. Note that incidentally, if this condition is satisfied, then already the expansions of $x(n + 1)$ and $y(n + 1)$ are regular in ϵ, that is, the singularity does not arise in the first place. Thus, the singularity confinement test on the discrete system (15) leads to a condition on the function f, although Kahan's method applied to the original system (5) would not have led to success. So although the system, with z replaced by some rescaled function $az+b$, has the Painlevé property, only for the specific value $a = \alpha - \beta$ do we find some kind of singularity confinement.

The case $\sigma = -1$. Again, inserting the expressions (16) into the system (15) we obtain expansions

$$x(n+1) = \frac{2(2+\alpha+\beta) + f(n)(\alpha-\beta+f(n)+2f(1+n))}{2\epsilon f(n)} + O(1),$$

$$y(n+1) = -\frac{2(2+\alpha+\beta) + f(n)(\alpha-\beta+f(n)+2f(1+n))}{2\epsilon f(n)} + O(1).$$

In the next step,

$$x(n+2) = \frac{1}{2}(f(n+2) - 2)$$
$$+ \epsilon\frac{(2+\alpha+\beta)(f(n+2)-2) - f(n)(2+2\beta+2f(n+1)+f(n+2))}{2(2+\alpha+\beta)+f(n)^2+f(n)(\alpha-\beta+2f(n+1))},$$

$$y(n+2) = \frac{1}{2}(f(n+2) + 2)$$
$$+ \epsilon\frac{f(n)(2+2\alpha-f(n+1)-f(n+2)) - (2+\alpha+\beta)(2+f(n+2))}{2(2+\alpha+\beta)+f(n)^2+f(n)(\alpha-\beta+2f(n+1))}.$$

We are now essentially in the situation of the case $\sigma = 1$. After one more step we find

$$(2f(n)\left((\alpha-\beta-f(n+2)-2f(n+1))f(n+2)-2(2+\alpha+\beta)\right)\epsilon)\, x(n+3)$$
$$= (2(2+\alpha+\beta) + f(n)(\alpha-\beta+f(n)+2f(n+1)))\cdot$$
$$(2+f(n+2))(\alpha-\beta+f(n+2)-f(n+3)) + O(1),$$
$$(2f(n)\left(-2(2+\alpha+\beta)+(\alpha-\beta-f(n+2)-2f(n+1))f(n+2)\right))\, y(n+3)$$
$$= (2(2+\alpha+\beta) + f(n)(\alpha-\beta+f(n)+2f(n+1)))\cdot$$
$$(2-f(n+2))(\alpha-\beta+f(n+2)-f(n+3)) + O(1).$$

Here are two possibilities for which we find singularity confinement. Either,

$$\alpha - \beta + f(n+2) - f(n+3) = 0,$$

but this is essentially the same condition already encountered in the case $\sigma = 1$. One the other hand, we could have

$$\alpha + \beta + 2 = 0, \qquad \alpha - \beta + f(n) + 2f(n+1) = 0.$$

Again, for all singularities of this type to be confined we require these equations to be satisfied identically, which results in

$$f(n) = c \cdot \left(-\frac{1}{2}\right)^n - \frac{2}{3}(\alpha + 1),$$

where c is an arbitrary constant. In this case, however, only singularities of the type $\sigma = -1$ are confined whereas in the case $f(n) = (\alpha - \beta)n + \gamma$, singularities of both types are confined.

4. More General Non-Standard Discretization

As we have seen in the previous section, singularity confinement in the system (15) forces $f(n)$ to be a linear function. We will therefore investigate if this condition leads to a unique discretization of a re-scaled system (5). A more general class of non-standard discretization than Kahan's method was also applied to the famous Lotka–Volterra system of population dynamics [9]. For a quadratic vector field in a set of dependent variables $x_1, \ldots, x_n$, the discretization with step size h is given by

$$\dot{x}_i \to \frac{\tilde{x}_i - x_i}{h}, \qquad x_i \to ax_i + \tilde{a}\tilde{x}_i, \qquad x_i x_j \to bx_i x_j + cx_i \tilde{x}_j + d\tilde{x}_i x_j + e\tilde{x}_i \tilde{x}_j,$$

where $\tilde{x}_i$ denotes the forward shift of x_i and $a + \tilde{a} = 1$ and $b + c + d + e = 1$. Kahan's method is the case when $a = \tilde{a} = \frac{1}{2}$ as well as $c = d = \frac{1}{2}$ and $b = e = 0$.

We discretize the system (5) by re-writing it as an autonomous system

$$\begin{aligned}
\dot{x} &= y^2 + zx + \alpha, \\
\dot{y} &= -x^2 - zy - \beta, \\
\dot{z} &= 1,
\end{aligned}$$

where $\dot{x} = \frac{dx}{ds}$ denotes the derivative with respect to some parameter s, thus treating z as a dependent variable. We obtain the system

$$\frac{\tilde{x} - x}{h} = a_1 y^2 + b_1 \tilde{y}^2 + c_1 y\tilde{y} + d_1 zx + e_1 z\tilde{x} + f_1 \tilde{z}x + g_1 \tilde{z}\tilde{x} + \alpha,$$

$$\frac{\tilde{y} - y}{h} = -a_2 x^2 - b_2 \tilde{x}^2 - c_2 x\tilde{x} - d_2 zy - e_2 z\tilde{y} - f_2 \tilde{z}y - g_2 \tilde{z}\tilde{y} - \beta,$$

$$\frac{\tilde{z} - z}{h} = 1.$$

$$(17)$$

Due to the given constraints, this system now depends on 10 independent parameters. One requirement for integrability is the time-reversibility of the discrete map, that is, the equations should define a bi-rational relation $(x, y) \leftrightarrow (\tilde{x}, \tilde{y})$, meaning that one can solve the system for $(\tilde{x}, \tilde{y})$ as a rational expression in (x, y) and vice versa. This condition implies $a_1 = b_1 = a_2 = b_2 = 0$ and $c_1 = c_2 = 1$. Furthermore, we assume that discretization is symmetric in y and x so we suppose that $d_1 = d_2$, $e_1 = e_2$, $f_1 = f_2$ and $g_1 = g_2$. Furthermore, as we saw above that the constant multiplying the term zpq in the Hamiltonian (4) need not be equal to unity in order for the system to have the Painlevé property, we will also lift the restriction $d + e + f + g = 1$. From the last equation in (17) we obtain $z = nh$, so $\tilde{z} = (n+1)h$. Inserting this into the first two equations and letting $h = 1$, after a re-labelling of the parameters we obtain the system

$$x(n+1) - x(n) = y(n)y(n+1) + anx(n) + bnx(n+1)$$
$$+ cx(n) + dx(n+1) + \alpha,$$
$$y(n+1) - y(n) = -x(n)x(n+1) - (any(n) + bny(n+1)$$
$$+ cy(n) + dy(n+1)) - \beta,$$

which yields a bi-rational relation between $x(n)$, $y(n)$ and $x(n+1)$, $y(n+1)$,

$$x(n+1)$$
$$= \frac{\beta y(n) + (-1 + c + an)y(n)^2 - (1 + d + bn)(\alpha + (1 + c + an)x(n))}{(-1 + d + bn)(1 + d + bn) - x(n)y(n)},$$

$$y(n+1)$$
$$= \frac{\alpha x(n) + (1 + c + an)x(n)^2 - (-1 + d + bn)(\beta + (-1 + c + an)y(n))}{(-1 + d + bn)(1 + d + bn) - x(n)y(n)}.$$

$$(18)$$

The system has a movable singularity when, in the common denominator,

$$x(n)y(n) = (-1 + d + bn)(1 + d + bn).$$

Again, this defines a 1-parameter curve of singularities in the x-y-plane. To perform the singularity confinement test, we consider only the cases where x and y are rational functions of n,

$$x(n) = d + bn - \sigma + \varepsilon, \quad y(n) = d + bn + \sigma + \varepsilon, \quad \sigma \in \{+1, -1\}. \quad (19)$$

The case $\sigma = 1$. Inserting (19) into (18) we obtain expressions of the form

$$x(n+1) = \frac{(1 + d + bn)(-2c + 2d - 2an + 2bn + \alpha - \beta)}{2(d + bn)\epsilon} + O(1),$$

$$y(n+1) = \frac{(-1 + d + bn)(2c - 2d + 2an - 2bn - \alpha + \beta)}{2(d + bn)\epsilon} + O(1),$$

which are singular around $\epsilon = 0$. For the second iteration, we obtain the expressions

$$x(n+2) = \frac{(-1 + a + c + an)(-1 + d + bn)}{1 + d + bn} + O(\epsilon),$$

$$y(n+2) = \frac{(1 + a + c + an)(1 + d + bn)}{-1 + d + bn} + O(\epsilon),$$

both of which are regular expansions in ϵ. However, in the next step the singularity will return if in the common denominator of $(x(n+3),\ y(n+3))$,

$$(-1 + d + b(n+2))(1 + d + b(n+2)) - x(n+2)y(n+2),$$

vanishes to leading order in ϵ, i.e. if

$$(-1+d+b(n+2))(1+d+b(n+2))-(-1+a+c+an)(1+a+c+an) = 0.$$

For this condition to be identically satisfied, we collect the terms of distinct powers of n,

$$4b^2 + 4bd + d^2 - a^2 - c^2 - 2ac = 0,$$
$$4b^2 + 2bd - 2a^2 - 2ac = 0,$$
$$b^2 - a^2 = 0.$$

120   *G. Filipuk & T. Kecker*

Apart from the autonomous case where $a = b = 0$, there are four cases which satisfy these conditions:

$$
\begin{aligned}
\text{(i)} \quad & b = a, \quad c = a + d, \\
\text{(ii)} \quad & b = a, \quad c = -a, \quad d = -2a, \\
\text{(iii)} \quad & b = -a, \quad c = a - d, \\
\text{(iv)} \quad & b = -a, \quad c = -a, \quad d = 2a.
\end{aligned}
\tag{20}
$$

In the different cases, the singularity can be avoided if the coefficients of the terms of order ϵ^{-1} in the expressions for $x(n+3), y(n+3)$ vanish identically. In the cases (i) and (ii), these expressions are of the form

$$
\begin{aligned}
x(n+3) &= \frac{(2a - \alpha + \beta)}{\epsilon} R_1(n) + O(1), \\
y(n+3) &= \frac{(2a - \alpha + \beta)}{\epsilon} R_2(n) + O(1),
\end{aligned}
$$

where we do not write out the rational functions $R_1(n)$, $R_2(n)$ due to their lengthy expressions. So, in both cases (i) and (ii) we find the additional constraint

$$
a = \frac{\alpha - \beta}{2}.
$$

Case (i), where d is arbitrary, corresponds to the case already encountered in Section 3 with $f(n) = (\alpha - \beta)n + \gamma$, $(\gamma = d)$, i.e.

$$
\begin{aligned}
&\left(-4 + (n(\alpha - \beta) + \gamma)^2 - 4x(n)y(n)\right) x(n+1) \\
&\quad = 4\beta y(n) - (2 + n(\alpha - \beta) + \gamma)(2\alpha + (2 + (n+1)(\alpha - \beta) + \gamma)x(n)) \\
&\qquad + 2(-2 + (n+1)(\alpha - \beta) + \gamma)y(n)^2, \\
&\left(-4 + (n(\alpha - \beta) + \gamma)^2 - 4x(n)y(n)\right) y(n+1) \\
&\quad = 4\alpha x(n) - (-2 + n(\alpha - \beta) + \gamma)(2\beta + (-2 + (n+1)(\alpha - \beta) + \gamma)y(n)) \\
&\qquad + 2(2 + (n+1)(\alpha - \beta) + \gamma)x(n)^2.
\end{aligned}
\tag{21}
$$

Case (ii), where all parameters are fixed, is essentially the same as the above system with $\gamma = -2(\alpha - \beta)$, that is we obtain the system,

$$
\begin{aligned}
((-2 &+ (n-2)(\alpha - \beta))(2 + (n-2)(\alpha - \beta)) - 4x(n)y(n))\,x(n+1) \\
&= 4\beta y(n) - (2 + (n-2)(\alpha - \beta))(2\alpha + (2 + (n-1)(\alpha - \beta))x(n)) \\
&\quad + 2(-2 + (n-1)(\alpha - \beta))y(n)^2, \\
((-2 &+ (n-2)(\alpha - \beta))(2 + (n-2)(\alpha - \beta)) - 4x(n)y(n))\,y(n+1) \\
&= 4\alpha x(n) - (2 - (n-2)(\alpha - \beta))(-2\beta + (2 - (n-1)(\alpha - \beta))y(n)) \\
&\quad + 2(2 + (n-1)(\alpha - \beta))x(n)^2.
\end{aligned}
$$

$$(22)$$

We remark that in the cases (iii) and (iv) the singularity is present in the expressions for $x(n+3), y(n+3)$, and cannot be avoided, so the singularities are not confined in these cases.

The case $\sigma = -1$. Inserting (19) with $\sigma = -1$ into the system (18) we obtain the expansions

$$
x(n+1) = \frac{2 + \alpha + \beta + (d + bn)(4c + 2d + 4an + 2bn + \alpha - \beta)}{2(d + bn)\epsilon} + O(1),
$$

$$
y(n+1) = -\frac{2 + \alpha + \beta + (d + bn)(4c + 2d + 4an + 2bn + \alpha - \beta)}{2(d + bn)\epsilon} + O(1).
$$

After the second step, the expansions are regular in ϵ,

$$
\begin{aligned}
x(n+2) &= -1 + a(n+1) + c + O(\epsilon), \\
y(n+2) &= 1 + a(n+1) + c + O(\epsilon).
\end{aligned}
$$

Again, in the third step the singularity will reappear if one of the four conditions (20) is satisfied. However, the coefficients of the order ϵ^{-1} terms in $x(n+3), y(n+3)$ vanish identically if and only if $2a + b = 0$, $4c + 2d + \alpha - \beta = 0$ and $2 + \alpha + \beta = 0$. These conditions, however, are not compatible with any of the conditions (20), so the singularities are not confined.

5. Algebraic Entropy of the Discrete Systems

We determine the algebraic entropy of the discrete systems encountered in Sections 3 and 4. Starting with some rational functions

$x(n), y(n)$ of degree 1, such as $x(n) = r \cdot n$, $y(n) = s \cdot n$, we compute the degrees of the iterates under the discrete mappings (21) and (22), which we have done using Mathematica® (www.wolfram.com), obtaining a sequence of integers $\langle d_n \rangle$ in each case. These computations rely on Mathematica's ability of fully reducing the rational expressions. In the generic case, with no singularity confinement, we find that the degrees double in every iteration leading to the sequence $\langle d_n \rangle = 1, 2, 4, 8, 16, 32, \ldots$, as expected.

For the system (21), where we have singularity confinement, with a generic value for γ the first terms in the sequence of degrees are,

$$\langle d_n \rangle = 1, 2, 4, 8, 15, 28, 52, 96, 177, 326, 600, 1104, 2031, \ldots,$$

i.e. we see that when cancellations occur, the degrees are lower than in the generic case. The first terms of the sequence are consistent with the recursive law

$$d_n = 2d_{n-1} - d_{n-4}, \quad d_0 = 1, \quad d_n = 0 \text{ for } n < 0.$$

The characteristic equation for this recursive formula is $\lambda^4 - 2\lambda^3 + 1 = 0$, has largest real root $\lambda \approx 1.839\ldots$, meaning that the sequence exhibits exponential growth $d_n \sim \lambda^n$. The algebraic entropy of the system with the prescribed parameters can thus be computed to be

$$e_{\mathrm{alg}} = \log \lambda \approx 0.609\ldots,$$

which is positive but smaller than $\log 2 \approx 0.693\ldots$ in the generic case.

In the case of system (22) the sequence of degrees is slightly different:

$$\langle d_n \rangle = 1, 2, 4, 8, 13, 24, 44, 82, 151, 278, 512, 942, 1733\ldots, \tag{23}$$

which still exhibits exponential growth, obeying a similar recursion to the one above, but with some additional cancellations occurring in some of the steps. So although Kahan's method leads to a "good" discretization with singularity confinement and a reduction in algebraic entropy compared to the generic case, it does not fix all the constants in the system to maximize the number of cancellations occurring in the iteration of the discrete map.

6. Conclusion

Discretization of non-autonomous integrable equations such as the Painlevé equations is a non-unique process which in general does not lead to discrete integrable systems. A class of non-standard methods, of which Kahan's method is a special case, has been applied to a certain system associated with the fourth Painlevé equation. Using singularity confinement test we have identified the cases within this class that lead to a reduced, though non-zero, algebraic entropy, compared with the generic case. It is seen that Kahan's method leads to a system with this property, however, a more general class of discretization contains a system with slightly more cancellations occurring in the iteration of the discrete map. Although the discrete systems turn out to be non-integrable, this provides an interesting example of a discrete system with singularity confinement where the behavior of the solution at a singularity depends on the independent variable.

Acknowledgments

The results leading to this chapter were obtained during visits by TK to the Mathematical Institute of the Polish Academy of Science (IMPAN) and the University of Warsaw, hosted by GF. The visits were funded by the LMS (London Mathematical Society) and WCNM (Warszawskie Centrum Nauk Matematycznych), for which we would like to acknowledge their generous support. GF acknowledges the support of the National Science Center (Poland) via grant OPUS 2017/25/B/BST1/00931. GF also acknowledges the support of the Alexander von Humboldt Foundation.

References

1. E. Celledoni, R. I. McLachlan, B. Owren, and G. R. W. Quispel, Geometric properties of Kahan's method. *J. Phys. A*, **46**(2013), 025201, 12p.
2. E. Celledoni, R. I. McLachlan, D. I. McLaren, B. Owren, and G. R. W. Quispel, Integrability properties of Kahan's method. *J. Phys. A*, **47**(2014), 365202, 20p.

3. G. Filipuk and T. Kecker, Kahan discretisation of a cubic Hamiltonian system. In: *Mathematical and Numerical Aspects of Dynamical System Analysis*, J. Awrejcewicz, M. Kaźmierczak, J. Mrozowski, P. Olejnik (eds.), DSTA Conference Proceedings Vol. 2, Łódź, 2017, pp. 185–192.

4. B. Gambier, Sur les équations différentielles du second ordre et du premier degré dont l'intégrale générale est a points critiques fixes. *Acta Math.*, **33**(1910), 1–55.

5. R. G. Halburd and R. J. Korhonen, Three approaches to detecting discrete integrability. *Comput. Methods Funct. Theory*, **19**(2019), (2) 299–313.

6. R. G. Halburd and R. J. Korhonen, Finite-order meromorphic solutions and the discrete Painlevé equations. *Proc. London Math. Soc.*, **94**(2007), 443–474.

7. R. G. Halburd and R. J. Korhonen, Meromorphic solutions of difference equations integrability and the discrete Painlevé equations. *J. Phys. A*, **40**(2007), 1–38.

8. J. Hietarinta and C. Viallet, Singularity confinement and chaos in discrete systems. *Phys. Rev. Lett.*, **82**(1998), 325–328.

9. A. Hone and K. Towler, Non-standard discretisation of biological models, *Nat. Comput.*, **14**(2015), 39–48.

10. R. Hirota and K. Kimura, Discretization of the Euler top. *J. Phys. Soc. Japan*, **69**(2000), 627–630.

11. R. Hirota and K. Kimura, Discretization of the Lagrange top. *J. Phys. Soc. Japan*, **69**(2000), 3193–3199.

12. W. Kahan and R. C. Li, Unconventional schemes for a class of ordinary differential equations. *J. Comput. Phys.*, **134**(1997), 316–331.

13. T. Kecker, A cubic Hamiltonian system with meromorphic solutions. *Comput. Methods Funct. Theory*, **16**(2016), 307–317.

14. S. Kovalevskaya, Sur le problème de la rotation d'un corps solide autour d'un point fixe. *Acta Math.*, **12**(1889), 177–232.

15. P. Painlevé, Mémoire sur les équations différentielles dont l'intégrale générale est uniformé. *Bull. Soc. Math. France*, **28**(1900), 201–261.

16. G. R. W. Quispel, J. A. G. Roberts, and C. J. Thompson, Integrable mappings and soliton equations I. + II. *Phys. Lett. A*, **126** (1988), 419–421; *Physica D*, **34**(1989), 183–192.

17. A. Ramani, B. Grammaticos and J. Hietarinta, Discrete versions of the Painlevé equations. *Phys. Rev. Lett.*, **67**(1991), 1829–1832.

© 2022 World Scientific Publishing Europe Ltd.
https://doi.org/10.1142/9781800611368_0005

Chapter 5

Notes on the Zeros of the Solutions of Airy's Non-homogeneous Equation

Federico Zullo

DICATAM, University of Brescia
Via Valotti 9, 25133, Brescia, Italy
federico.zullo@unibs.it

We present some observations on the distribution of the zeros of solutions of the non-homogeneous Airy equation. We show the existence of a principal family of solutions, with simple zeros, and particular solutions, characterized by a double zero in a given position of the complex plane. A recursion, describing the distribution of the zeros is introduced and the limits of its applicability are discussed. The results can be considered a generalization of previous works on the distribution of the zeros for the solutions of the corresponding homogeneous equation.

1. Introduction

It is well known that the knowledge of the distribution of the zeros in the complex plane for transcendental entire functions of finite order gives the possibility to get a representation of such functions through the Weierstrass–Hadamard factorization theorem. In general, for entire functions defined by the solutions of suitable differential or difference equations, the problem of the localization of the corresponding zeros may be very demanding. For second-order homogeneous differential equations the problem has been investigated under different perspectives and many results are collected

125

in the literature [4,7]. For entire solutions of non-homogeneous differential equations, there are many withstanding problems about the distribution and localization of the corresponding solutions. In previous works, [16,17], we examined the distribution of the zeros of the general solution of the homogeneous Airy differential equation through the introduction of two suitable parameters into the equation. A limiting process on a parametric recursion has been shown to give the location of the zeros and the dependence of the zeros by varying the parameters has been investigated. In this work, we consider the solutions of a non-homogeneous Airy equation obtained by adding a constant term to the homogeneous equation. As we will see, this simple extension considerably changes the structure of the solutions regarding the distribution of the zeros.

As a last remark, we would like to underline that the non-homogeneous Airy equation possesses several applications in mathematical physics [9]. For example, it has a strict relation with the second member of Burger's hierarchy [6]:

$$\psi_t + (\psi_{xx} + \psi^3 + 3\psi\psi_x)_x = 0. \tag{1}$$

Under the self-similarity transformation

$$\psi = \frac{1}{\eta(3t)^{1/3}} f(z), \quad z \doteq \frac{x}{\eta(3t)^{1/3}} - \frac{b}{\eta^3}, \quad \eta^3 = a, \tag{2}$$

equation (1) becomes

$$\frac{d^3 f}{dz^3} + 3\left(\frac{df}{dz}\right)^2 + 3f\frac{d^2 f}{dz^2} + 3f^2\frac{df}{dz} = (az + b)\frac{df}{dz} + af. \tag{3}$$

Integrating once, one gets

$$\frac{d^2 f}{dz^2} + 3\frac{df}{dz}f + f^3 = k + (az + b)f(z), \tag{4}$$

where k is the integration constant. The previous equation will be considered in the next sections (see in particular equation (46)). Some of the solutions of (4) have been considered in [6] for the description of liquids with gas bubbles.

2. The Non-homogeneous Equation and a Particular Family of Solutions

The non-homogeneous Airy differential equations

$$\frac{d^2y(z)}{dz^2} = zy(z) \pm \frac{1}{\pi}, \tag{5}$$

and their particular solutions, denoted by Hi(z) and Gi(z) (the Scorer functions), have been considered, among others, by Scorer [13], Olver [11], MacLeod [9], Gil *et al.* [2,3]. In particular, the function Hi(z) possesses the integral representation

$$\mathrm{Hi}(z) = \frac{1}{\pi} \int_0^\infty \exp\left(zt - \frac{t^3}{3}\right) dt, \tag{6}$$

and is a solution of (5) with the $+$ sign, whereas Gi(z) has the representation

$$\mathrm{Gi}(z) = \frac{1}{\pi} \int_0^\infty \sin\left(zt + \frac{t^3}{3}\right) dt, \tag{7}$$

and is a solution of (5) with the $-$ sign. The corresponding initial conditions are given by [10]

$$\mathrm{Gi}(0) = \frac{1}{2}\mathrm{Hi}(0) = \frac{1}{3^{7/6}\Gamma(2/3)}, \quad \mathrm{Gi}'(0) = \frac{1}{2}\mathrm{Hi}'(0) = \frac{1}{3^{5/6}\Gamma(1/3)}. \tag{8}$$

The two functions are related by the equation $\mathrm{Gi}(z) + \mathrm{Hi}(z) = \mathrm{Bi}(z)$, where $\mathrm{Bi}(z)$ is one of the two fundamental solutions of the homogeneous Airy equation [10]. Both Gi and Hi are entire functions with an infinite number of zeros in the complex plane: many properties of the zeros of these two functions and their derivatives can be found in [3]. Also, Hi decays at infinity on the negative real axis, whereas Gi decays on the positive real axis and is always less than 1 in magnitude on the real axis.

To not keep the $\pm$ signs, let us insert a constant c instead of $\pm\frac{1}{\pi}$ in (5):

$$\frac{d^2y(z)}{dz^2} = zy(z) + c. \tag{9}$$

It is clear that with a rescaling of the dependent variable it is possible to fix the value of the constant c to be any suitable value. The general solution of equation (9) can be written as

$$y(z) = C_1 \mathrm{Ai}(z) + C_2 \mathrm{Bi}(z) + c\pi \mathrm{Hi}(z), \tag{10}$$

where C_1 and C_2 are two arbitrary constants. Ai, Bi and Hi are entire with order equal to $3/2$, so equation (9) possesses entire solutions with order of growth equal to $\frac{3}{2}$: since this number is not an integer, all solutions of (9) possess an infinite number of zeros [14]. These zeros are movable, in the sense that if the initial conditions change, the positions of the zeros change. The position of one zero can be considered a constant of integration. A second constant is the value of the derivative of the function at this zero. This is true also for the homogeneous equation corresponding to equation (5) (see [16,17]), but in this case there is a significant difference. It is known that, further to the trivial solution, entire solutions of linear, second-order, homogeneous differential equations

$$\frac{d^2 y(z)}{dz^2} = A(z)\frac{dy(z)}{dz} + B(z)y(z), \tag{11}$$

where $A(z)$ and $B(z)$ are entire functions, have all their zeros simple. This result comes directly from the local analysis of equation (11) around any zero z_0. In the non-homogeneous case, one has the possibility to have double zeros. Given any $p \in \mathbb{C}$, equation (9) possesses always a solution with a double zero in p. Indeed, it holds the following

Proposition 2.1. *Given $p \in \mathbb{C}$, equation (9) possesses just one solution $\tau_d(z,p)$, proportional to c, with a double zero in $z = p$. This solution is defined by the series*

$$\tau_d(z,p) = c\sum_{n=2} e_n(z-p)^n, \quad e_2 = \frac{1}{2}, \quad e_3 = 0,$$

$$e_4 = \frac{p}{24}, \quad e_{n+2} = \frac{pe_n + e_{n-1}}{(n+1)(n+2)}. \tag{12}$$

For p real, $\tau_d(z,p)$ can also be represented as

$$\tau_d(z,p) = c\pi\left(\mathrm{Bi}(z)\int_p^z \mathrm{Ai}(x)dx - \mathrm{Ai}(z)\int_p^z \mathrm{Bi}(x)dx\right). \tag{13}$$

The second part of this proposition is also given in [3], the first part can be obtained by considering the Taylor series of $y(z)$ with a double zero in $z = p$ around this point. As we will see, the solutions $y(z)$ possessing a zero in $z = 0$ (double or simple) will play a central role in the rest of the chapter. In the case $p = 0$, the recursion in (12) can be solved and the corresponding series is a generalized hypergeometric series. More explicitly, one has

$$\tau_d(z, 0) = c \sum_{n=0}^{\infty} \frac{3^n n!}{(3n+2)!} z^{3n+2} = \frac{c}{2} z^2 {}_1F_2\left(1; \frac{4}{3}, \frac{5}{3}; \frac{z^3}{9}\right). \tag{14}$$

The hypergeometric function ${}_1F_2\left(1; \frac{4}{3}, \frac{5}{3}; \frac{z^3}{9}\right)$ is actually related to the Lommel functions $s_{\mu,\nu}(z)$. The Lommel functions are particular solutions of the non-homogeneous Bessel equations

$$z^2 \frac{d^2 y(z)}{dz^2} + z \frac{dy(z)}{dz} + (z^2 - \nu^2) y(z) = z^{\mu+1} \tag{15}$$

determined by the behavior $y(z) \sim \frac{z^{\mu+1}}{(1+\mu)^2 - \nu^2}(1 + O(z^2))$ around $z = 0$ when $(\mu + n)^2 \neq \nu^2$, $n \in \mathbb{N}$. One has [15]

$$s_{\mu,\nu}(z) = \frac{z^{\mu+1}}{(\mu+1)^2 - \nu^2} {}_1F_2\left(1; \frac{\mu - \nu + 3}{2}, \frac{\mu + \nu + 3}{2}; -\frac{1}{4} z^2\right). \tag{16}$$

The values of (μ, ν) corresponding to the parameters $(\frac{4}{3}, \frac{5}{3})$ in (14) are $(\mu, \nu) = (0, 1/3)$. Further, $s_{0,v}(z)$ possesses a nice integral representation, given by

$$s_{0,v}(z) = \frac{1}{1 + \cos(\pi v)} \int_0^\pi \sin(z \sin(t)) \cos(vt) dt. \tag{17}$$

Indeed, by inserting into the left-hand side of (15) the integral (17) and multiplying by $(1 + \cos(\pi v))$ one gets

$$(1 + \cos(\pi v)) \left(z^2 \frac{d^2 y(z)}{dz^2} + z \frac{dy(z)}{dz} + (z^2 - v^2) y(z)\right)$$

$$= \int_0^\pi \cos(vt) \left(\sin(z \sin(t))(\cos(t)^2 z^2 - v^2) + z \cos(z \sin(t)) \sin(t)\right) dt$$

$$= -\left(\cos(vt) \cos(z \sin(t)) \cos(t) z + \sin(z \sin(t)) \sin(vt) v\right)\big|_0^\pi$$

$$= z(1 + \cos(\pi v)). \tag{18}$$

The integral on the right-hand side of (17) has also the right behavior around $z = 0$, i.e. $\sim \frac{z}{1-\nu^2}(1 + O(z^2))$, showing the equivalence stated in (17). From this integral representation it follows another formula, that will be important in the following. The formula is

$$
s_{0,\nu}(z) = \frac{1}{1 + \cos(\pi\nu)}
$$
$$
\int_0^1 \sin(zt) \frac{\cos(\nu \arcsin(t)) + \cos(\nu\pi - \nu \arcsin(t))}{\sqrt{1 - t^2}} dt, \quad (19)
$$

where the range of arcsin is $(-\pi/2, \pi/2)$. At this point, it is useful to recall a Theorem of Polya about the zeros of entire functions defined by integrals like (19). One has

Theorem 2.2 ([12]). *Suppose that the function $f(t)$ is positive and not decreasing in $(0, 1)$. Then the function of z defined by*

$$
V(z) = \int_0^1 \sin(zt) f(t) dt, \quad (20)
$$

possesses only real zeros. Further, if $f(t)$ grows steadily, these zeros are simple and the intervals $(k\pi, (k + 1)\pi)$, $k = 1, 2, \ldots$ contain the positive zeros of $V(z)$, each interval containing just one zero.

By growing steadily it is meant that the function is not piecewise constant with a finite number of rational points of discontinuity in $(0, 1)$. We can apply the Theorem of Polya to formula (19). It is possible to show indeed that the function multiplying $\sin(zt)$ in (19) is positive and increasing for $|\nu| < 1$ and $t \in (0, 1)$. Putting together equations (14), (16) and (19) we get the following

Proposition 2.3. *The function $\tau_d(z, 0)$ defined by the series (14) has a double zero in $z = 0$. All other zeros are simple and are located on the three rays $(e^{i\pi}, e^{\pm i\pi/3})$. The modulus of the simple zeros are contained in the intervals*

$$
\left[\left(\frac{3}{2}\pi k \right)^{\frac{2}{3}}, \left(\frac{3}{2}\pi(k + 1) \right)^{\frac{2}{3}} \right], \quad k = 1, 2, \ldots \quad (21)
$$

This global result can be compared with the asymptotic distribution of the zeros. Indeed, since ${}_1F_2\left(1; \frac{3-\nu}{2}, \frac{3+\nu}{2}; -\frac{z^2}{4} \right) \sim \frac{C_\nu}{z^{3/2}} \cos(z - \frac{3\pi}{4})$ [8], where C_ν is a constant, it follows that the modulus of the zeros is asymptotically approximated by

$$|z_k| \sim \left(\frac{3}{2}\pi \left(k + \frac{1}{4} \right) \right)^{\frac{2}{3}}, \quad k = 1, 2, \ldots \tag{22}$$

Formula (22) gives quite accurate results also for small values of k: indeed If $z_{\mathrm{app},k}$ and $z_{e,k}$ are the approximated (through formula (22)) with exact values for the modulus of the zeros of $\tau_d(z,0)$, then numerically we find that the relative errors are bounded by the following inequalities:

$$\left| 1 - \frac{z_{app,k}}{z_{e,k}} \right| < 0.01, \text{ for } k > 3. \tag{23}$$

Remark 2.4. It can be interesting to remark that Polya's Theorem and equation (19) imply that the generalized hypergeometric function $_1F_2 \left(1; \frac{3-\nu}{2}, \frac{3+\nu}{2}; z \right)$ has only real negative simple zeros for $|\nu|<1$. These zeros are contained in the intervals

$$- \left[\frac{\pi^2 (k+1)^2}{4}, \frac{\pi^2 k^2}{4} \right], \quad k = 1, 2, \ldots \tag{24}$$

3.   The Principal Family

So far, we characterized the zeros of the particular solution $\tau_d(z,0)$ of equation (9) by showing that this solution possesses just one double zero, the other being simple. Actually, as it will be shown, this property is shared by all solutions $\tau_d(z,p)$, for any $p \in \mathbb{C}$. The particular solutions $\tau_d(z,p)$ of equation (9) are characterized by the presence of a double zero in the arbitrary point $z = p$. As we said, the position of one zero can be considered a constant of integration. The other constant is the value of the derivative of the function at this zero. The functions $\tau_d(z,p)$ however have a derivative equal to 0 in $z = p$ (for any p). The set of functions $\tau_d(z,p)$ then represents a family of particular solutions of equation (9). This point will be further clarified in this section. Besides the family $\tau_d(z,p)$, equation (9) possesses a *principal family* of solutions. Proposition (2.1) can be easily extended to the case of a solution with a simple zero in an arbitrary point $z = q$. Indeed, one has

Proposition 3.1. *Given $q \in \mathbb{C}$, equation (9) possesses a solution $\tau(z,q,\alpha)$ with a simple zero in $z = q$. This solution is defined by the series*

$$\tau(z,q,\alpha) = \sum_{n=1} f_n(z-q)^n, \quad f_1 = \alpha, \ f_2 = \frac{c}{2}, \ f_3 = \frac{q\alpha}{6},$$

$$f_{n+2} = \frac{q f_n + f_{n-1}}{(n+1)(n+2)}. \tag{25}$$

For q real, $\tau(z,q,\alpha)$ can also be represented as

$$\tau(z,q,\alpha) = \pi\alpha\left(\mathrm{Bi}(z)\mathrm{Ai}(q) - \mathrm{Ai}(z)\mathrm{Bi}(q)\right)$$
$$+ c\pi\left(\mathrm{Bi}(z)\int_q^z \mathrm{Ai}(x)dx - \mathrm{Ai}(z)\int_q^z \mathrm{Bi}(x)dx\right). \tag{26}$$

In the previous formulae, q and α are arbitrary parameters and so the series (25), which converges everywhere in the complex plane, can be considered the general solution of equation (9). At this point, a note of caution may be worth understanding the characteristics of this representation of the general solution and its potential intersections with the representation (12). It is clear that if $\alpha = 0$, the series (25) represents a solution with a double zero in $z = q$, like the series (12) with $p = q$. We recall however that each element of the set of solutions (12) $\tau_d(z,p)$ possesses, further to the double zero in $z = p$, only simple zeros. Let $\{\Omega_k(p)\}_{k=1,2,\dots}$ be the set of these simple zeros. If $q = \Omega_n(p)$ for some n and p and $\alpha = \tau_d'(\Omega_n(p))$, from the existence and uniqueness theorem it follows that the two series (12) and (25) coincide. In all other cases, the function $\tau(z,q,\alpha)$ defined by the series (25) must have only simple zeros since it cannot be represented by any element of the family $\tau_d(z,p)$.

The fact that the set of solutions of (9) possessing only simple zeros is not empty follows from the observation that if it would be empty, the set of particular solutions (12) would represent the general solution of equation (9), that is impossible since in (12) there is just one free parameter. So we arrived at the conclusion that equation (9) possesses two families of solutions: the particular families given by the solutions having just one double zero in $z = p$, all other zeros being simple, and the principal family, given by the solutions of equation (9) having only simple zeros. This result is summarized in the following:

Proposition 3.2. *Let $p \in \mathbb{C}$ be arbitrary and let $\tau_d(z,p)$ be the generic element of the set of particular solutions of equation (9) defined by the series (12). Let $\{\Omega_k(p)\}_{k=1,2,\dots}$ be the set of the zeros of*

$\tau_d(z, p)$. *Then, if $q \neq \Omega_n(p)$ for some n and p and $\alpha \neq \tau_d'(\Omega_k(p))$, the series* (25) *gives a generic element of the principal family of solutions of equation* (9). *Further, any element of the principal family possesses only simple zeros.*

Remark 3.3. In the case p and q are real, then from proposition (3.1) it follows that the function defined by the series (25) possesses a double zero in $z = p$ if and only if α is the common value of the following expressions:

$$\alpha = \frac{c \int_q^p \mathrm{Ai}(x)dx}{\mathrm{Ai}(q)} = \frac{c \int_q^p \mathrm{Bi}(x)dx}{\mathrm{Bi}(q)}. \tag{27}$$

The characterization of the two families of solutions can be further clarified by looking at the logarithmic derivative of $y(z)$, i.e. to $u(z) = \frac{y'(z)}{y(z)}$. This function is meromorphic and, from (9), it solves the following differential equation

$$\frac{d^2u(z)}{dz^2} + 3u(z)\frac{du(z)}{dz} + u(z)^3 = 1 + zu(z). \tag{28}$$

Clearly, $u(z)$ possesses the Painlevé property, but still it is instructive to apply the Painlevé test to $u(z)$. Here, we will use the standard terms that can be found in modern literature about the Painlevé test (see e.g. [5]). The dominant balance gives singularities of the type $c_0(z - p)^{-1}$, but, seeking the resonances, one discovers that there are two families of solutions:

- One family characterized by $c_0 = 1$ and with resonance polynomial given by $(r - 1)(r + 1)$.
- The other characterized by $c_0 = 2$ and with resonance polynomial given by $(r + 1)(r + 2)$.

The two resonances $r = -1$ correspond to the arbitrariness of the position of the pole for $u(z)$, i.e. the arbitrariness of the position of the zero $z = p$ for $y(z)$. For the other resonances, one has:

- In the first family, $c_0 = 1$ implies that the zero in $z = p$ of $y(z)$ is simple, whereas the resonance $r = 1$ indicates that the coefficient of $(z - p)^0$ in the Laurent expansion of $u(z)$ is the other arbitrary constant describing the solutions of the second-order equation (46). This is the principal family of solutions [1].

- In the second family, $c_0 = 2$ implies that the zero in $z = p$ of $y(z)$ is double, whereas the resonance $r = -2$ is negative: this indicates that the second family is a particular solution of equation (46) [1].

As we shall see, the coefficients of the Laurent expansion of the logarithmic derivative of $y(z)$ are explicitly connected with the zeros of $y(z)$: this connection was crucial for the characterization of the distribution of the zeros of the solutions of the homogeneous Airy equation in [16,17]. The same line of reasoning will be adopted in the next section for the non-homogeneous case. Before that, as will be explained, we need to slightly modify equation (9) and insert two more parameters.

4.   A Mapping Among Solutions

As we have shown in the previous sections, it is possible to pick a solution of equation (9) by assigning the position of one zero and the value of the first derivative of the function at this zero. If the solution of equation (9) has a zero at $z = z_0$, with z_0 an arbitrary point, by a shift $z \to z + z_0$ and a rescaling of z, we get the following equation for $y(z)$:

$$\frac{d^2y(z)}{dz^2} = (az + b)y(z) + c. \tag{29}$$

A solution of equation (29) with a zero in $z = 0$ corresponds to a solution of equation (9) with a zero in $z_0 = \frac{b}{\eta^2}$, where $a = \eta^3$. Instead of looking at solutions of equation (9) with zeros in $z = z_0$, with z_0 arbitrary, we can look at the particular solution of equation (29) with a zero in $z = 0$ for an arbitrary value of the parameter b. This second point of view will be adopted in the rest of the chapter.

    As we have seen, the particular solution $\tau_d(z, 0)$ (12) of equation (9) possesses just one double zero in $z = 0$, with all other zeros simple. Equivalently, for $b = 0$ there is a solution of equation (29) possessing a double zero in $z = 0$, the other zeros being simple. We would like to prove that this property is shared by all the particular solutions $\tau_d(z, p)$ (12) of equation (9). This is equivalent to saying that, for any $b \neq 0$, equation (29) possesses a solution with a double zero in $z = 0$, with all the other zeros simple. We can give two arguments to justify the presence of just one double zero. For clarity, let us call the solution of equation (29) with a double zero in $z = 0$ $c\Xi(z, a, b)$, where $\Xi(z, a, b)$ is defined by the series

$$\Xi(z, a, b) = \sum_{n=2} d_n z^n, \quad d_2 = \frac{1}{2}, \quad d_3 = 0, \quad d_4 = \frac{b}{24},$$

$$d_{n+2} = \frac{b d_n + a d_{n-1}}{(n+1)(n+2)}. \tag{30}$$

By multiplying equation (29) by $y'(z)$, integrating and taking into account that $\Xi(0, a, b) = 0$ and $\Xi'(0, a, b) = 0$, we find, for $z \in \mathbb{R}$

$$\left(\Xi'(z, a, b)\right)^2 = (az + b)\Xi^2(z, a, b) + 2c\Xi(z, a, b) - a \int_0^z \Xi^2(z, a, b) dz. \tag{31}$$

In the case we have another double zero in $z = p$, $p \in \mathbb{R}$, one should have

$$\int_0^p \Xi^2(z, a, b) dz = 0, \tag{32}$$

which is impossible. Note that we tacitly assumed Ξ to be real on the real axis and this is the case if the coefficients appearing in equation (29) are real.

For the second argument, we observe that the solutions of equation (29) with a double zero in $z = 0$ have a nice homogeneity property with respect to the parameters. Indeed, for any $\lambda \in \mathbb{C}^*$ one has:

$$\Xi(\lambda^{-1}z, \lambda^3 a, \lambda^2 b) = \lambda^2 \Xi(z, a, b). \tag{33}$$

For $b = 0$, from (14) we have

$$\Xi(z, a, 0) = \sum_{n=0} \frac{(3a)^n n!}{(3n+2)!} z^{3n+2} = \frac{z^2}{2} {}_1F_2\left(1; \frac{4}{3}, \frac{5}{3}; \frac{az^3}{9}\right). \tag{34}$$

Now we ask how the zeros of $\Xi(z, a, 0)$ will move in the complex plane if we take a small value of b. We remember that the zeros of $\Xi(z, a, 0)$, for $a = 1$, are distributed on the rays $(e^{i\pi}, (e^{\pm i\pi/3})$. Changing this value of a corresponds to a scaling (for $|a| \neq 1$) and global rotation (for $\arg(a) \neq 0$) of their positions. If we expand $\Xi(z, a, b)$ in a series of b, each coefficient being a function of z and a, we get

$$\Xi(z, a, b) = \Xi_0(z, a) + b\Xi_1(z, a) + O(b^2), \tag{35}$$

where $\Xi_0(z, a) = \Xi(z, a, 0)$ is given by (34). The function $\Xi_1(z, a)$ solves the differential equation

$$\frac{d^2\Xi_1(z, a)}{dz^2} = az\Xi_1(z, a) + \Xi_0(z, a), \tag{36}$$

with the initial conditions $\Xi_1(0, a) = 0$, $\Xi_1'(0, a) = 0$. It is possible to show that the corresponding solution is defined by the series

$$\Xi_1(z, a) = \sum_{n=0}^{\infty} a^n \left(\frac{3^{n+1}(n+1)!}{(3n+4)!} - \frac{1}{3^{n+1}n!} \prod_{k=1}^{n+1} \frac{1}{3k+1} \right) z^{3n+4}, \tag{37}$$

that can be resummed to the difference of two generalized hypergeometric functions:

$$\Xi_1(z, a) = \frac{z^4}{8} {}_1F_2\left(1; \frac{5}{3}, \frac{7}{3}; \frac{az^3}{9}\right) - \frac{z^4}{12} {}_1F_2\left(1; 2, \frac{7}{3}; \frac{az^3}{9}\right). \tag{38}$$

If $\xi_{1,k}$ are the zeros of $\Xi(z, a, b)$ and $\xi_{k,0}$ are those of $\Xi(z, a, 0)$, at first order in b we have

$$\xi_{1,k} - \xi_{0,k} = -b\frac{\Xi_1(z_{0,k}, a)}{\Xi_0'(z_{0,k}, a)}, \tag{39}$$

or, in terms of hypergeometric functions:

$$\xi_{1,k} - \xi_{0,k} = -\frac{5b}{9a} \frac{3\,{}_1F_2\left(1; \frac{5}{3}, \frac{7}{3}; \frac{az_{0,k}^3}{9}\right) - 2\,{}_1F_2\left(1; 2, \frac{7}{3}; \frac{az_{0,k}^3}{9}\right)}{{}_1F_2\left(2; \frac{7}{3}, \frac{8}{3}; \frac{az_{0,k}^3}{9}\right)}. \tag{40}$$

Note that $\xi_{1,k} - \xi_{0,k}$ depends only on the value of $\xi_{0,k}^3$: the difference is the same for all the triples of zeros lying on the circle of radius $|\xi_{0,k}|$. All the zeros move in the same direction and, further, each triple of zeros on a given circle moves the same amount in that direction, as if they belong to a rigid body. The denominator of (39) is always different from zero since, from Proposition (2.3), the zeros of Ξ_0 are simple. The numerator is an entire function of $\xi_{0,k}$. It follows that, for any k, the ratio $\frac{\Xi_1(z_{0,k},a)}{\Xi_0'(z_{0,k},a)}$ is bounded by a suitable constant M_k. Then, for any k, it is possible to choose a constant b_k such that, for $|b| < |b_k|$, the zeros $\xi_{1,n}$, $n < k$, are all simple. This implies that, for suitable small values of b, the zeros of $\Xi(z,a,b)$ are simple (apart from the double zero in $z = 0$). This line of reasoning is independent on the value of a, since, as we said, the zeros of $\Xi_0(z,a,0)$ are only scaled and/or rotated with respect to those of $\Xi(z,1,0)$. Now we can go back to the homogeneity property (33). Indeed, we have seen that the zeros of $\Xi(z,\frac{a}{\lambda^3},0)$ are simple for any value of the ratio a/λ^3. Also, the zeros of $\Xi(z,\frac{a}{\lambda^3},b)$ are simple for sufficiently small values of b. From the homogeneity property (33), it follows that the zeros of $\Xi(z,\frac{a}{\lambda^3},b)$ coincide with those of $\Xi(\lambda^{-1}z,a,\lambda^2 b)$ for any value of λ, i.e. coincide, with a proper rescaling and rotation, with the zeros of $\Xi(z,a,\lambda^2 b)$. Since λ is arbitrary we get that the zeros of $\Xi(z,a,b)$ are all simple (apart $z = 0$) for any value of a and b.

Up to now we investigated the distribution of the zeros of the solutions of equation (9) with a double zero in $z = p$. In the rest of the chapter, we are going to investigate the distribution of the zeros of the solutions of equation (9) possessing only simple zeros (equivalently, the solution of equation (29) for suitable values of the parameters a and b). To distinguish between the two types of solutions, we make the following definition:

Definition 4.1. $X(z,a,b,c)$ are the solutions of (29) having a simple zero in an arbitrary point of the complex plane, all the other zeros being simple. These zeros are denoted by $\chi_k(a,b,c)$. $c\Xi(z,a,b)$ are the solutions of (29) having one double zero in an arbitrary point of the complex plane. The other zeros are denoted by $\xi_k(a,b)$.

It will be useful to consider the following remark:

Remark 4.2. If $y_0(z,a,b,c)$ is any particular solution of equation (29), then another solution is given by $\mathrm{T}_{A,B}y_0$, where $\mathrm{T}_{A,B}$ is defined by

$$\mathrm{T}_{A,B}y_0(z,a,b,c) = Ay_0\left(z - B, a, b + aB, \frac{c}{A}\right) \tag{41}$$

with A and B two arbitrary constants.

From Proposition (3.2) and the discussion just before that proposition, it is possible to write the following charts for the action of the operator $T_{A,B}$ on the two families of solutions $X(z,a,b,c)$ and $\Xi(z,a,b)$:

$$\text{1)}\ \ A=\frac{c\Xi'(\xi_n,a,b)}{X'(\chi_m,a,b,c)},\ B=\xi_n-\chi_m \quad\Longrightarrow\quad \begin{cases} T_{A,B}X(z,a,b,c)=c\Xi(z,a,b)\\ X(\chi_m,a,b,c)=0\\ \Xi(\xi_n,a,b)=0 \end{cases}$$

$$\text{2)}\ \ A=\frac{\tilde{X}'(\chi_n,a,b,c)}{X'(\chi_m,a,b,c)},\ B=\chi_n-\chi_m \quad\Longrightarrow\quad \begin{cases} T_{A,B}X(z,a,b,c)=\tilde{X}(z,a,b,c)\\ X(\chi_m,a,b,c)=0\\ \tilde{X}(\chi_n,a,b,c)=0 \end{cases}$$

The transformation (1) brings the element of the family $X(z,a,b,c)$ having a zero in $z=\xi_m$ to an element of the family $\Xi(z,a,b)$ having a zero in $z=\xi_n$: if this zero is simple, then $A\neq 0$ and the function $\Xi(z,a,b)$ has a double zero in some other point of the complex plane. If ξ_n is the double zero, then $A=0$. The transformation $T_{0,B}$ eliminates all the terms not proportional to c in $X(z,a,b,c)$. We recall that the element of the family $X(z,a,b,c)$ having a zero in an arbitrary point $z=\chi_m$ is linear in c and in the first derivative of X evaluated in χ_m, as can be seem also from Proposition 3.1).

The transformation (2) brings the element of the family $X(z,a,b,c)$ having a zero in $z=\xi_m$ to an element $\tilde{X}(z,a,b)$ of the same family. This new solution has a simple zero in $z=\chi_n$.

It is also possible to get a transformation from $\Xi(z,a,b,c)$ to the same family: the value of A is redundant in this case since the starting solution is proportional to c and there is just the parameter B surviving in equation (41).

In the next Proposition it will be useful to answer the following question: is it possible, for a suitable choice of A and B in the transformation (2), to get the identity transformation? If we start from the solution possessing a zero in $z=\xi_m$, then the value of B must be equal to $\xi_n-\xi_m$, where ξ_n is another zero of the *same* solution. Also, the value of A must be equal to the ratio of the first derivative evaluated at $z=\xi_n$ over the first derivative evaluated at $z=\xi_m$. Since ξ_n is an arbitrary zero, actually there is an infinite, numerable set of choices for the values of A and B. In this case, $T_{A,B}$ is the identity transformation and we can write the following

Proposition 4.3. *If $X(z, a, b, c)$ is the solution of equation (29) having a zero in $z = \xi_m$, then one has*

$$X(z, a, b, c) = A_{n,m} X\left(z - (\xi_n - \xi_m), a, b + a(\xi_n - \xi_m), c/A_{n,m}\right), \tag{42}$$

where $A_{n,m} = \frac{X'(\xi_n, a, b, c)}{X'(\xi_m, a, b, c)}$ and ξ_n is any other zero of $X(z, a, b, c)$.

As we have seen, a solution of equation (29) with a zero in $z = 0$ corresponds to a solution of equation (9) with a zero in $z_0 = \frac{b}{\eta^2}$, where $a = \eta^3$. Also, the characterization of the solutions of equation (9) with a zero in an arbitrary point z_0 of the complex plane corresponds to the characterization of the particular solution of equation (29) having a zero in $z = 0$ for arbitrary values of a and b. Proposition (4.3) gives a quasi-periodic property of the zeros of this particular solution. Indeed, from (4.3) we get the following:

Remark 4.4. Let $S(z, a, b, c)$ be a particular solution of equation (29) having a simple zero in $z = 0$, with all the other zeros simple. Let $\{\xi_k^0(a, b, c)\}$ denote the set of the zeros of $S(z, a, b, c)$. Then from Proposition (4.3) it follows that, for any choice of ξ_n^0 one has

$$\left\{\xi_k^0\left(a, b + a\xi_n^0(a, b, c), c/A_n\right)\right\} = \left\{\xi_k^0(a, b, c) - \xi_n^0(a, b, c)\right\}, \tag{43}$$

where $A_n = \frac{S'(\xi_n, a, b, c)}{S'(0, a, b, c)}$.

We note that for $c = 0$, formula (43) gives the quasi-periodicity of the zeros of the solutions of the homogeneous Airy equation, as described in [16,17], whereas for $c = 0$ and $a = 0$ one gets the periodicity of the trigonometric functions.

5. A Cubic Recursion for the Zeros

To further characterize the zeros of the function $S(z, a, b, c)$, we can look at the logarithmic derivative of this function. Indeed, since $S(z, a, b, c)$ is an entire function with order of growth equal to $3/2$, from the Weierstrass–Hadamard factorization theorem, [14] we get

$$S(z, a, b, c) = S'(0, a, b, c)z e^{\beta z} \prod_{k=1} \left(1 - \frac{z}{\xi_k}\right) e^{\frac{z}{\xi_k}}, \qquad (44)$$

where $\beta = \frac{c}{2S'(0,a,b,c)}$ and hereafter we omit the apex "0" from the zeros ξ_n^0 for ease of readability. The logarithmic derivative of S, $u = \frac{S'}{S}$ is a meromorphic function and, from the previous product formula, possesses the (global) Mittag Leffler representation

$$u(z, a, b, c) = \frac{1}{z} + \beta + \sum_{\xi_n \neq 0} \frac{1}{z - \xi_n} + \frac{1}{\xi_n}. \qquad (45)$$

From the differential equation (29), it is possible to show that u solves the following second-order, cubic, differential equation

$$\frac{d^2 u(z)}{dz^2} + 3u(z)\frac{du(z)}{dz} + u(z)^3 = a + (az + b)u(z). \qquad (46)$$

We can expand $u(z, a, b, c)$ in a (local) Laurent series around $z = 0$, getting

$$u(z, a, b, c) = \frac{1}{z} + \sum_{n=0} c_n z^n. \qquad (47)$$

From the differential equation (46), the coefficients $c_n(a, b, c)$ solve the following recursion:

$$(n + 4)(n + 2)c_{n+2} = bc_n + ac_{n-1} - 3\sum_{k=0}^{n+1} c_k c_{n+1-k}$$
$$+ - 3\sum_{k=0}^{n}(k + 1)c_{k+1}c_{n-k} - \sum_{k=0}^{n}\sum_{j=0}^{k} c_j c_{k-j} c_{n-k}, \quad n \geq 1, \quad (48)$$

where $c_0 = \beta, c_1 = \frac{b}{3} - c_0^2, c_2 = c_0^3 - \frac{1}{4}bc_0 + \frac{a}{4}$. The compatibility between the expansion (47) and the representation (45) gives the following expression relating the coefficients c_n to the zeros ξ_k:

$$c_n = -\sum_{\xi_k \neq 0} \frac{1}{\xi_k^{n+1}}, \quad n = 1, 2... \qquad (49)$$

In the case $c = 0$ (i.e. for the solutions of the homogeneous Airy equation), the previous recursion is not more cubic but quadratic in the coefficients c_n. In that case, formula (48) can be inverted, in the sense that given all the coefficients c_n it is possible, in principle, to calculate all the zeros ξ_k [16,17]. In this case, we can repeat the same arguments as given in [16] and we refer to that work for more details. The result is the following:

Proposition 5.1. Let $c_n(a, b + ax, \beta)$ *be the sequence of polynomials defined by the recursion (48) with* $b \to b + ax$. *Then, if the distance between successive zeros is decreasing and* ξ_1 *is the zero of the function* $S(z, a, b, c)$ *closest to the origin, a subset of the zeros of* $S(z, a, b, c)$ *are given by*

$$\xi_{k+1} = \xi_k + \lim_{n \to \infty} \frac{c_n(a, b + a\xi_k, \beta/S'(\xi_k))}{c_{n+1}(a, b + a\xi_k, \beta/S'(\xi_k))}, \quad \xi_0 = 0. \tag{50}$$

The subset of the zeros are those on the semi-axis containing both 0 and ξ_1 in the direction of ξ_1. We are assuming also that there is just one zero closest to the origin. On the contrary, with more than one zero closest to the origin lying on the circle of radius $|\xi_1|$, the right-hand side of (50) would oscillate indefinitely. The constraint that the distance between successive zeros must be decreasing is crucial: for $c = 0$ (i.e. $\beta = 0$ in the recursion (50)) it has been shown [16] that, for suitable choice of the constants a and b, this property holds true and the recursion (50) can be effectively used to calculate the corresponding zeros. In the case $c \neq 0$, the situation is different: for small values of c we expect that this property may hold true again, but, by increasing c, we have a different picture. Also, another zero can appear on the semi-axis opposite to that containing 0 and ξ_1. This rich and interesting portrait will be further considered in a next work. Here, we will give just few examples of applications of the recursion (50). We know from [16,17] that, for $c = 0$, $a < 0$ and $b < 0$, $S(z, a, b, 0)$ has an infinite number of zeros on the positive real axis. We first consider small values of the parameter c: for definiteness, we take $(a, b, c) = (-1, -1, -0.1)$. Then, we will consider a value of c comparable to those of a and b: we will take $(a, b, c) = (-1, -1, -1)$. In both the examples we fix the value of $S'(0, a, b, c)$ to be equal to 1 (the distribution of the zeros depends only on the ratio between c and $S'(0, a, b, c)$). In the last case, the zero closest to $z = 0$ is real and is equal to

$$\xi_1 = 1.4230603\ldots \tag{51}$$

The next real zero is about equal to $\xi_2 = 3.53175\ldots$, meaning that the condition given in Proposition (5.1) about the distance of successive zeros is not satisfied, since $|\xi_2 - \xi_1| > |\xi_1 - \xi_0|$: *in this case the recursion (50) gives the zero that is closest to* ξ_1, i.e. one gets

$$\xi_2 = 0 = \xi_1 + \lim_{n \to \infty} \frac{c_n(-1, -1 - \xi_1, -1/2)}{c_{n+1}(-1, -1 - \xi_1, -1/2)}, \tag{52}$$

that is a 2-periodic sequence. On the contrary, in the first case, i.e. for $(a, b, c) = (-1, -1, -0.1)$, the conditions of Proposition (5.1) are satisfied: the zero closest to $\xi_0 = 0$ is

$$\xi_1 = 2.0977152\ldots \tag{53}$$

The next two zeros, found with the recursion (50) itself, are given by

$$\xi_2 = 3.7233151\ldots, \quad \xi_3 = 5.0507149\ldots \tag{54}$$

The previous zeros have been obtained by taking the ratio of c_n/c_{n+1} up to $n = 80$: with larger number of n it would be possible to get more significant digits in the calculation of the zeros.

6. Conclusions

We hope to have highlighted some of the very interesting mathematical structures defined by the general solutions of equation (9) in the complex plane. The possibility to look at a particular solution of equation (29) to characterize the general solution of equation (9) is surely useful and may be relevant in future potential developments of this work. The dynamics of the zeros by varying the parameters in equation (29), in the spirit of [17], is an issue that will be investigated in future works. The relation between nonlinear recurrence relations, like (48), and the zeros of entire functions, seems to be another point deserving more consideration in the literature.

Acknowledgments

The support of University of Brescia, INdAM-GNFM and INFN is gratefully acknowledged. Also, the author wishes to thank Prof. O. Ragnisco for his advice and comments.

References

1. R. Conte, A. P. Fordy, and A. Pickering, A perturbative Painlevé approach to nonlinear differential equations. *Physica D*, **69**(1993), 33–58.

2. A. Gil, J. Segura, N. M. Temme, On nonoscillating integrals for computing inhomogeneous Airy functions, *Mathematics of Computation*, **70**(235) (2000), 1183—1194.

3. A. Gil, J. Segura, and N. M. Temme, On the zeros of the Scorer functions, *J. Approxi. Theory*, **120**(2020), 253–266.

4. E. Hille, *Ordinary Differential Equations in the Complex Domain*, John Wiley & Sons, New York, 1976.

5. A. N. W. Hone, Painleve tests, singularity structure and integrability, in Integrability, A. V. Mikhailov (ed.), *Lect. Notes Physics* 767. Springer, Berlin, 2009, pp. 245–277.

6. N. A. Kudryashov, D. I. Sinelshchikov, Analytical and numerical studying of the perturbed Korteweg-de Vries equation. *J. Mathemat. Phy.*, **55**(2014), 103504. doi: 10.1063/1.4897445.

7. I. Laine, *Nevanlinna Theory and Complex Differential Equations*. W. de Gruyter, Berlin, Germany, 1993.

8. Y. Lin and R. Wong, *Asymptotics of Generalized Hypergeometric Functions, in Frontiers in Orthogonal Polynomials and q-Series*, M. Z. Nashed and X. Li (eds.), World Scientific, Singapore, 2018.

9. A. J. MacLeod, Computation of inhomogeneous Airy functions. *J. Comput. Appl. Mathe.*, **53**(1994), 109–116.

10. NIST Digital Library of Mathematical Functions. Available at https://dlmf.nist.gov/9.12, F. W. J. Olver, A. B. Olde Daalhuis, D. W. Lozier, B. I. Schneider, R. F. Boisvert, C. W. Clark, B. R. Miller, and B. V. Saunders, (eds.).

11. F. W. J. Olver, *Asymptotics and Special Functions*. Academic Press, New York. Reprinted in 1997 by A. K. Peters.

12. Polya, G, Über die Nullstellen gewisser ganzer Funktionen, *Mathematische Zeitschrift*, **2**(1918), 352–383.

13. R. S. Scorer, Numerical evaluation of integrals of the form $I = \int_{x_1}^{x_2} f(x)e^{i\phi(x)}dx$ and the tabulation of the function $\mathrm{Gi}(z) = \sin(uz + \frac{1}{3}u^3)du$. *Quart. J. Mech. Appl. Math.*, **3**(1950), 107–112.

14. E. C. Titchmarsh, *The Theory of Functions*, Oxford University Press, London, 1939.

15. G. N. Watson, *A Treatise on the Theory of Bessel Functions*, Cambridge University Press, Cambridge, 1944.

     F. Zullo

16. F. Zullo, *On the Solutions of the Airy Equation and Their Zeros, in Complex Differential and Difference Equations*, G. Filipuk, A. Lastra, S. Michalik, Y. Takei and H. Żoładek (ed.), De Gruyter Proceedings in Mathematics, De Gruyter, Berlin/Boston, 2020.
17. F. Zullo, On the dynamics of the zeros of solutions of the Airy equation, *Mathematics and Computers in Simulation*, **176**(2020), 312–318.

© 2022 World Scientific Publishing Europe Ltd.
https://doi.org/10.1142/9781800611368_0006

Chapter 6

Bäcklund Transformations: A Tool to Study Abelian and Non-Abelian Nonlinear Evolution Equations

Sandra Carillo[*,†] and Cornelia Schiebold[‡,§]

*Basic and Applied Sciences Department (SBAI), Sapienza University of
Rome, Rome, Italy
†Gr. Roma1, IV, Mathematical Methods in NonLinear Physics
National Institute for Nuclear Physics (I.N.F.N.), Rome, Italy
‡Department of Natural Sciences, Engineering, and Mathematics,
Mid Sweden University, Sundsvall, Sweden
§cornelia.schiebold@miun.se

Bäcklund transformations, well known to represent a key tool in the
study of nonlinear evolution equations, are shown to allow the construction of a net of nonlinear links, termed *Bäcklund chart*, connecting Abelian as well as non-Abelian equations. In particular, Bäcklund
transformations are applied to reveal algebraic properties enjoyed by
nonlinear evolution equations they connect. The present study concerns
third-order nonlinear evolution equations, termed KdV-type, which are
all connected to the KdV equation. The Abelian wide Bäcklund chart
connecting these nonlinear evolution equations is recalled. Then, the
links, originally established in the case of Abelian equations, are shown
to conserve their validity when non-Abelian counterparts are considered. In addition, the non-commutative case reveals a richer structure
since there may be more than a single non-Abelian counterpart of the
same Abelian equation. Reduction from the non-commutative to the
commutative case allows to show the connection of the KdV equation
with KdV eigenfunction equation, in the *scalar* case. The main result
presented refers to the KdV eigenfunction equation: some explicit solutions it admits are constructed on application of an invariance property

proved via Bäcklund transformations. These, to the best of the authors' knowledge, new solutions represent an example of the powerfulness of the method devised. Matrix solutions of the mKdV equations, recently obtained, are mentioned in the closing remarks to stress the powerfulness of the method.

1. Introduction

The crucial role played by Bäcklund transformations in investigating *soliton equations* is well known as testified by [1,7,32,52,55,56] referring only to some well-known books on the subject. Third-order nonlinear evolution equations, termed KdV-type, are studied. Indeed, both in the Abelian case [9,10] as well as in the non-Abelian one [11,14,16,17], a wide net of nonlinear evolution equations turns out to be connected to the Korteweg–deVries (KdV) equation or, respectively, to its non-commutative counterpart. The net of links, termed Bäcklund chart, allows to reveal many interesting properties enjoyed by the nonlinear evolution equations it connects. Specifically, all algebraic properties which are preserved under Bäcklund transformations can be transferred from one equation to all the others within the Bäcklund chart. The nonlinear evolution equations under investigation, further to the KdV equation, are the potential Korteweg–deVries (pKdV), the modified Korteweg–deVries (mKdV), and the KdV eigenfunction (KdV eig.). In addition, in the commutative case [9,10] also the Dym equation is included in the Bäcklund chart. On the other hand, in the non-commutative one, as pointed out in [17], there are two different mKdV equations which are connected to each other via two further nonlinear evolution equations: both of them reducing to the KdV eigenfunction equation when the non-commutativity condition is removed.

Section 2 is devoted to briefly recall the definition of Bäcklund transformation together with some remarkable properties. In the subsequent Section 3, the Bäcklund chart connecting Abelian KdV-type equations is provided. In particular, the Bäcklund chart in [9,10] is recalled. In Section 4, the nontrivial invariance exhibited by the KdV eigenfunction equation is used to construct some of its explicit solutions. These solutions are independent of time, but show an explicit dependence on the space variable x. The extension of the Bäcklund chart, induced by the Möbius invariance enjoyed by the

KdV singularity equation, is given in Section 5. Section 6 reconsiders the non-Abelian Bäcklund chart, constructed in [11,14,16]. The last section briefly discusses further remarkable results Bäcklund transformations allow to achieve. Matrix solutions, generalization of results to hierarchies of nonlinear evolution equations are mentioned.

2. Bäcklund Transformations

This section is devoted to a brief review on Bäcklund transformations[a] aiming to provide the definitions needed in the following; thus, given two nonlinear evolution equations of the type

$$u_t = K(u), \quad v_t = G(v), \tag{1}$$

the following definition, see [24], is adopted.

Definition 1. Given two evolution equations,

$$u_t = K(u), \ K : M_1 \to TM_1, \ u \colon (x,t) \in \mathbb{R}^n \times \mathbb{R} \to u(x,t) \in \mathbb{R}^m \subset M_1,$$
$$v_t = G(v), \ G : M_2 \to TM_2, \ v \colon (x,t) \in \mathbb{R}^n \times \mathbb{R} \to v(x,t) \in \mathbb{R}^m \subset M_2,$$

then B (u , v) = 0 represents a Bäcklund transformation between them whenever given two solutions of such equations, say, respectively, $u(x,t)$ and $v(x,t)$, such that

$$B(u(x,t), v(x,t))|_{t=0} = 0, \tag{2}$$

it follows that,

$$B(u(x,t), v(x,t))|_{t=\bar{t}} = 0, \quad \forall \bar{t} > 0 \ , \ \forall x \in \mathbb{R}. \tag{3}$$

As a usual choice, when soliton solutions are considered, it is assumed $M := M_1 \equiv M_2$ and, in addition, the generic *fiber* $T_u M$, at $u \in M$, is identified with M itself.[b]

---

[a]The interested reader is addressed to [52,55] and references therein, where it is possible to find the general definition of Bäcklund transformation in implicit form as well as a detailed study on the subject.

[b]It is generally assumed that M is the space of functions $u(x,t)$ which, for each fixed t, belong to the Schwartz space S of *rapidly decreasing functions* on $\mathbb{R}^n$, i.e. $S(\mathbb{R}^n) := \{f \in C^\infty(\mathbb{R}^n) : \|f\|_{\alpha,\beta} < \infty, \forall \alpha, \beta \in \mathbb{N}_0^n\}$, where $\|f\|_{\alpha,\beta} := sup_{x \in \mathbb{R}^n} |x^\alpha D^\beta f(x)|$, and $D^\beta := \partial^\beta / \partial x^\beta$; and throughout this Chapter, $n = 1$.

As a consequence, solutions of such two equations are linked via the Bäcklund transformation which establishes a correspondence between them: it can be graphically represented as follows:

$$\boxed{u_t = K(u)} \overset{B}{\text{---}} \boxed{v_t = G(v)} \tag{4}$$

which shows that the Bäcklund transformation relates the two non-linear evolution equations.

3. Abelian Bäcklund Chart

The net of links connecting the many different nonlinear evolution equations can be summarized in the Bäcklund chart in [17]. Indeed, the construction is directly related to results in [9,28] further developed in [11,12,14,16], where an analogous Bäcklund chart is constructed in the case of non-commutative nonlinear evolution equations. A comparison between the Abelian and the non-Abelian Bäcklund chart is analyzed in [17]. The links among the various KdV-type equations are summarized in the Bäcklund chart (Fig. 1) where, in turn, the linked nonlinear evolution equations are as follows:

$$u_t = u_{xxx} + 6uu_x \qquad\qquad\qquad \text{(KdV)},$$

$$v_t = v_{xxx} - 6v^2 v_x \qquad\qquad\qquad \text{(mKdV)},$$

$$w_t = w_{xxx} - 3\frac{w_x w_{xx}}{w} \qquad\qquad\qquad \text{(KdV eig.)},$$

$$\varphi_t = \varphi_x\{\varphi; x\}, \text{ where } \{\varphi; x\} := \left(\frac{\varphi_{xx}}{\varphi_x}\right)_x - \frac{1}{2}\left(\frac{\varphi_{xx}}{\varphi_x}\right)^2 \text{ (KdV sing.)},$$

$$s^2 s_t = s^2 s_{xxx} - 3ss_x s_{xx} + \frac{3}{2}s_x{}^3 \qquad\qquad \text{(int. sol. KdV)},$$

$$\rho_t = \rho^3 \rho_{\xi\xi\xi} \qquad\qquad\qquad\qquad \text{(Dym)},$$

$$\boxed{\text{KdV}(u)} \overset{(a)}{\text{---}} \boxed{\text{mKdV}(v)} \overset{(b)}{\text{---}} \boxed{\text{KdV eig.}(w)} \overset{(c)}{\text{---}} \boxed{\text{KdV sing.}(\varphi)} \overset{(d)}{\text{---}} \boxed{\text{int. sol KdV}(s)} \overset{(e)}{\text{---}} \boxed{\text{Dym}(\rho)}$$

Fig. 1.   KdV-type equations Bäcklund chart.

that is, in order, the KdV, the mKdV, and the KdV eigenfunction (KdV eig.), the Korteweg–deVries interacting soliton (int. sol. KdV), the Korteweg–deVries singularity manifold (KdV sing.)[c] and the Dym equations. Respectively, in the Bäcklund chart, (a), (b), (c), (d), (e) denote the following Bäcklund transformations

$$\text{(a) } u + v_x + v^2 = 0, \qquad \text{(b) } v - \frac{w_x}{w} = 0, \qquad (5)$$

$$\text{(c) } w^2 - \varphi_x = 0, \qquad \text{(d) } s - \varphi_x = 0, \qquad (6)$$

and

$$\text{(e) } \bar{x} := D^{-1}s(x), \ \rho(\bar{x}) := s(x), \quad \text{where } D^{-1} := \int_{-\infty}^{x} d\xi, \qquad (7)$$

so that $\bar{x} = \bar{x}(s, x)$ and, hence, $\rho(\bar{x}) := \rho(\bar{x}(s, x))$. The transformation (e) is termed *reciprocal transformation* since it interchanges the role of the dependent and independent variables.[d]

4. The KdV Eigenfunction Equation: Invariance Properties and Solutions' Construction

The *KdV eigenfunction* equation (KdV eig.) is included in a wide study by Konopelchenko in [38] where, among many other ones, it is proved to be integrable via the *inverse spectral transform* (IST) method. Indeed, this equation was firstly derived in a founding article of the IST method [47] and also [61], later further investigated in [38,43] wherein a wide variety of nonlinear evolution equations have been studied. Nevertheless, the KdV eigenfunction equation does not appear in subsequent classification studies of integrable nonlinear evolution equations, such as [5,44,45,50,62], until very recently when, in [4], linearizable nonlinear evolution equations have been classified.

---

[c]Throughout the chapter, the equation, introduced by Weiss in [63], is termed *Korteweg–deVries singularity manifold equation (KdV sing.)* following Fuchssteiner's notation according to [27,28]; however, the same equation is also known as UrKdV or Schwarzian KdV [64].

[d]See, for instance, [55] where reciprocal transformations are defined and applications are provided. The transformation (e) is analyzed in [10,28] where it is shown to represent a Bäcklund transformation between the extended manifold consisting of both the dependent and the independent variables.

4.1. *Invariance*

This section is concerned only about an invariance property enjoyed by the KdV eigenfunction equation. It can be trivially checked to be *scaling invariant* since on substitution of $\alpha w, \forall \alpha \in \mathbb{C}$, to w it remains unchanged. In addition, according to [9] (see prop. 4 therein) the following proposition shows further nontrivial invariances.

Proposition 1. *The KdV eigenfunction equation* $w_t = w_{xxx} - 3\dfrac{w_x w_{xx}}{w}$ *is invariant under the transformation*

$$I : \hat{w}^2 = \frac{ad - bc}{(cD^{-1}(w^2) + d)^2} w^2,$$

$$a, b, c, d \in \mathbb{C} \quad s.t. \ ad - bc \neq 0, \quad \text{where } D^{-1} := \int_{-\infty}^{x} d\xi. \quad (8)$$

Note that D^{-1} is well defined since the so-called *soliton solutions* are looked for in the Schwartz space $S(\mathbb{R}^n)$.

The proof, according to [9], is based on the invariance under the Möbius group of transformations

$$M : \ \hat{\varphi} = \frac{a\varphi + b}{c\varphi + d}, \quad a, b, c, d \in \mathbb{C} \quad \text{such that} \quad ad - bc \neq 0,$$

$$(9)$$

enjoyed by the KdV singularity manifold equation (KdV sing.). Indeed, the invariance I is induced by the Bäcklund chart in Fig. 2, which shows how the invariance under the Möbius group of transformations M, combined with the Bäcklund transformations B and $\widehat{B}$,

$$
\begin{array}{ccc}
\boxed{w_t = w_{xxx} - 3\dfrac{w_x w_{xx}}{w}} & \xrightarrow{\ B\ } & \boxed{\varphi_t = \varphi_x\{\varphi; x\}} \\[2ex]
\updownarrow \ I & & \updownarrow \ M \\[2ex]
\boxed{\hat{w}_t = \hat{w}_{xxx} - 3\dfrac{\hat{w}_x \hat{w}_{xx}}{\hat{w}}} & \xrightarrow{\ \widehat{B}\ } & \boxed{\hat{\varphi}_t = \hat{\varphi}_x\{\hat{\varphi}; x\}}
\end{array}
$$

Fig. 2. Induced invariance Bäcklund chart.

given in (6*c*), i.e. respectively:

$$\text{B}: \quad w^2 - \varphi_x = 0 \quad \text{and} \quad \widehat{\text{B}}: \quad \hat{w}^2 - \hat{\varphi}_x = 0 \ ,$$

induces the invariance I. Via the invariance I, it is possible to construct solutions of the KdV eigenfunction equation.

4.2. *Explicit solutions: An example*

In this subsection, an example of solutions admitted by the KdV eigenfunction equation is constructed on the basis of the invariance presented in the previous subsection. Indeed, it is easily checked that $w(x,t) = k_1, \forall k_1 \in \mathbb{R}$ represents a solution of the KdV eigenfunction equation

$$w_t = w_{xxx} - 3\frac{w_x w_{xx}}{w} \ . \tag{10}$$

When, in the Möbius group the parameters are set to be

$$a = d = 0, \quad b = 1, \quad c = -1, \tag{11}$$

the invariance I indicates that also

$$\hat{w}(x,t) = (k_1^2 x + k_2)^{-1} \ , \tag{12}$$

$\forall k_2 \in \mathbb{R}$, represent solutions of the KdV eigenfunction equation.

Further solutions can be obtained in the same way. Remarkably, also in the non-commutative case solutions can be constructed.

5. Extension of the Abelian Bäcklund Chart

Note that the whole Bäcklund chart in Fig. 1 can be extended, as indicated in Fig. 3.

$$\boxed{\text{KdV}(u)}\overset{(a)}{-}\boxed{\text{mKdV}(v)}\overset{(b)}{-}\boxed{\text{KdV eig.}(w)}\overset{(c)}{-}\boxed{\text{KdV sing.}(\varphi)}\overset{(d)}{-}\boxed{\text{int. sol KdV}(s)}\overset{(e)}{-}\boxed{\text{Dym}(\rho)}$$

$$AB_1 \updownarrow \qquad AB_2 \updownarrow \qquad AB_3 \updownarrow \qquad M \updownarrow \qquad AB_4 \updownarrow \qquad AB_5 \updownarrow$$

$$\boxed{\text{KdV}(\tilde{u})}\overset{(a)}{-}\boxed{\text{mKdV}(\tilde{v})}\overset{(b)}{-}\boxed{\text{KdV eig.}(\tilde{w})}\overset{(c)}{-}\boxed{\text{KdV sing.}(\tilde{\varphi})}\overset{(d)}{-}\boxed{\text{int. sol KdV}(\tilde{s})}\overset{(e)}{-}\boxed{\text{Dym}(\tilde{\rho})}$$

Fig. 3.   Abelian KdV-type hierarchies Bäcklund chart: induced invariances.

That is, since the KdV singularity manifold equation (KdV sing.) is invariant under the Möbius group of transformations (9), all the auto-Bäcklund transformations $AB_k, k = 1\ldots 5$ follow. Note that AB_1 and AB_2 are, respectively, the well-known KdV and the mKdV auto-Bäcklund transformations [7,28,46]. The invariance of the KdV eigenfunction equation is I$\equiv AB_3$ [9] and, according to [28], auto-Bäcklund transformations of the int. sol. KdV and Dym equations are also obtained, denoted as AB_4, and AB_5. In detail, the listed auto-Bäcklund transformations follow:

$$AB_1 : \ (u + \hat{u}) + \frac{1}{2}\left(D^{-1}(u + \hat{u})\right)^2 = 0, \tag{13}$$

$$AB_2 : \ (v + \hat{v})_x - (v + \hat{v})^2 = 0, \tag{14}$$

$$AB_3 \equiv \text{I} : \ \hat{w}^2 = \frac{ad - bc}{(cD^{-1}(w^2) + d)^2} w^2, \tag{15}$$

$$\text{M} : \ \hat{\varphi} = \frac{a\varphi + b}{c\varphi + d}, \quad a, b, c, d \in \mathbb{C} \text{ such that } ad - bc \neq 0, \tag{16}$$

$$AB_4 : \ \hat{s} = \frac{aD^{-1}s + b}{cD^{-1}s + d}, \tag{17}$$

$$AB_5 : \ \rho(\bar{x}) \to \hat{\rho}(\hat{x}) = \frac{(c\bar{x} + d)^2}{ad - bc}\rho(\bar{x}), \quad \bar{x} \to \hat{x} = \frac{a\bar{x} + b}{c\bar{x} + d}, \tag{18}$$

where $a, b, c, d \in \mathbb{C}$ such that $ad - bc \neq 0$, the operator D^{-1} as well as the independent and dependent variables $\bar{x}$ and $\rho(\bar{x})$ in the Dym equation, related via the reciprocal transformation (e) to the int. sol KdV equation, are given in (7).

6.   Non-Abelian Bäcklund Chart

In this section, attention is focused on non-Abelian equations. Specifically, according to [2,21,42], this section is concerned with nonlinear evolution equations in which the unknown is an operator on a Banach space. These, for short, are termed *operator equations*. Crucial to the present study, both in the Abelian as well as in the non-Abelian setting, is that the algebraic properties of interest are invariant under Bäcklund transformations. To stress the distinction between scalar and operator unknown functions, respectively, scalar quantities are denoted by lower-case letters while upper-case letters

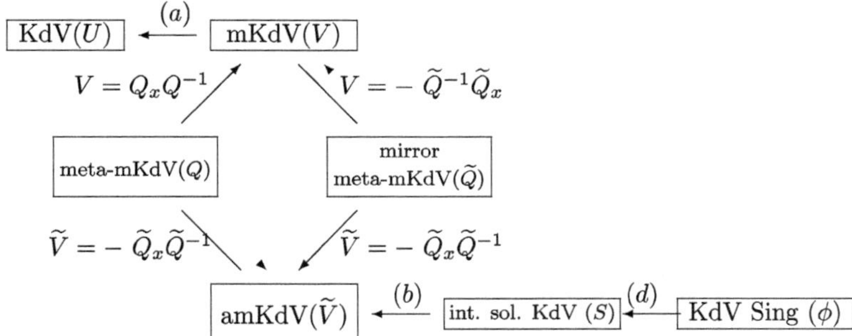

Fig. 4. KdV-type equations Bäcklund chart: the non-Abelian case.

are used referring to operators. Taking into account the results in [3,11,14,16], a Bäcklund chart which connects operator KdV-type equations, can be constructed: it is depicted in Fig. 4.

In Fig. 4, the third-order nonlinear operator evolution equations are presented, where, as pointed out, all unknowns are denoted via capital case letters. The KdV-type operator equations are, in turn,

$$U_t = U_{xxx} + 3\{U, U_x\} \qquad \text{(KdV)},$$

$$V_t = V_{xxx} - 3\{V^2, V_x\} \qquad \text{(mKdV)},$$

$$Q_t = Q_{xxx} - 3Q_{xx}Q^{-1}Q_x \qquad \text{(meta-mKdV)},$$

$$\widetilde{Q}_t = \widetilde{Q}_{xxx} - 3\widetilde{Q}_x\widetilde{Q}^{-1}\widetilde{Q}_{xx} \qquad \text{(mirror meta-mKdV)},$$

$$\widetilde{V}_t = \widetilde{V}_{xxx} + 3[\widetilde{V}, \widetilde{V}_{xx}] - 6\widetilde{V}\widetilde{V}_x\widetilde{V} \qquad \text{(amKdV)},$$

$$\phi_t = \phi_x\{\phi; x\}, \text{where} \quad \{\phi; x\} = \left(\phi_x^{-1}\phi_{xx}\right)_x$$
$$- \frac{1}{2}\left(\phi_x^{-1}\phi_{xx}\right)^2 \qquad \text{(KdV sing.)},$$

$$S_t = S_{xxx} - \frac{3}{2}\left(S_x S^{-1} S_x\right)_x \qquad \text{(int. sol KdV)},$$

where the Bäcklund transformations (a), (b), (c) linking the KdV with the mKdV, the alternative modified Korteweg–deVries (amKdV) with the int.sol KdV and the latter with the KdV sing. are the non-commutative counterparts of those in the commutative

                     S. Carillo & C. Schiebold

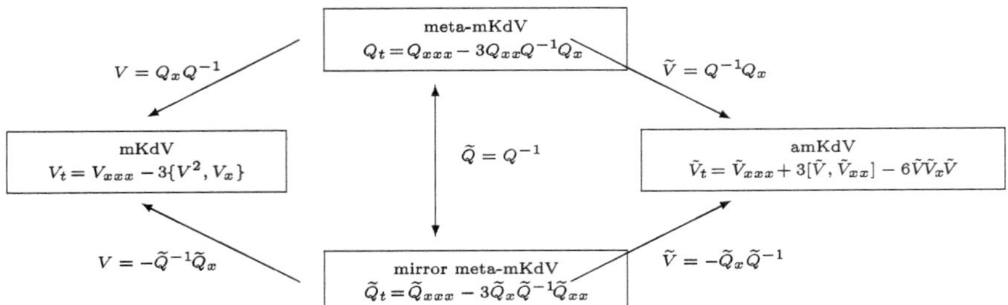

Fig. 5. Non-Abelian modified KdV equations: a *closer* look.

Bäcklund chart (see Figs. 1 and 3).

$$U = -\,(V^2 + V_x), \tag{a}$$

$$\widetilde{V} = \frac{1}{2}S^{-1}S_x, \tag{b}$$

$$S = \phi_x\,. \tag{c}$$

Notably, the two equations, which in [16] are termed meta-mKdV and mirror meta-mKdV coincide with the KdV eigenfunction equation when the non-commutativity condition is removed. Similarly, also the mKdV and amKdV equations when commutativity is assumed reduce to the usual mKdV equation and, therefore, the *box* in Fig. 4 collapses to a line and the Bäcklund chart in Fig. 1 can be recovered. The links are depicted in the following Fig. 5, which explains the situation.

In particular, it can be noted that, when the non-commutativity condition is removed, the amKdV and mKdV equations coincide both with the commutative mKdV equation. Furthermore, composition of the transformations $V = Q_xQ^{-1}$ and $\widetilde{V} = -\,Q^{-1}Q_x$ reduce to $V \to -V$, a trivial sign invariance admitted by the mKdV equation; correspondingly, the two forms of modified KdV equations (amKdV and mKdV) coincide with the (Abelian) mKdV equation.

7. Conclusions and Perspectives

This closing section aims to give a brief overview on two different lines of further developments in the present research project. On the one hand, the determination of explicit solutions, and on the other hand, the extension of the Bäcklund chart from nonlinear evolution equations to the corresponding hierarchies are considered. Two different lines of developments are currently under investigation: the first one closely related to applicative results while the second one is more concerned with algebraic properties of soliton equations.

- A special case which is of applicative interest concerns Matrix solutions of soliton equations. This is the special case when the operator is finite dimensional so that it admits a matrix representation. Thus, the aim is to emphasize the importance of Bäcklund transformations also when solutions admitted by non-Abelian soliton equations are looked for. Solutions admitted by the matrix equations are a subject of interest in the literature. The study presented, based on previous results [11,12] further developed in [18,19], takes into account multisoliton solutions of the matrix KdV equation obtained by Goncharenko [31], via a generalization of the Inverse Scattering Method. Theorem 3 in [12] generalizes Goncharenko's multisoliton solutions which follow as special ones. Solutions of matrix mKdV equation are obtained in [18], where some 2×2 and some 3×3 examples are presented; in [19] the solution formula, which in [12] is obtained in the general operator case, is discussed in the case of a $d \times d, d \in \mathbb{N}$. Further solutions are produced to give an idea of the results in this direction and currently under investigation [20]. Further matrix solutions are obtained in [23,31,35,40,58–60].
- Some further observations deserve attention. First of all, one of the properties of Bäcklund transformations crucial in the present research project, which involves not only the two authors but also further collaborators, is the notion of *recursion operator*. Indeed the existence of a hereditary recursion operator admitted by a nonlinear evolution equation is a remarkable algebraic property [25]. Such a property, on the one hand, allows to construct a whole *hierarchy* of nonlinear evolution equations associated, for instance, to the KdV equation. On the other hand, the algebraic properties which characterize a hereditary recursion operator are preserved under Bäcklund transformations. Hence, since the KdV equation admits a hereditary recursion operator, the Bäcklund chart in Fig. 1 indicates the way to construct the recursion operators of

all the nonlinear evolution equations it connects. In addition, such a Bäcklund chart can be naturally extended to the whole hierarchies of all the nonlinear evolution equations therein [28], in the Abelian case. The corresponding non-Abelian Bäcklund chart is constructed in [11,12] and further extended in [14,16]. It is worth mentioning that even the case of the non-Abelian Burgers equation [13,15,33,34,36,39] shows a richer structure with respect to the corresponding Abelian case.

- Furthermore, a $2 + 1$-dimensional [48] Bäcklund chart which links the Kadomtsev–Petviashvili (KP), the modified Kadomtsev-Petviashvili (mKP) and other $2+1$-dimensional soliton equations, such as the KP singularity manifold equation and the $2 + 1$-dimensional version of the Dym equation. The link, obtained by Rogers [51], between the KP and the $2+1$-dimensional Dym equation indicates the way to the construction of solutions of the Dym equation in $2+1$-dimensions. Thus, the study of $2+1$-dimensional non-Abelian equations seems of interest.

- In [48], it is shown that the Bäcklund chart in Fig. 3 can be regarded as a *constrained* version of the KP Bäcklund chart in [49]. In addition, as shown in [28], the Hamiltonian and bi-Hamiltonian structure admitted by the KdV equation, since these properties are preserved under Bäcklund transformations [25], all the nonlinear evolution KdV-type equations in the Bäcklund chart, in Fig. 3, are all proved to admit a bi-Hamiltonian structure [26,29,30,41].

- As discussed in [8], a Bäcklund chart connects the Caudrey–Dodd–Gibbon–Sawata-Kotera and Kaup–Kupershmidt hierarchies [22,37,57]. All the involved equations are 5th-order nonlinear evolution equations; notably, the Bäcklund chart linking them all shows an impressive resemblance to the one connecting KdV-type equations. Again, such a Bäcklund chart can be extended to the corresponding whole hierarchies [10,53].

Acknowledgments

The financial support of G.N.F.M.-I.N.d.A.M., I.N.F.N. and SAPIENZA University of Rome, Italy, is gratefully acknowledged. C. Schiebold thanks SAPIENZA University of Rome, Italy, and, in particular, Dipartimento Scienze di Base e Applicate per l'Ingegneria, for kind hospitality.

References

1. M. J. Ablowitz and H. Segur, Solitons and the inverse scattering transform. *SIAM Stud. Appl. Math.*, **4**(1981).
2. H. Aden and B. Carl, On realizations of solutions of the KdV equation by determinants on operator ideals. *J. Math. Phys.*, **37**(1996), 1833–1857.
3. C. Athorne and A. Fordy, Generalised KdV and MKdV equations associated with symmetric spaces. *J. Phys. A: Math. Gen.*, **20**(1987), 1377–1386.
4. P. Basarab-Horwath and F. Güngör, Linearizability for third order evolution equations. *J. Math. Phys.*, **58**(2017), 081507.
5. F. Calogero and A. Degasperis, A modified Korteweg-deVries equation. *Inv. Probl.*, **1**(1985), 57–66.
6. F. Calogero and S. De Lillo, The Burgers equation on the semiline with general boundary conditions at the origin. *J. Math. Phys.*, **32**(1) (1991), 99–105.
7. F. Calogero and A. Degasperis, Spectral transform and solitons I. *Studies in Mathematics and its Application*, Vol. 13. North Holland, Amsterdam, 1980.
8. S. Carillo, Nonlinear evolution equations: Bäcklund transformations and Bäcklund charts. *Acta Applicandae Math.*, **122**(1) (2012), 93–106.
9. S. Carillo, KdV-Type Equations Linked via Bäcklund transformations: Remarks and perspectives. *Appl. Numer. Math.*, **141**(2019), 81–90.
10. S. Carillo and B. Fuchssteiner, The abundant symmetry structure of hierarchies of nonlinear equations obtained by reciprocal links. *J. Math. Phys.*, **30**(1989), 1606–1613.
11. S. Carillo and C. Schiebold, Non-commutative KdV and mKdV hierarchies via recursion methods. *J. Math. Phys.*, **50**(2009), 073510.
12. S. Carillo and C. Schiebold, Matrix Korteweg-deVries and modified Korteweg–deVries hierarchies: Non-commutative soliton solutions. *J. Math. Phys.*, **52**(2011), 053507.
13. S. Carillo and C. Schiebold, On the recursion operator for the non-commutative Burgers hierarchy. *J. Nonlin. Math. Phys.*, **19**(1) (2012), 1250003, p. 11.
14. S. Carillo, M. Lo Schiavo and C. Schiebold, Bäcklund Transformations and Non Abelian Nonlinear Evolution Equations: A novel Bäcklund chart. *SIGMA*, **12**(2016), 087, p. 17.
15. S. Carillo, M. Lo Schiavo and C. Schiebold, Recursion Operators admitted by non-Abelian Burgers equations: Some remarks. *Math. Comp. Simul.*, **147C**(2018), 40–51.

16. S. Carillo, M. Lo Schiavo, E. Porten and C. Schiebold, A novel non-commutative KdV-type equation, its recursion operator, and solitons. *J. Math. Phys.*, **59**(3) (2018), 3053–3060.

17. S. Carillo, M. Lo Schiavo and C. Schiebold, *Abelian versus non-Abelian Bäcklund charts: Some remarks. Evolut. Equat. Control Theory*, **8**(2019), 43–55.

18. S. Carillo, M. Lo Schiavo and C. Schiebold, Matrix solitons solutions of the modified Korteweg-deVries equation. In: *Nonlinear Dynamics of Structures, Systems and Devices.* W. Lacarbonara, B. Balachandran, J. Ma, J. Tenreiro Machado, G. Stepan. (eds.) Springer, Cham, 2020, pp. 75–83.

19. S. Carillo and C. Schiebold, Construction of soliton solutions of the matrix modified Korteweg-deVries equation, In: *Nonlinear Dynamics of Structures, Systems and Devices,* W. Lacarbonara *et al.* (eds.) Springer, Cham, 2021.

20. S. Carillo, M. Lo Schiavo and C. Schiebold, N-soliton matrix mKdV solutions: A step towards their classification, Preprint 2020.

21. B. Carl and C. Schiebold, Nonlinear equations in soliton physics and operator ideals. *Nonlin.*, **12**(1999), 333–364.

22. P. J. Caudrey, R. K. Dodd and J. D. Gibbon, A new hierarchy of Korteweg-deVries equations. *Proc. Roy. Soc. London,* **A 351**(1976), 407–422.

23. X. Chen, Y. Zhang, J. Liang and R. Wang, The N-soliton solutions for the matrix modified Korteweg-deVries equation via the Riemann-Hilbert approach. *Eur. Phys. J. Plus,* **135**(2020), 574.

24. A. S. Fokas and B. Fuchssteiner, Bäcklund transformation for hereditary symmetries. *Nonlin. Anal., Theory Methods Appl.,* **5**(4) (1981), 423-432.

25. B. Fuchssteiner, Application of hereditary symmetries to nonlinear evolution equations. *Nonlin. Anal., Th. Meth. Appl.,* **3**(6) (1979), 849–862.

26. B. Fuchssteiner, The Lie algebra structure of degenerate hamiltonian and bi-Hamiltonian systems. *Progr. Theor. Phys.,* **68**(1982), 1082–1104.

27. B. Fuchssteiner, Solitons in interaction. *Progr. Theor. Phys.,* **78**(5) (1987), 1022–1050.

28. B. Fuchssteiner and S. Carillo, Soliton structure versus singularity analysis: Third order completely integrable nonlinear equations in 1+1 dimensions. *Phys. A,* **154**(1989) 467–510.

29. B. Fuchssteiner and S. Carillo, The action-angle transformation for soliton equations. *Physica A,* **166**(1990), 651–676.

30. B. Fuchssteiner and W. Oevel, The bi-Hamiltonian structure of some nonlinear fifth- and seventh-order differential equations and recursion

formulas for their symmetries and conserved covariants. *J. Math. Phys,* **23**(3) (1982), 358–363.

31. V. M. Goncharenko, Multisoliton solutions of the matrix KdV equation. *Theor. Math. Phys.,* **126**(2001), 81–91.

32. C. Gu, H. Hu and Z. Zhou, *Darboux transformations in integrable systems.* Springer, Dordrecht, 2005, ISBN 1-4020-3087-8.

33. M. Gürses, A. Karasu and V.V. Sokolov, On construction of recursion operators from Lax representation. *J. Math. Phys.,* **40**(1999), 6473–6490.

34. M. Gürses, A. Karasu and R. Turhan, On non-commutative integrable Burgers equations. *J. Nonlinear Math. Phys.,* **17**(2010), 1–6.

35. M. Hamanaka, Noncommutative solitons and quasideterminants. *Phys. Scripta,* **89**(2014), 038006.

36. M. Hamanaka and K. Toda, Noncommutative Burgers equation. *J. Phys. A: Math. Gen.,* **36**(2003), 11981.

37. S. Kawamoto, An exact transformation from the Harry Dym equation to the modified KdV equation. *J. Phys. Soc. Japan,* **54**(1985), 2055–2056.

38. B.G. Konopelchenko, Soliton eigenfunction equations: the IST integrability and some properties. *Rev. Math. Phys.,* **2**(4) (1990), 399–440.

39. B.A. Kupershmidt, On a group of automorphisms of the noncommutative Burgers hierarchy. *J. Nonlinear Math. Phys.,* **12**(4) (2005), 539–549.

40. D. Levi, O. Ragnisco and M. Bruschi. Continuous and discrete matrix Burgers' hierarchies. *Il Nuovo Cimento,* **74B**(1983), 33–51.

41. F. Magri, A simple model of the integrable Hamiltonian equation. *J. Math. Phys.,* **19**(1978), 1156–1162.

42. V. A. Marchenko, Nonlinear equations and operator algebras. *Mathematics and its Applications (Soviet Series),* Vol. 17. D. Reidel Publishing Co., Dordrecht, 1988.

43. A. G. Meshkov and V. V. Sokolov, Integrable evolution equations with a constant separant. *Ufimsk. Mat. Zh.,* **4**(3) (2012), 104–154.

44. A. V. Mikhailov, A. B. Shabat and R. I. Yamilov, The symmetry approach to the classification of non-linear equations. Complete lists of integrable systems. *Russian Math. Surv.,* **42**(4) (1987), 1–63.

45. A. V. Mikhailov, V. S. Novikov and J. P. Wang, On classification of integrable nonevolutionary equations. *Stud. Appl. Math.,* **118**(2007), 419–457.

46. R. M. Miura (ed), *Bäcklund transformations, I.S.T. Method and Their Applications, Lecture Notes in Math.,* Vol. 515. Springer, Berlin, 1976.

47. R. M. Miura, C. S. Gardner and M. D. Kruskal, Korteweg-deVries equation and generalizations. II. Existence of conservation laws and constants of motion. *J. Math. Phys.,* **9**(8) (1968), 1204–1209.

48. W. Oevel and S. Carillo, Squared eigenfunction symmetries for soliton equations: Part I, *J. Math. Anal. and Appl.*, **217**(1) (1998), 161–178.

49. W. Oevel and S. Carillo, Squared eigenfunction symmetries for soliton equations: Part II, *J. Math. Anal. Appl.*, **217**(1) (1998), 179–199.

50. P. J. Olver and V. V. Sokolov, Integrable evolution equations on nonassociative algebras. *Comm. Math. Phys.*, **193**(1998), 245–268.

51. C. Rogers, The Harry Dym equation in 2+1-dimensions: a reciprocal link with the Kadomtsev-Petviashvili equation. *Phys. Lett.*, **120A**(1987), 15–18.

52. C. Rogers and W. F. Ames, *Nonlinear Boundary Value Problems in Science and Engineering*. Academic Press, Boston, 1989.

53. C. Rogers and S. Carillo, On reciprocal properties of the Caudrey-Dodd-Gibbon and Kaup-Kupershmidt hierarchies. *Phys. Scripta*, **36**(1987), 865–869.

54. C. Rogers and M. C. Nucci, On reciprocal Bäcklund transformations and the Korteweg-deVries hierarchy. *Phys. Scripta*, **33**(1986), 289–292.

55. C. Rogers and W. F. Shadwick. Bäcklund transformations and their applications. *Mathematics in Science and Engineering*. Vol. 161. Academic Press, Inc., New York-London, 1982.

56. C. Rogers and W. K. Schief, *Bäcklund and Darboux Transformations: Geometry and Modern Applications in Soliton Theory*. Cambridge University Press, Cambridge, 2002.

57. A. K. Sawada and A. T. Kotera, A method for finding N-soliton solutions of the KdV and KdV-like equations. *J. Progr. Theor. Phys.*, **51**(1974), 1355–1367.

58. A. L. Sakhnovich, L. A. Sakhnovich and I. Ya. Roitberg, Inverse problems and nonlinear evolution equations. Solutions, Darboux matrices and Weyl-Titchmarsh functions. *Studies in Mathematics*, Vol. 47. De Gruyter, Berlin, 2013.

59. C. Schiebold, Noncommutative AKNS systems and multisoliton solutions to the matrix sine-Gordon equation. *Discr. Cont. Dyn. Systems Suppl.*, **2009**(2009), 678–690.

60. C. Schiebold, Matrix solutions for equations of the AKNS system. In: *Nonlinear Systems and their Remarkable Mathematical Structures*, N. Euler, (ed.). CRC Press, Boca Raton, FL, USA, 2018, Chapter B.5, pp. 256–293.

61. S. I. Svinolupov and V. V. Sokolov, Evolution equations with nontrivial conservative laws. *Funct. Anal. Its Appl.*, **16**(4) (1982), 317–319.

62. J. P. Wang, A list of $1 + 1$ dimensional integrable equations and their properties. *J. Nonlinear Math. Phys.*, **9**(1) (2002), 213–233.
63. J. Weiss, On classes of lntegrable systems and the Painlevé property. *J. Math. Phys.*, **25**(1984), 13–24.
64. G. Wilson, On the quasi-hamiltonian formalism of the KdV equation. *Phys. Lett. A*, **132**(1988), 445–450.

© 2022 World Scientific Publishing Europe Ltd.
https://doi.org/10.1142/9781800611368_0007

Chapter 7

Voros Coefficients and the Topological Recursion for the Hypergeometric Differential Equation of Type $(1, 1, 3)$

Yumiko Takei

National Institute of Technology, Ibaraki College
866 Nakane, Hitachinaka, Ibaraki 312-8508, Japan
ytakei@ibaraki-ct.ac.jp

In my joint papers with Iwaki and Koike we found an intriguing relation between the Voros coefficients in the exact WKB analysis and the free energy in the topological recursion introduced by Eynard and Orantin in the case of the confluent family of the Gauss hypergeometric differential equations. In this chapter we discuss its generalization to the case of the hypergeometric differential equation of type $(1, 1, 3)$.

1. Introduction

The exact WKB analysis is a method of analyzing differential equations with a small parameter $\hbar$ (or a large parameter $\eta = \hbar^{-1}$). It was first developed for second-order ordinary differential equations and then has been extended to higher order ordinary differential equations. In this chapter, we discuss the hypergeometric differential equation of type $(1, 1, 3)$

$$\left[2t^2\hbar^3\frac{d^3}{dx^3} + (2x - 3t^2)\hbar^2\frac{d^2}{dx^2} - (2x - t^2 - 2\lambda_{\infty,0} - 2\lambda_{\infty,1})\hbar\frac{d}{dx}\right.$$

$$\left. -2\lambda_{\infty,0}\right]\psi = 0. \tag{1}$$

This equation is obtained from the class of hypergeometric systems of two variables studied in [15] by fixing the second variable. A Voros coefficient is defined as a contour integral of the logarithmic derivative of WKB solutions. The explicit form of Voros coefficients enables us to describe parametric Stokes phenomena. The explicit forms of Voros coefficients are first computed for the Weber equation [16,17], and now known for many equations such as the Whittaker equation [13], the Kummer equation [2], the Gauss hypergeometric equation [1], the hypergeometric equation of type $(1,4)$ [10], and so on.

On the other hand, the topological recursion introduced by Eynard and Orantin [7] is a generalization of the loop equations that the correlation functions of the matrix model satisfy. For a Riemann surface Σ and meromorphic functions x and y on Σ, it produces an infinite tower of multidifferential $W_{g,n}(z_1,\ldots,z_n)$ on Σ and free energies F_g. A triplet (Σ, x, y) is called a spectral curve and $W_{g,n}(z_1,\ldots,z_n)$ is called a correlation function.

The quantization scheme for the spectral curve connects WKB solutions with the topological recursion ([4,5,8] etc.). More precisely, it is found that WKB solutions can be constructed by correlation functions for the spectral curve when the spectral curve satisfies the "admissibility condition" in the sense of [4, Definition 2.7]. WKB solutions can be constructed by correlation functions also for spectral curves not satisfying the admissibility condition, in particular, for elliptic curves [9] and for any degree 2 spectral curve [6,14]. It means that the recursive relations satisfied by the coefficients of WKB solutions have some relationship with the topological recursion. Furthermore, in the case of the confluent family of the Gauss hypergeometric differential equations and the hypergeometric differential equations of types $(1,4)$ and $(2,3)$, we found that the Voros coefficients are expressed as the difference of the free energy of the spectral curve obtained as the classical limit of the equations [11,12,18].

Taking these situations into account, we consider a similar problem for the hypergeometric differential equations of type $(1, 1, 3)$. The main results are the following:

(1) We show that the Voros coefficients are expressed as a difference value of the free energy.
(2) We also get the explicit forms of the free energy and Voros coefficients.

In what follows, we explain the precise claim of these results and sketch some proofs.

The chapter is organized as follows: In Section 2, we introduce WKB solutions and Voros coefficients. In Section 3, we introduce the topological recursion. In Section 4, we explain the quantization introduced by Bouchard and Eynard [4]. In Section 5, we state our main results for the hypergeometric differential equation of type $(1, 1, 3)$.

2. WKB Solutions and Voros Coefficients

We consider the following third-order ordinary differential equation with a small parameter $\hbar \neq 0$

$$P\left(x, \hbar\frac{d}{dx}\right)\psi = \left\{\hbar^3\frac{d^3}{dx^3} + p(x)\hbar^2\frac{d^2}{dx^2} + q(x, \hbar)\hbar\frac{d}{dx} + r(x, \hbar)\right\}\psi = 0.$$

$$(2)$$

Here,

$$p(x) = \frac{2x - 3t^2}{2t^2}, \tag{3}$$

$$q(x, \hbar) = -\frac{2x - t^2 - 2\lambda_{\infty,0} - 2\lambda_{\infty,1}}{2t^2} + \frac{\nu_{\infty,0} + \nu_{\infty,1} - 2}{t^2}\hbar, \tag{4}$$

$$r(x, \hbar) = -\frac{\lambda_{\infty,0}}{t^2} + \frac{\nu_{\infty,0} - 1}{t^2}\hbar, \tag{5}$$

and $t, \lambda_{\infty,j} \neq 0$, $\nu_{\infty,j}$ $(j \in \{0, 1\})$ are parameters. We call this equation the hypergeometric differential equation of type $(1, 1, 3)$. We consider (2) as a differential equation on the Riemann sphere $\mathbb{P}^1$ with regular or irregular singular points.

2.1. *WKB solutions*

For (2), we construct a formal solution, called a WKB solution, of the form

$$\psi(x,\hbar) = \exp\left[\int^x S(x,\hbar)dx\right].$$ (6)

We substitute this into (2) and have the following equation:

$$\hbar^3\left(\tfrac{d^2}{dx^2}S(x,\hbar) + 3S(x,\hbar)\tfrac{d}{dx}S(x,\hbar) + S(x,\hbar)^3\right)$$
$$+ p(x)\hbar^2\left(\tfrac{d}{dx}S(x,\hbar) + S(x,\hbar)^2\right) + \hbar q(x,\hbar)S(x,\hbar) + r(x,\hbar) = 0.$$ (7)

This is a counterpart of the Riccati equation in the second-order case. We seek for a formal series solution of the form

$$S(x,\hbar) = \hbar^{-1}S_{-1}(x) + S_0(x) + \hbar S_1(x) + \cdots = \sum_{m=-1}^{\infty} \hbar^m S_m(x).$$ (8)

By substituting (8) into (7), and equating like powers of both sides with respect to $\hbar$, we obtain

$$S_{-1}^3 + p(x)S_{-1}^2 + q_0(x)S_{-1} + r_0(x) = 0,$$ (9)

$$\left(3S_{-1}^2 + 2p(x)S_{-1} + q_0(x)\right)S_0 + 3S_{-1}\frac{dS_{-1}}{dx} + p(x)\frac{dS_{-1}}{dx}$$
$$+ q_1(x)S_{-1} + r_1(x) = 0,$$ (10)

and

$$\left(3S_{-1}^2 + 2p(x)S_{-1} + q_0(x)\right)S_{m+1} + \sum_{\substack{i+j+k=m-1\\i,j,k\geq 0}} S_i S_j S_k$$
$$+ 3\sum_{j=0}^{m-1} S_{m-j-1}S_j + 3S_m\frac{dS_{-1}}{dx} + 3S_{-1}\frac{dS_m}{dx} + \frac{d^2 S_{m-1}}{dx^2}$$
$$+ p(x)\sum_{j=0}^{m} S_{m-j}S_j + p(x)\frac{dS_m}{dx} + q_1(x)S_m = 0 \quad (m \geq 0).$$ (11)

Equation (9) has three solutions, and once we fix one of them, we can determine S_m for $m \geq 0$ uniquely and recursively by (10) and (11). $S_m(x)$ for $m \geq 0$ are functions on the Riemann surface defined by $S_{-1}(x)$.

2.2. *Voros coefficients*

In this subsection, we define Voros coefficients of the hypergeometric differential equation of type $(1,1,3)$.

By straightforward computations, we obtain asymptotic behaviors of characteristic roots of the quantum $(1,1,3)$ curve; these are solutions of $P(x,\xi) = 0$,

$$\begin{cases} \xi^{(0)}(x,\underline{\lambda}) = -\dfrac{\lambda_0}{x} - \dfrac{\lambda_0(t^2 + 2\lambda_1)}{2x^2} + O(x^{-3}), \\[2ex] \xi^{(1)}(x,\underline{\lambda}) = 1 - \dfrac{\lambda_1}{x} + O(x^{-2}), \\[2ex] \xi^{(2)}(x,\underline{\lambda}) = -\dfrac{x}{t^2} + \dfrac{1}{2} + O(x^{-1}). \end{cases} \tag{12}$$

Here, $\underline{\lambda} = (\lambda_{\infty,0}, \lambda_{\infty,1})$. Let $\gamma_{\infty,j}$ ($j \in \{0,1\}$) be a path on the Riemann surface of $S_{-1}(x)$ which starts from $x = \infty$ on the third sheet (i.e. $\xi(x,\underline{\lambda}) = \xi^{(2)}(x,\underline{\lambda})$ on the sheet), turn around a turning point, and returns to $x = \infty$ on the $(j+1)$th sheet (i.e. $\xi(x,\underline{\lambda}) = \xi^{(j)}(x,\underline{\lambda})$ on the sheet). A turning point is defined by a point at which two characteristic roots coincide.

Remark 1. We will consider x as a branched covering map $x(z)$ from $\mathbb{P}^1$ to itself in what follows. We can verify that $\gamma_{\infty,j}$ ($j \in \{0,1\}$) is defined as the image by x of a path from ∞ to j on the z-plane.

Then, we define Voros coefficients of the hypergeometric differential equation of type $(1,1,3)$

$$V(\underline{\lambda}, t, \nu_{\infty,0}, \nu_{\infty,1}; \hbar) = \int_{\gamma_{\infty,j}} \Big(S(x,\hbar) - \hbar^{-1}S_{-1}(x) - S_0(x) \Big) dx$$
$$(j \in \{0,1\}). \tag{13}$$

3. Topological Recursion

3.1. *Topological recursion*

In this section, we briefly explain the global topological recursion. It was shown in [3] that it is indeed equivalent to the usual local formulation of the topological recursion introduced by Eynard and Orantin [7] when the ramification points are all simple.

Let us start from the definition of genus 0 spectral curves (see [7] for general definition of spectral curves).

Definition 1. A spectral curve of genus 0 is a pair $(x(z), y(z))$ of non-constant rational functions on $\mathbb{P}^1$, such that their exterior differentials dx and dy never vanish simultaneously.

Let R be the set of ramification points of $x(z)$, i.e. R consists of zeros of $dx(z)$ of any order and poles of $x(z)$ whose orders are greater than or equal to two (here we consider x as a branched covering map from $\mathbb{P}^1$ to itself). We need to introduce some notation to define the topological recursion.

Definition 2. Let $A \subseteq_k B$ if $A \subseteq B$ and $|A| = k$.

Definition 3. Let $\mathcal{S}(t)$ be the set of set partitions of an ensemble t.

Then, we define the recursive structure:

Definition 4 ([4, Definition 3.4]). *Let $\{W_{g,n}\}$ be an arbitrary collection of symmetric multidifferentials on $(\mathbb{P}^1)^n$ with $g \geq 0$ and $n \geq 1$. Let $k \geq 1$, $t = \{t_1, \ldots, t_k\}$ and $z = \{z_1, \ldots, z_n\}$. Then, we define*

$$\mathcal{R}^{(k)} W_{g,n+1}(t; z)$$

$$:= \sum_{\mu \in \mathcal{S}(\mathbf{t})} \sum_{\sqcup_{i=1}^{l(\mu)} J_i = \mathbf{z}} \sideset{}{'}\sum_{\sum_{i=1}^{l(\mu)} g_i = g + l(\mu) - k} \left\{ \prod_{i=1}^{l(\mu)} W_{g_i, |\mu_i| + |J_i|}(\mu_i, J_i) \right\}. \tag{14}$$

The first summation in (14) is over set partitions of $\mathbf{t}$, $l(\mu)$ is the number of subsets in the set partition μ. The third summation in (14) is over all $l(\mu)$-tuple of non-negative integers $(g_1, \ldots, g_{l(\mu)})$ such that $\sum_{i=1}^{l(\mu)} g_i = g + l(\mu) - k$. $\sqcup$ denotes the disjoint union, and the prime $'$ on the summation symbol in (14) means that we exclude terms with $(g_i, |\mu_i| + |J_i|) = (0, 1)$ $(i = 1, \ldots, l(\mu))$ (so that $W_{0,1}$ does not appear) from the sum. We also define

$$\mathcal{R}^{(0)} W_{g,n+1}(\mathbf{z}) := \delta_{g,0} \delta_{n,0}, \tag{15}$$

where $\delta_{i,j}$ is the Kronecker delta symbol.

We now define the topological recursion.

Definition 5 ([4, Definition 3.6]). *Eynard–Orantin's correlation function $W_{g,n}(z_1, \ldots, z_n)$ for $g \geq 0$ and $n \geq 1$ is defined as a multi-differential on $(\mathbb{P}^1)^n$ using the recurrence relation*

$$
\begin{aligned}
W_{g,n+1}\, &(z_0, z_1, \ldots, z_n) := \\
&\sum_{r \in R} \operatorname{Res}_{z=r} \left\{ \sum_{k=1}^{r-1} \sum_{\beta(z) \subseteq_k \tau'(z)} (-1)^{k+1} \frac{w^{z-\alpha}(z_0)}{E^{(k)}(z; \beta(z))} \right. \\
&\qquad\qquad \left. \times \mathcal{R}^{(k+1)} W_{g,n+1}(z, \beta(z); z_1, \ldots, z_n) \right\}
\end{aligned}
\tag{16}
$$

for $2g + n \geq 2$ with initial conditions

$$
W_{0,1}(z_0) := y(z_0)dx(z_0), \quad W_{0,2}(z_0, z_1) = B(z_0, z_1) := \frac{dz_0 dz_1}{(z_0 - z_1)^2}.
\tag{17}
$$

Here, we set $W_{g,n} \equiv 0$ for a negative g and

$$
E^{(k)}(z; t_1, \ldots, t_k) := \prod_{i=1}^{k} (W_{0,1}(z) - W_{0,1}(t_i)),
\tag{18}
$$

$$
w^{z-\alpha}(z_0) := \left(\frac{1}{z_0 - z} - \frac{1}{z_0 - \alpha} \right) dz_0.
\tag{19}
$$

The second and third summations in (16) together mean that we are summing over all subsets of $\tau'(z)$, where $\tau'(z)$ is a set of preimages of $x(z)$ except for the original point z. α is an arbitrary base point on $\mathbb{P}^1$, but it can be checked (see [3]) that the definition is actually independent of the choice of base point α.

See [7] for basic properties of $W_{g,n}$.

3.2. *Definition of free energies*

The free energy F_g $(g \geq 0)$ is defined for the spectral curve, and is one of the most important objects in Eynard–Orantin's theory.

Definition 6 ([7, **Definition 4.3**]). *For $g \geq 2$, the gth free energy F_g is defined by*

$$F_g := \frac{1}{2 - 2g} \sum_{r \in R} \mathrm{Res}_{z=r} \left[\Phi(z) W_{g,1}(z) \right] \quad (g \geq 2), \tag{20}$$

where $\Phi(z)$ is a primitive of $y(z)dx(z)$. The free energies F_0 and F_1 are also defined, but in a different manner (see [7, §4.2.2 and §4.2.3] for the definition).

4. Quantization of Spectral Curves

We treat the quantization introduced by Bouchard and Eynard [4].

4.1. *Results of bouchard and eynard*

Here, we only consider the defining equation of the spectral curve

$$P(x,y) = p_0(x)y^3 + p_1(x)y^2 + p_2(x)y + p_3(x) = 0, \tag{21}$$

where $p_j(x)$ are polynomials of x and we assume $p_0(x) \neq 0$. We assume that a spectral curve $(x(z), y(z))$ satisfies (21). Then we introduce some notations.

Definition 7. Let us rewrite (21) as

$$P(x,y) = \sum_{i,j \in A} \alpha_{i,j} x^i y^j = 0, \tag{22}$$

where

$$A = \{n \in \mathbb{N} \mid 0 \leq n \leq \max\{3, \deg(p_j(x)) \mid j = 0, 1, 2, 3\}, \alpha_{i,j} \neq 0\}.$$

Then the Newton polygon Δ is the convex hull of the set A.

Definition 8. For $m = 2, 3$, we define the following meromorphic function on $\mathbb{P}^1$

$$P_m(x,y) = \sum_{k=1}^{m-1} p_{m-1-k}(x)y^k. \tag{23}$$

Definition 9 ([4, **Definition 2.7**]). *We say that a spectral curve is admissible if:*

1. *Its Newton polygon Δ has no interior point;*
2. *If the origin $(x, y) = (0,0) \in \mathbb{C}^2$ is on the curve $\{P(x,y) = 0 \subset \mathbb{C}^2\}$, then the curve is smooth at this point.*

We assume that our spectral curve $(x(z), y(z))$ is admissible. Then the following theorem holds according to [4].

Theorem 1 ([4, Lemma 5.14]). *Let β_i $(1 \leqq i \leqq n)$ be some of the simple poles of $x(z)$ and*

$$D(z; \underline{\nu}) = [z] - \sum_{i=1}^{n} \nu_i [\beta_i] \tag{24}$$

be a divisor on $\mathbb{P}^1$, where ν_i $(1 \leq i \leq n)$ are complex numbers satisfying $\sum_{i=1}^{n} \nu_i = 1$ and $\underline{\nu} = (\nu_1, \ldots, \nu_n)$. Let $W_{g,n}(z_1, \ldots, z_n)$ be the correlation functions of a spectral curve $(x(z), y(z))$ defined from (16). Then,

$$\psi(x; \underline{\nu}, \hbar) = \exp\left[\frac{1}{\hbar} \int_{\zeta_1 \in D(z; \underline{\nu})} W_{0,1}(\zeta_1) + \frac{1}{2!} \int_{\zeta_1 \in D(z; \underline{\nu})} \int_{\zeta_2 \in D(z; \underline{\nu})} \right.$$
$$\left(W_{0,2}(\zeta_1, \zeta_2) - \frac{dx(\zeta_1)\, dx(\zeta_2)}{(x(\zeta_1) - x(\zeta_2))^2} \right)$$
$$+ \sum_{m=1}^{\infty} \hbar^m \left\{ \sum_{\substack{2g+n-2=m \\ g \geq 0,\, n \geq 1}} \frac{1}{n!} \int_{\zeta_1 \in D(z; \underline{\nu})} \cdots \int_{\zeta_n \in D(z; \underline{\nu})} \right.$$
$$\left. \left. W_{g,n}(\zeta_1, \ldots, \zeta_n) \right\} \right]\Bigg|_{z=z(x)}, \tag{25}$$

with the integration divisor (24) is a WKB-type formal solution of

$$\left[D_1 D_2 \frac{p_0(x)}{x^{\lfloor \alpha_3 \rfloor}} D_3 + D_1 \frac{p_1(x)}{x^{\lfloor \alpha_2 \rfloor}} D_2 + \frac{p_2(x)}{x^{\lfloor \alpha_1 \rfloor}} D_1 + \frac{p_3(x)}{x^{\lfloor \alpha_0 \rfloor}} \right.$$
$$\left. - \hbar C_1 D_1 \frac{x^{\lfloor \alpha_2 \rfloor}}{x^{\lfloor \alpha_1 \rfloor}} - \hbar C_2 \frac{x^{\lfloor \alpha_1 \rfloor}}{x^{\lfloor \alpha_0 \rfloor}} \right] \psi = 0, \tag{26}$$

where

$$\alpha_m = \inf\{a \mid (a, m) \in \Delta\} \quad (m = 0, 1, 2, 3),$$

$$D_i = \hbar \frac{x^{\lfloor \alpha_i \rfloor}}{x^{\lfloor \alpha_{i-1} \rfloor}} \frac{d}{dx} \quad (i = 1, 2, 3),$$

$$C_k = \sum_{i=1}^{n} \nu_i \left(\lim_{z \to \beta_i} \frac{P_{k+1}(x(z), y(z))}{x(z)^{\lfloor \alpha_{3-k} \rfloor + 1}} \right) \quad (k = 1, 2).$$

4.2. *Quantum $(1, 1, 3)$ curve*

In what follows, we restrict ourselves to the case of the $(1, 1, 3)$ curve

$$P(x, y) = 2t^2 y^3 + (2x - 3t^2)y^2 - (2x - t^2 - 2\lambda_{\infty,0} - 2\lambda_{\infty,1})y - 2\lambda_{\infty,0} = 0, \tag{27}$$

where $\lambda_{\infty,0}, \lambda_{\infty,1} \neq 0$ and t are parameters. We parametrize (27) as

$$\begin{cases} x = x(z) = \dfrac{-2t^2 z^3 + 3t^2 z^2 - (t^2 + 2\lambda_{\infty,0} + 2\lambda_{\infty,1})z + 2\lambda_{\infty,0}}{2z(z - 1)}, \\[2ex] y = y(z) = z \end{cases} \tag{28}$$

and regard it as a spectral curve in the sense of Definition 1.

We choose

$$\begin{aligned} D(z; \underline{\nu}) &= [z] - \nu_{\infty,0}[0] - \nu_{\infty,1}[1] - (1 - \nu_{\infty,0} - \nu_{\infty,1})[\infty] \\ &= \nu_{\infty,0}([z] - [0]) + \nu_{\infty,1}([z] - [1]) \\ &\quad + (1 - \nu_{\infty,0} - \nu_{\infty,1})([z] - [\infty]) \end{aligned} \tag{29}$$

as the divisor for the quantization, where $\nu_{\infty,0}, \nu_{\infty,1}$ are parameters. Then, Theorem 1 gives the quantum curve of the $(1, 1, 3)$ curve (quantum $(1, 1, 3)$ curve):

$$\begin{aligned} \Big[2t^2 \hbar^3 \frac{d^3}{dx^3} + (2x - 3t^2)\hbar^2 \frac{d^2}{dx^2} - (2x - t^2 - 2\hat{\lambda}_{\infty,0} - 2\hat{\lambda}_{\infty,1} - 4\hbar)\hbar \frac{d}{dx} \\ - 2\hat{\lambda}_{\infty,0} - 2\hbar \Big] \psi = 0. \end{aligned} \tag{30}$$

Here, we used the notation

$$\hat{\lambda}_{\infty,j} = \lambda_{\infty,j} - \nu_{\infty,j}\hbar \quad (j \in \{0, 1\}). \tag{31}$$

5. Main Results for the Hypergeometric Differential Equation of Type $(1, 1, 3)$

5.1. *Relations between the Voros coefficient and the free energy*

In this subsection, we describe the Voros coefficient of the quantum $(1, 1, 3)$ curve in terms of the free energy of the $(1, 1, 3)$ curve with a parameter shift. Let

$$F(\underline{\lambda}, t; \hbar) = \sum_{g=0}^{\infty} \hbar^{2g-2} F_g(\underline{\lambda}, t) \tag{32}$$

be the generating function of the free energies of the $(1, 1, 3)$ curve. Then, the following relations hold:

Theorem 2.

$$V^{(\infty,0)}(\underline{\lambda}, t, \underline{\nu}; \hbar) = F\left(\hat{\lambda}_{\infty,0} + \hbar, \hat{\lambda}_{\infty,1}, t; \hbar\right) - F\left(\hat{\lambda}_{\infty,0}, \hat{\lambda}_{\infty,1}, t; \hbar\right)$$
$$-\frac{1}{\hbar}\frac{\partial F_0}{\partial \lambda_{\infty,0}} + \left(\nu_{\infty,0} - \frac{1}{2}\right)\frac{\partial^2 F_0}{\partial \lambda_{\infty,0}{}^2} + \nu_{\infty,1}\frac{\partial^2 F_0}{\partial \lambda_{\infty,0}\lambda_{\infty,1}},$$
$$\tag{33}$$

$$V^{(\infty,1)}(\underline{\lambda}, t, \underline{\nu}; \hbar) = F\left(\hat{\lambda}_{\infty,0}, \hat{\lambda}_{\infty,1} + \hbar, t; \hbar\right) - F\left(\hat{\lambda}_{\infty,0}, \hat{\lambda}_{\infty,1}, t; \hbar\right)$$
$$-\frac{1}{\hbar}\frac{\partial F_0}{\partial \lambda_{\infty,1}} + \left(\nu_{\infty,1} - \frac{1}{2}\right)\frac{\partial^2 F_0}{\partial \lambda_{\infty,1}{}^2} + \nu_{\infty,0}\frac{\partial^2 F_0}{\partial \lambda_{\infty,0}\lambda_{\infty,1}}.$$
$$\tag{34}$$

(Recall $\hat{\lambda}_{\infty,j} = \lambda_{\infty,j} - \nu_{\infty,j}\hbar$ $(j \in \{0, 1\})$ as we have introduced in (31).)

We can prove Theorem 2 similarly to the case of the Gauss equation ([12, Theorem 3.1]).

Then, we get the three-term difference equation that the generating function of the free energies satisfies.

Theorem 3. *The free energy* (32) *satisfies the following difference equations.*

$$F(\lambda_{\infty,0} + \hbar, \lambda_{\infty,1}, t; \hbar) - 2F(\lambda_{\infty,0}, \lambda_{\infty,1}, t; \hbar) + F(\lambda_{\infty,0} - \hbar, \lambda_{\infty,1}, t; \hbar)$$
$$= \log(\lambda_{\infty,0}) + 2\log(t), \tag{35}$$
$$F(\lambda_{\infty,0}, \lambda_{\infty,1} + \hbar, t; \hbar) - 2F(\lambda_{\infty,0}, \lambda_{\infty,1}, t; \hbar) + F(\lambda_{\infty,0}, \lambda_{\infty,1} - \hbar, t; \hbar)$$
$$= \log(\lambda_{\infty,1}) + 2\log(t). \tag{36}$$

To prove Theorem 3, we need the following identity.

Lemma 1.

$$V^{(\infty,0)}(\underline{\lambda}, t, \nu_{\infty,0} - 1, \nu_{\infty,1}; \hbar) - V^{(\infty,0)}(\underline{\lambda}, t, \nu_{\infty,0}, \nu_{\infty,1}, \hbar)$$
$$= \log\left\{ 1 - \frac{(\nu_{\infty,0} - 1)\hbar}{\lambda_{\infty,0}} \right\}, \tag{37}$$
$$V^{(\infty,1)}(\underline{\lambda}, t, \nu_{\infty,0}, \nu_{\infty,1} - 1; \hbar) - V^{(\infty,1)}(\underline{\lambda}, t, \nu_{\infty,0}, \nu_{\infty,1}; \hbar)$$
$$= \log\left\{ 1 - \frac{(\nu_{\infty,1} - 1)\hbar}{\lambda_{\infty,1}} \right\}. \tag{38}$$

Using the following lemma, we can prove Lemma 1.

Lemma 2. *The WKB solution of the quantum* $(1,1,3)$ *curve satisfies*

$$\frac{d}{dx}\psi(x; \nu_{\infty,0}, \nu_{\infty,1}, \hbar) = \psi(x; \nu_{\infty,0} - 1, \nu_{\infty,1}, \hbar), \tag{39}$$
$$\left(-\hbar\frac{d}{dx} + 1 \right)\psi(x; \nu_{\infty,0}, \nu_{\infty,1}, \hbar) = \psi(x; \nu_{\infty,0}, \nu_{\infty,1} - 1, \hbar). \tag{40}$$

Proof of Theorem 3. Substituting $(\nu_{\infty,0}, \nu_{\infty,1}) = (1, 0)$ into (37), we have

$$V^{(\infty,0)}(\underline{\lambda}, t, 1, 0; \hbar) = V^{(\infty,0)}(\underline{\lambda}, t, 0, 0; \hbar). \tag{41}$$

On the other hand, (33) with $(\nu_{\infty,0}, \nu_{\infty,1}) = (1, 0)$ and $(\nu_{\infty,0}, \nu_{\infty,1}) = (0, 0)$ gives

$$V^{(\infty,0)}(\underline{\lambda}, t, 1, 0; \hbar) = F(\lambda_{\infty,0}, \lambda_{\infty,1}, t; \hbar) - F(\lambda_{\infty,0} - \hbar, \lambda_{\infty,1}, t; \hbar)$$
$$- \frac{1}{\hbar}\frac{\partial F_0}{\partial \lambda_{\infty,0}} + \frac{1}{2}\frac{\partial^2 F_0}{\partial \lambda_{\infty,0}{}^2}, \tag{42}$$

$$V^{(\infty,0)}(\underline{\lambda}, t, 0, 0; \hbar) = F(\lambda_{\infty,0} + \hbar, \lambda_{\infty,1}, t; \hbar) - F(\lambda_{\infty,0}, \lambda_{\infty,1}, t; \hbar)$$
$$- \frac{1}{\hbar}\frac{\partial F_0}{\partial \lambda_{\infty,0}} - \frac{1}{2}\frac{\partial^2 F_0}{\partial \lambda_{\infty,0}{}^2}. \tag{43}$$

Then, the desired equality (35) follows immediately from (41), (42), (43) and

$$F_0 = -\frac{1}{8}\{6\lambda_{\infty,0}{}^2 + 4\lambda_{\infty,0}\lambda_{\infty,1} + 6\lambda_{\infty,1}{}^2 - 4\lambda_{\infty,0}\lambda_{\infty,1}\log\left(-t^4\right)$$
$$- (\lambda_{\infty,0}+\lambda_{\infty,1})t^2 - 4\lambda_{\infty,0}{}^2\log\left(\lambda_{\infty,0}t^2\right) - 4\lambda_{\infty,1}{}^2\log\left(\lambda_{\infty,1}t^2\right)\}.$$

We obtain (36) similarly. $\qquad\qquad\square$

5.2. *The explicit forms of the free energy and Voros coefficient*

As a corollary, we obtain explicit formulas for the coefficients of the free energy and Voros coefficients. In this subsection, we finally provide the explicit forms for the free energy and Voros coefficients.

Theorem 4. (i) *For $g \geq 2$, the gth free energy of the $(1,1,3)$ curve has the following expression:*

$$F_g(\underline{\lambda}, t) = \frac{B_{2g}}{2g(2g-2)}\left\{\frac{1}{\lambda_{\infty,0}{}^{2g-2}} + \frac{1}{\lambda_{\infty,1}{}^{2g-2}}\right\}, \qquad (44)$$

where $\{B_n\}_{n\geq 0}$ designates the Bernoulli number defined by

$$\frac{w}{e^w - 1} = \sum_{n=0}^{\infty} B_n \frac{w^n}{n!}. \qquad (45)$$

(ii) The Voros coefficients for the quantum $(1,1,3)$ curve are explicitly given by

$$V^{(\infty,0)}(\underline{\lambda}, t, \underline{\nu}; \hbar) = \sum_{m=1}^{\infty} \hbar^m \frac{1}{m(m+1)} \frac{B_{m+1}(\nu_{\infty,0})}{\lambda_{\infty,0}{}^m}, \qquad (46)$$

$$V^{(\infty,1)}(\underline{\lambda}, t, \underline{\nu}; \hbar) = \sum_{m=1}^{\infty} \hbar^m \frac{1}{m(m+1)} \frac{B_{m+1}(\nu_{\infty,1})}{\lambda_{\infty,1}{}^m}, \qquad (47)$$

where $\{B_n(t)\}_{n\geq 0}$ is the Bernoulli polynomial defined through the generating function as

$$\frac{we^{Xw}}{e^w - 1} = \sum_{n=0}^{\infty} B_n(X) \frac{w^n}{n!}. \qquad (48)$$

One of the differences between this case and the case of the Gauss equation is that $W_{g,n}$ depends on t. However, from the following lemma we can prove F_g $(g \geq 1)$ does not depend on t. Therefore, we can prove Theorem 4 in a similar manner to the case of the Gauss equation.

Lemma 3. *The following relations hold for $g \geq 2$:*

$$\frac{\partial F_g}{\partial t} = 0. \tag{49}$$

To prove this theorem, we use the following two lemmas.

Lemma 4. *For the free energy for the (1,1,3) curve*

$$\frac{\partial F_g}{\partial t} = -\operatorname{Res}_{z=\infty} t(z^2 - z)W_{g,1}(z) \quad (g \geq 1) \tag{50}$$

holds.

Lemma 5. *For the (1,1,3) equation and the (1,1,3) curve, the following relations hold:*

$$\operatorname{Res}_{z=\infty} t(z^2 - z) \sum_{m=-1}^{\infty} \hbar^m S_m(x(z))dx(z) = -\frac{\lambda_{\infty,1}t}{\hbar} + \frac{\lambda_{\infty,0} + \lambda_{\infty,1}}{t}, \tag{51}$$

$$\operatorname{Res}_{z=\infty} t(z^2 - z) \sum_{\substack{g \geq 0, n \geq 2 \\ (g,n) \neq (0,2)}} \frac{\hbar^{2g-2+n}}{(n-1)!} \int_{\zeta_2=\infty}^{\zeta_2=z} \cdots$$
$$\int_{\zeta_n=\infty}^{\zeta_n=z} W_{g,n}(z, \zeta_2, \ldots, \zeta_n)$$
$$= \sum_{g \geq 1} \hbar^{2g} C_{g,2}(\underline{\lambda}, t, \underline{\nu}), \tag{52}$$

where $C_{g,2}(\underline{\lambda}, t, \underline{\nu})$ $(g \geq 1)$ are constant with respect to $\hbar$.

Proof of Lemma 4. Because

$$\frac{\partial y(z)}{\partial t} \cdot dx(z) - \frac{\partial x(z)}{\partial t} \cdot dy(z) = t(2z-1)\, dz = -\frac{1}{2\pi i} \int_{\zeta \in \gamma} t(\zeta^2 - \zeta)B(z, \zeta) \tag{53}$$

holds, the variation formula [7, Theorem 5.1] implies (50). Here, γ is a circle around ∞ in a counterclockwise manner. $\qquad \square$

Proof of Lemma 5. From, (10), $S_0(x)$ is expressed in $S_{-1}(x)$. Therefore, we obtain

$$S_0(x(z)) = \frac{\nu_{\infty,0} + \nu_{\infty,1}}{t^2 z} + \frac{\nu_{\infty,1}}{t^2 z^2}$$
$$+ \frac{\nu_{\infty,1}t^2 + (\nu_{\infty,0} + \nu_{\infty,1} + 1)(\lambda_{\infty,0} + \lambda_{\infty,1})}{t^4 z^3} \qquad (54)$$
$$+ o(z^{-4}) \quad (z \to \infty).$$

We can also show

$$S_m(x(z)) = o(z^{-4}) \quad (z \to \infty) \qquad (55)$$

for $m \geq 1$. Then, by a straightforward computation, we can verify (51) directly. Because $W_{g,n}(z_1, \ldots, z_n)$ are holomorphic at $z_i = \infty$ ($1 \leq i \leq n$) for $2g - 2 + n \geq 1$, we find that

$$W_{g,n}(z, z_2, \ldots, z_n) \sim \frac{dz_2}{z_2{}^2} \left(C_{g,n}^{(2)}(z, z_3, \ldots, z_n, \underline{\lambda}, t, \underline{\nu}) + o(1/z_2) \right)$$
$$(z_2 \to \infty),$$

and for $i = 2, \ldots, n - 2$

$$C_{g,n}^{(i)}(z, z_{i+1}, \ldots, z_n, \underline{\lambda}, t, \underline{\nu})$$
$$\sim \frac{dz_{i+1}}{z_{i+1}{}^2} \left(C_{g,n}^{(i+1)}(z, z_{i+2}, \ldots, z_n, \underline{\lambda}, t, \underline{\nu}) + o(1/z_{i+1}) \right) \quad (z_{i+1} \to \infty),$$
$$C_{g,n}^{(n-1)}(z, z_n, \underline{\lambda}, t, \underline{\nu}) \sim \frac{dz_n}{z_n{}^2} \left(C_{g,n}^{(n)}(z, \underline{\lambda}, t, \underline{\nu}) + o(1/z_n) \right) \quad (z_n \to \infty),$$
$$C_{g,n}^{(n)}(z, \underline{\lambda}, t, \underline{\nu}) \sim \frac{dz}{z^2} \left(C_{g,n}(\underline{\lambda}, t, \underline{\nu}) + o(1/z) \right) \quad (z \to \infty).$$

Then, since lower order terms $o(1/z_i)$ vanish in the limit $z_i \to \infty$, we obtain

$$\int_{\zeta_2=\infty}^{\zeta_2=z_2} W_{g,n}(z, \zeta_2, \ldots, \zeta_n) \sim \int_{\zeta_2=\infty}^{\zeta_2=z_2} \frac{C_{g,n}(\underline{\lambda}, t, \underline{\nu}) d\zeta_2 \cdots d\zeta_n}{z^2 \zeta_2{}^2 \zeta_3{}^2 \cdots \zeta_n{}^2} dz$$
$$= -\frac{C_{g,n}(\underline{\lambda}, t, \underline{\nu}) d\zeta_3 \cdots d\zeta_n}{z^2 z_2 \zeta_3{}^2 \cdots \zeta_n{}^2} dz \qquad (56)$$
$$(z, \zeta_2, \ldots, \zeta_n \to \infty).$$

Therefore,

$$\int_{\zeta_2=\infty}^{\zeta_2=z_2} \cdots \int_{\zeta_n=\infty}^{\zeta_n=z_n} W_{g,n}(z,\zeta_2,\ldots,\zeta_n)\Bigg|_{z_2=\cdots=z_n=z} \tag{57}$$
$$\sim \left(\frac{(-1)^{n+1}C_{g,n}(\underline{\lambda},t,\underline{\nu})}{z^{n+1}} + \cdots \right) dz$$

holds. Multiplying both sides of the equation by $t(z^2 - z)$ and calculating residues, we get (52). $\qquad\square$

Proof of Lemma 3. By taking $(\nu_{\infty,0},\nu_{\infty,1}) = (0,0)$ in Theorem 1, we obtain

$$\log \psi|_{x=x(z)} = \sum_{m=-1}^{\infty} \hbar^m \left\{ \sum_{\substack{2g+n-2=m \\ g\geq 0,\, n\geq 1}} \frac{1}{n!} \int_{\infty}^{z} \cdots \int_{\infty}^{z} \left(W_{g,n}(z_1,\ldots,z_n) \right. \right.$$
$$\left. \left. -\delta_{g,0}\delta_{n,2}\frac{dx(z_1)\,dx(z_2)}{(x(z_1)-x(z_2))^2} \right) \right\}. \tag{58}$$

It follows from this equation that

$$\sum_{g\geq 0} \hbar^{2g-1}\left(-\operatorname{Res}_{z=\infty} t(z^2 - z)W_{g,1}(z)\right)$$

$$= -\operatorname{Res}_{z=\infty} t(z^2 - z) \sum_{m=-1}^{\infty} \hbar^m S_m(x(z))dx(z)$$

$$+ \operatorname{Res}_{z=\infty} t(z^2 - z) \int_{\infty}^{z} \left(W_{0,2}(z,z_2) - \frac{dx(z)\,dx(z_2)}{(x(z)-x(z_2))^2} \right)$$

$$+ \operatorname{Res}_{z=\infty} t(z^2 - z) \sum_{\substack{g\geq 0,\, n\geq 2 \\ (g,n)\neq(0,2)}} \frac{\hbar^{2g-2+n}}{(n-1)!} \int_{\infty}^{z} \cdots \int_{\infty}^{z} W_{g,n}(z,z_2,\ldots,z_n)$$

$$= -\frac{\lambda_{\infty,1}t}{\hbar} + \frac{\lambda_{\infty,0}+\lambda_{\infty,1}}{t} + \operatorname{Res}_{z=\infty} t(z^2 - z) \int_{\infty}^{z} \left(W_{0,2}(z,z_2) \right.$$

$$\left. -\frac{dx(z)\,dx(z_2)}{(x(z)-x(z_2))^2} \right) + \sum_{g\geq 1} \hbar^{2g} C_{g,2}(\underline{\lambda},t,\underline{\nu}).$$

$$\tag{59}$$

Here, we have used Lemma 5 in the last equality. Because the left-hand side of this equation is written by

$$\sum_{g \geq 0} \hbar^{2g-1} \left(- \operatorname{Res}_{z=\infty} t(z^2 - z) W_{g,1}(z) \right)$$
$$= -\hbar^{-1} \operatorname{Res}_{z=\infty} t(z^2 - z) W_{0,1}(z) + \sum_{g \geq 1} \hbar^{2g-1} \frac{\partial F_g}{\partial t}, \tag{60}$$

we compare the odd degree terms with respect to $\hbar$ of both sides. Then, we find that there is no odd degree term whose order with respect to $\hbar$ is greater than or equal to one in the right-hand side. It means that (49) holds. $\qquad \Box$

Acknowledgment

The author is grateful to Prof. Takashi Aoki, Prof. Sampei Hirose, Prof. Kohei Iwaki, Prof. Shingo Kamimoto, Prof. Takahiro Kawai, Prof. Nobuki Takayama, Prof. Yoshitsugu Takei and Prof. Mika Tanda for helpful discussions and communications.

References

1. T. Aoki and M. Tanda, Parametric Stokes phenomena of the Gauss hypergeometric differential equation with a large parameter. *J. Math. Soc. Japan*, **68**(2016), 1099–1132.
2. T. Aoki, T. Takahashi and M. Tanda, Exact WKB analysis of confluent hypergeometric differential equations with a large parameter. *RIMS Kôkyûroku Bessatsu*, **B 52**(2014), 165–174.
3. V. Bouchard and B. Eynard, Think globally, compute locally. *J. High Energy Phys.*, (2013), doi:10.1007/JHEP02(2013)143.
4. V. Bouchard and B. Eynard, Reconstructing WKB from topological recursion *J. de l'Ecole polytechnique — Mathematiques*, **4**(2017), pp. 845–908.
5. O. Dumitrescu and M. Mulase, Quantum curves for Hitchin fibrations and the Eynard-Orantin theory. *Lett. Math. Phys.*, **104**(2014), 635–671.
6. B. Eynard and E. Garcia-Failde, From topological recursion to wave functions and PDEs quantizing hyperelliptic curves, arXiv:1911.07795.

7. B. Eynard and N. Orantin, Invariants of algebraic curves and topological expansion. *Comm. Numb. Theory Phys.*, **1**(2007), pp. 347–452; arXiv:math-ph/0702045.

8. S. Gukov and P. Sułkowski A-polynomial, B-model, and quantization. *JHEP*, **2012**(2012), 70.

9. K. Iwaki, 2-parameter τ-function for the first Painlevé equation — Topological recursion and direct monodromy problem via exact WKB analysis, arXiv:1902.06439.

10. K. Iwaki and T. Koike, On the computation of Voros coefficients via middle convolutions. *Kôkyûroku Bessatsu*, **B52**(2014), 55–70.

11. K. Iwaki, T. Koike and Y.-M. Takei, Voros coefficients for the hypergeometric differential equations and Eynard-Orantin's topological recursion. Part I: For the weber equation, arXiv:1805.10945.

12. K. Iwaki, T. Koike and Y.-M. Takei, Voros coefficients for the hypergeometric differential equations and Eynard-Orantin's topological recursion: Part II: For confluent family of hypegeometric equations. *J. Integr. Syst.*, **3**(2019), 1–46.

13. T. Koike and Y. Takei, On the Voros coefficient for the Whittaker equation with a large parameter – Some progress around Sato's conjecture in exact WKB analysis. *Publ. RIMS*, Kyoto Univ. **47**(2011), 375–395.

14. O. Marchal and N. Orantin, Isomonodromic deformations of a rational differential system and reconstruction with the topological recursion: The $\mathfrak{sl}_2$ case, arXiv:1901.04344.

15. K. Okamoto and H. Kimura, On particular solutions of the Garnier systems and the hypergeometric functions of several variables. *Quart. J. Math.*, **37**(1986), 61–80.

16. H. Shen and H. J. Silverstone, Observations on the JWKB treatment of the quadratic barrier. In: *Algebraic Analysis of Differential Equations from Microlocal Analysis to Exponential Asymptotics*, Springer, 2008, pp. 237–250.

17. Y. Takei, Sato's conjecture for the Weber equation and transformation theory for Schrödinger equations with a merging pair of turning points. *RIMS Kôkyurôku Bessatsu*, **B 10**(2008), 205–224.

18. Y.-M. Takei, Voros coefficients for a class of the hypergeometric differential equations associated with the degeneration of the 2-dimensional Garnier system and the topological recursion; arXiv:2005.08957.

© 2022 World Scientific Publishing Europe Ltd.
https://doi.org/10.1142/9781800611368_0008

Chapter 8

Exact WKB Analysis of the Hypergeometric Differential Equation with a Simple Pole

Toshinori Takahashi*,§ and Mika Tanda†,‡

*Institute for Fundamental Sciences
Faculty of Science and Engineering, Setsunan University
17-8 Ikedanaka-machi, Neyagawa City, Osaka, 572-8508, Japan
†Department of Mathematical Sciences, Kwansei Gakuin University
2-1 Gakuen, Sanda 669-1337, Japan
§takahashi@math.kindai.ac.jp
‡tanda.m@kwansei.ac.jp

The Gauss hypergeometric differential equation with a large parameter deformed to a differential equation with a simple pole at the origin is considered. The relations between Kummer's solutions in the neighborhood of the regular singular point 1 and the Borel sums of the WKB solutions are established.

1. Introduction

The principal aim of this chapter is to investigate the relations between the fundamental solutions in the neighborhood of $x = 1$ of the Gauss hypergeometric differential equation with a large parameter η and the Borel sums of the WKB solutions of the following differential equation:

$$\left\{ -\frac{d^2}{dx^2} + \eta^2 \left(\frac{Q_0}{x} + \eta^{-2} \frac{Q_2}{x^2} \right) \right\} \psi = 0, \tag{1}$$

where

$$Q_0 = \frac{(\alpha - \beta)^2 x + 4\alpha\beta}{4(x-1)^2}, \quad Q_2 = -\frac{x^2 - x + 1}{4(x-1)^2} \tag{2}$$

and $\alpha, \beta \in \mathbb{C}$. Equation (1) is obtained by applying a suitable gauge transformation to the Gauss hypergeometric differential equation

$$x(1-x)\frac{d^2 w}{dx^2} + (c - (a+b+1)x)\frac{dw}{dx} - abw = 0 \tag{3}$$

with

$$a = \frac{1}{2} + \alpha\eta, \quad b = \frac{1}{2} + \beta\eta, \quad c = 1 \quad (\alpha, \beta \in \mathbb{C}). \tag{4}$$

Note that equation (1) has one simple pole at $x = 0$ in its potential function. It is well known that equation (3) has 24 standard solutions which are called Kummer's solutions [9]. We pick the solutions at $x = 1$ from Kummer's solutions:

$$u_2 = {}_2F_1(a, b, a + b + 1 - c; 1 - x),$$
$$u_6 = (1-x)^{c-a-b} {}_2F_1(c - a, c - b, c - a - b + 1; 1 - x),$$

where ${}_2F_1(a, b, c; x)$ denotes the hypergeometric function defined by the hypergeometric series:

$$_2F_1(a, b, c; x) = \sum_{n=0}^{\infty} \frac{(a)_n (b)_n}{(c)_n n!} x^n. \tag{5}$$

Here, $(a)_n = \Gamma(a + n)/\Gamma(a)$ for a complex number a. The solutions u_2 and u_6 converge if $|x - 1| < 1$ and define holomorphic functions.

On the other hand, we can construct formal solutions which are called WKB solutions of equation (1). Koike and Schäfke proved the Borel summability of WKB solutions of the equation with a polynomial potential [14], and Takei studied the Borel summability with another method [17]. If we take suitable normalization of the integrals in WKB solutions of (1), they are Borel summable under some assumptions by using these methods and the Borel sums of the WKB solutions define analytic solutions of (1). In [6], Aoki and the authors of the present chapter found the relation between the hypergeometric function (5) with

$$a = \alpha_0 + \alpha\eta, \quad b = \beta_0 + \beta\eta, \quad c = \gamma_0 + \gamma\eta \tag{6}$$

and the Borel sum of the WKB solution normalized at the origin (i.e. a singular point) of the hypergeometric equation with η. Moreover, computing monodromy matrices of the Borel resummed WKB solutions (cf. [13,18]) and the Voros coefficients which connect the WKB solutions with different ways of normalization (cf. [6,8,10,15,16,19]), the relation between Kummer's solutions on a neighborhood of the origin and the Borel resummed WKB solutions normalized at a turning point were given.

In this chapter, we give the relations which hold between (u_2, u_6) and the Borel resummed WKB solutions of (1) on a neighborhood of a simple pole, namely, the origin. In order to obtain the relations, we use a similar method used in [6], that is, we employ two special ways of normalization for WKB solutions normalized at a turning point and at $x = 1$, computing the Voros coefficient at $x = 1$ and the monodromy matrices of the Borel sums of the WKB solutions. We also find the relations between (u_2, u_6) with

$$a = 1/2 + \alpha\eta, \quad b = 1/2 + \beta\eta, \quad c = 1 + \gamma\eta \tag{7}$$

and the Borel resummed WKB solutions to the following equation:

$$\left\{ -\frac{d^2}{dx^2} + \eta^2 \left(\frac{Q_0}{x^2} + \eta^{-2}\frac{Q_2}{x^2} \right) \right\} \psi = 0 \tag{8}$$

with

$$Q_0 = \frac{(\alpha - \beta)^2 x^2 + 2(2\alpha\beta - \alpha\gamma - \beta\gamma)x + \gamma^2}{4(x - 1)^2}, \quad Q_2 = -\frac{x^2 - x + 1}{4(x - 1)^2}. \tag{9}$$

The comparison of the relations obtained in two cases (4) and (7) shows that those relations are connected via the limit $\gamma \to 0$.

2. Some Basics of Exact WKB Analysis of the HGDE with a Simple Pole

We put a large parameter η in the parameters a, b, c of (3) as (4). The explicit form of (3) becomes

$$x(1-x)\frac{d^2w}{dx^2}+(1-((\alpha+\beta)\eta+2)x)\frac{dw}{dx}-\left(\alpha\eta+\frac{1}{2}\right)\left(\beta\eta+\frac{1}{2}\right)w=0. \tag{10}$$

We introduce a new unknown function ψ as

$$w = x^{1/2}(1-x)^{((\alpha+\beta)\eta+1)/2}\psi. \tag{11}$$

Then we have (1). We assume that (α, β) is not contained in the following set $\tilde{E}_0$:

$$\tilde{E}_0 = \{(\alpha, \beta) \in \mathbb{C}^2 \mid \alpha \cdot \beta \cdot (\alpha - \beta) \cdot (\alpha + \beta) = 0\}. \tag{12}$$

In this case, there are two distinct zeros of Q_0/x which do not coincide with $0, 1, \infty$. The zero is called a simple turning point of (1), which is denoted by τ. Stokes curves emanating from a point a (cf. [13]) are defined by

$$\mathrm{Im} \int_a^x \sqrt{\frac{Q_0}{x}}\, dx = 0 \quad (a = \tau, 0). \tag{13}$$

In this case, Stokes geometry is degenerate (cf. [2,13]), that is, there is one of the Stokes curves which starts from the origin and flows into the singular points 1 or ∞. A Stokes region is, by definition, a connected component of the Riemann sphere obtained by excluding all the Stokes curves, τ, the origin, 1, and ∞. We define the sets $\tilde{E}_1$ and $\tilde{E}_2$ of pairs of the parameters (α, β) as follows:

$$\tilde{E}_1 = \{(\alpha, \beta) \in \mathbb{C}^2 \mid \mathrm{Re}\alpha \cdot \mathrm{Re}\beta = 0\}, \tag{14}$$

$$\tilde{E}_2 = \{(\alpha, \beta) \in \mathbb{C}^2 \mid \mathrm{Re}(\alpha - \beta) \cdot \mathrm{Re}(\alpha + \beta) = 0\}. \tag{15}$$

If (α, β) belongs to $\tilde{E}_1$, τ and the origin are connected by a Stokes curve. If (α, β) is contained in $\tilde{E}_2$, a Stokes curve forms a closed curve with a turning point or the origin as the base point. Throughout this chapter, we assume that (α, β) does not belong to both sets $\tilde{E}_1$ and $\tilde{E}_2$. Equation (1) has the formal power series solutions which are called the WKB solutions normalized at a (cf. [13]):

$$\psi_{a,\pm} = \frac{1}{\sqrt{S_{\text{odd}}}} \exp(\pm \int_a^x S_{\text{odd}}\, dx), \tag{16}$$

where $a = \tau, 0$ and S_{odd} denotes the odd-order part of the formal solution

$$S(x) = S_{\text{odd}} + S_{\text{even}} = \eta S_{-1}(x) + S_0(x) + \eta^{-1} S_1(x) + \cdots \tag{17}$$

in η^{-1} of the Riccati equation

$$\frac{dS}{dx} + S^2 = \eta^2 Q(x) \tag{18}$$

to (1). Note that $S_{-1} = \sqrt{Q_0/x}$ holds by substituting (17) into (18) and comparing the coefficients of η^2. If

$$\operatorname{Re} \int_a^x \sqrt{\frac{Q_0}{x}}\, dx < 0 \quad \left(\text{resp., } \operatorname{Re} \int_a^x \sqrt{\frac{Q_0}{x}}\, dx > 0 \right) \tag{19}$$

holds along a Stokes curve, $\psi_{a,+}$ (resp., $\psi_{a,-}$) is said to be recessive (resp., dominant) on the Stokes curve. The recessive (resp., dominant) WKB solution does not have (resp., has) the Stokes phenomena on the Stokes curve. Thus, the WKB solutions $\psi_{a,\pm}$ are convenient for describing connection formulas concerned with Stokes curves emanating from a (see [13] for the simple turning point $a = \tau$ and [12] for the simple pole $a = 0$).

On the other hand, we can take another natural normalization of WKB solutions: By using a method given in [1,13], we obtain

$$\operatorname*{Res}_{x=1} S_{\text{odd}}\, dx = \frac{\alpha + \beta}{2} \eta. \tag{20}$$

Here, we choose the branch of $\sqrt{Q_0/x}$ so that

$$\sqrt{\frac{Q_0}{x}} \sim \frac{\alpha + \beta}{2(x - 1)} \tag{21}$$

holds near $x = 1$. WKB solutions to (1) normalized at $x = 1$ are given by

$$\psi_\pm^{(1)} := \frac{(x - 1)^{\pm(\alpha+\beta)\eta/2}}{\sqrt{S_{\text{odd}}}} \exp\left(\pm \int_1^x \left(S_{\text{odd}} - \frac{(\alpha + \beta)\eta}{2(x - 1)} \right) dx \right). \tag{22}$$

The WKB solutions $\psi_{a,\pm}$ and $\psi_{\pm}^{(1)}$ are Borel summable under some assumptions and analytic solutions of (1) by using the method in [14,17]. If the WKB solution (16) is the recessive (resp., dominant) solution, then the WKB solution (22) at 1 is a recessive (resp., dominant) solution.

The sets $\tilde{\omega}_k$ $(k = 3, 4)$ of the parameters (α, β) are defined by

$$\tilde{\omega}_3 = \{(\alpha, \beta) \in \mathbb{C}^2 \mid 0 < \mathrm{Re}\alpha < \mathrm{Re}\beta\}, \tag{23}$$

$$\tilde{\omega}_4 = \{(\alpha, \beta) \in \mathbb{C}^2 \mid 0 < \mathrm{Re}\alpha + \mathrm{Re}\beta < \mathrm{Re}\beta\}. \tag{24}$$

Let G denote the group defined in the space of parameters $(\alpha, \beta) \in \mathbb{C}^2$. The group G is generated by the involutions $l = 0, 1, 2$:

$$\tilde{\iota}_0 : (\alpha, \beta) \mapsto (\beta, \alpha), \tag{25}$$

$$\tilde{\iota}_1 : (\alpha, \beta) \mapsto (-\alpha, -\beta), \tag{26}$$

$$\tilde{\iota}_2 : (\alpha, \beta) \mapsto (-\beta, -\alpha). \tag{27}$$

Moreover, we define open subsets $\tilde{\Pi}_k$ $(k = 3, 4)$ in $\mathbb{C}^2$ by

$$\tilde{\Pi}_k = \bigcup_{r \in G} r(\tilde{\omega}_k). \tag{28}$$

A Stokes graph can be classified by its order sequence $\tilde{n} = (\{b_j\}; n_1, n_2)$, where n_1 and n_2 are the number of Stokes curves that start from τ and flow into 1 and ∞, respectively, and $\{b_j\}$ designates the symbol for the Stokes curve emanating from the origin and flowing into b_j $(j = 1, \infty;\ b_1 = 1, b_\infty = \infty)$.

Proposition 1. *Let $\tilde{n} = (\{b_j\}; n_1, n_2)$ denote the order sequences of the Stokes graph $(j = 1, 2)$.*

(1) *If $(\alpha, \beta) \in \tilde{\Pi}_3$, then $\hat{n} = (\{b_1\}; 2, 1)$.*
(2) *If $(\alpha, \beta) \in \tilde{\Pi}_4$, then $\hat{n} = (\{b_\infty\}; 1, 2)$.*

Remark 1. A proof of this proposition will be given in our forthcoming paper.

Remark 2. We show typical configuration of Stokes curves for each case in Figs. 1(a) and 1(b). Here, triangle, black triangle, small circle and the wavy line designate turning points τ, the simple pole 0, one of the singular points 1 and the branch cut for $\sqrt{Q_0/x}$, respectively.

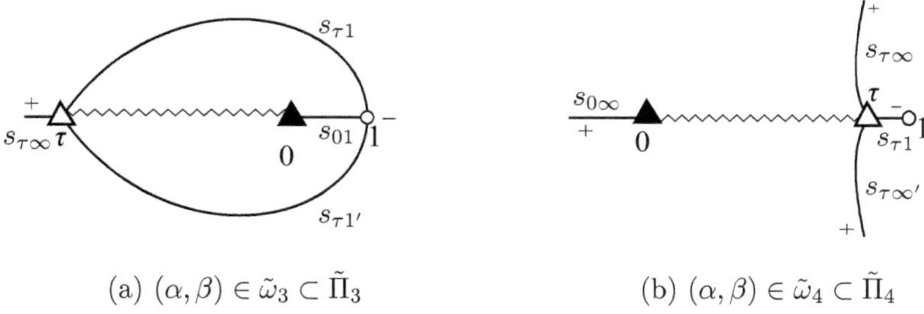

(a) $(\alpha, \beta) \in \tilde{\omega}_3 \subset \tilde{\Pi}_3$ (b) $(\alpha, \beta) \in \tilde{\omega}_4 \subset \tilde{\Pi}_4$

Fig. 1.   Stokes curves $\tilde{\omega}_3$ and for $\tilde{\omega}_4$.

There are three Stokes curves emanating from τ and one Stokes curve emanating from the simple pole 0. Let s_{aj} $(a = \tau, 0; j = 1, 1', \infty, \infty')$ be the Stokes curve shown in Fig. 1. The plus or minus signs on the Stokes curves s_{aj} show which WKB solutions $\psi_{a,\pm}$ and $\psi_{\pm}^{(1)}$ are dominant, i.e. for example, since the residue of $S_{\mathrm{odd}}\, dx$ at 1 is taken as (20), $\psi_{a,-}$ and $\psi_{-}^{(1)}$ (resp., $\psi_{a,+}$ and $\psi_{+}^{(1)}$) are the dominant (resp., recessive) solutions on $s_{a,1}, s_{a,1'}$ in both cases.

We consider the relations between (u_2, u_6) with the large parameter η and the Borel sums of $(\psi_{\tau,+}, \psi_{\tau,-})$ in the case where $(\alpha, \beta) \in \tilde{\omega}_k$. In both cases, the residue of $S_{\mathrm{odd}}\, dx$ at ∞ is given by

$$\operatorname*{Res}_{x=\infty} S_{\mathrm{odd}}\, dx = \frac{\alpha - \beta}{2}\eta. \tag{29}$$

3.   Voros Coefficient at $x = 1$ and Its Borel Sums

The WKB solutions $\psi_{\tau,\pm}$ normalized at τ and $\psi_{\pm}^{(1)}$ are connected as

$$\psi_{\pm}^{(1)} = \exp(\pm\hat{W}_1)\psi_{\tau,\pm}, \tag{30}$$

where $\hat{W}_1$ is

$$\hat{W}_1 = \int_1^{\tau}\left(S_{\mathrm{odd}} - \frac{(\alpha + \beta)\eta}{2(x-1)}\right) dx + \frac{(\alpha + \beta)\eta}{2(x-1)}\log(\tau - 1). \tag{31}$$

We write the formal series $\hat{W}_1$ of η^{-1} in the form

$$\hat{W}_1 = \sum_{j=-1}^{\infty} \eta^{-j} W_{1,j}. \tag{32}$$

We divide $\hat{W}_1$ into two parts:

$$\hat{W}_1 = W_1 + \hat{W}_{1,\leq 0}, \tag{33}$$

where we set

$$W_1 = \sum_{j=1}^{\infty} \eta^{-j} W_{1,j}, \quad \hat{W}_{1,\leq 0} = \eta W_{1,-1} + W_{1,0}. \tag{34}$$

These formal series can be rewritten as the following forms:

$$W_1 = \frac{1}{2} \int_{\Gamma_1} S_{\mathrm{odd},>0} \, dx, \tag{35}$$

$$\hat{W}_{1,\leq 0} = \lim_{x \to 1} \left(\int_x^\tau S_{\mathrm{odd},\leq 0} \, dx + \frac{(\alpha + \beta)\eta}{2(x-1)} \log(x-1) \right). \tag{36}$$

Here, Γ_1 is a path that runs from $x = 1$, encircles τ in a counterclockwise manner and returns to $x = 1$. Since the residue of $S_{\mathrm{odd},>0} \, dx$ at any regular singular point vanishes, W_1 is independent of the choice of a ($a = \tau, 0$) and the path connecting $x = 1$ and τ. Note that W_1 is invariant under the action of $\tilde{\iota}_l$ ($= 0, 1$). We call W_1 the Voros coefficient of (1) at $x = 1$. Since we take the branch of $\sqrt{Q_0/x}$ as (21) holds near $x = 1$, we obtain the explicit form of W_1 as follows:

Theorem 1. *If* $\mathrm{Re}(\alpha + \beta)$ *is positive, the explicit form of the Voros coefficient* W_1 *is the following form:*

$$W_1 = \sum_{n=2}^{\infty} \frac{B_n \eta^{1-n}}{n(n-1)} \left\{ (1 - 2^{1-n}) \left(\frac{1}{\alpha^{n-1}} + \frac{1}{\beta^{n-1}} \right) + \frac{1}{(\alpha + \beta)^{n-1}} \right\}. \tag{37}$$

Here, B_n *denotes the Bernoulli number:*

$$\frac{te^t}{e^t - 1} = \sum_{n=0}^{\infty} \frac{B_n}{n!} t^n, \tag{38}$$

where the first term B_1 *of* B_n *is* $\frac{1}{2}$.

Remark 3. The explicit form of Voros coefficient at $x = 1$ of (8) is obtained by Theorem 2.3 in [4].

The proof of **Theorem 1** is essentially the same as the proof of Theorem 2.3 in [4], however, the definition of the Voros coefficient is different and some careful computations are required. The calculation method of the Voros coefficient is described in [7]. Recently another derivation of the Voros coefficient for the hypergeometric differential equation is given in [11] by using the theory of topological recursion.

Next, we consider the explicit form of $\hat{W}_{1,\leq 0}$. Note that we should be more careful about the choice of branch of logarithm. For simplicity, we assume that α, β are real in this chapter. If we take the residue of $S_{\mathrm{odd}}\,dx$ at 1 and ∞ as (20) and (29), respectively, we obtain the following theorem:

Theorem 2. (I) *If* (α, β) *is contained in* $\tilde{\omega}_3$, *we have*

$$\hat{W}_{1,\leq 0} = -\alpha\eta \log \alpha - \beta\eta \log \beta + (\alpha + \beta)\eta \log(\alpha + \beta) + \frac{\alpha - \beta}{2}\eta\pi i. \tag{39}$$

(II) *If* (α, β) *is contained in* $\tilde{\omega}_4$, *we have*

$$\hat{W}_{1,\leq 0} = -\alpha\eta \log(-\alpha) - \beta\eta \log \beta + (\alpha + \beta)\eta \log(\alpha + \beta)$$
$$+ \frac{3\beta - \alpha}{2}\eta\pi i. \tag{40}$$

The Voros coefficient W_1 is Borel summable in $\tilde{\omega}_k$ $(k = 3, 4)$ by using the method in [14,17]. Let W_1^k be the Borel sums of W_1 in $\tilde{\omega}_k$. We can compute the explicit forms of W_1^k:

Theorem 3. *The Voros coefficient* W_1 *is the Borel summable in* $\tilde{\omega}_k$.
(I) *If* $(\alpha, \beta) \in \tilde{\omega}_3$, *the Borel sum* W_1^3 *of* W_1 *has the following form:*

$$W_1^3 = \log \frac{\sqrt{2\pi}\Gamma((\alpha + \beta)\eta)\alpha^{\alpha\eta}\beta^{\beta\eta}\eta^{\frac{1}{2}}}{\Gamma(\frac{1}{2} + \alpha\eta)\Gamma(\frac{1}{2} + \beta\eta)(\alpha + \beta)^{(\alpha+\beta)\eta-\frac{1}{2}}}.$$

(II) *If* $(\alpha, \beta) \in \tilde{\omega}_4$, *the Borel sum* W_1^4 *of* W_1 *has the following form:*

$$W_1^4 = \log \frac{\Gamma((\alpha + \beta)\eta)\Gamma(\frac{1}{2} - \alpha\eta)\beta^{\beta\eta}\eta^{\frac{1}{2}}}{\sqrt{2\pi}\Gamma(\frac{1}{2} + \beta\eta)(-\alpha)^{-\alpha\eta}(\alpha + \beta)^{(\alpha+\beta)\eta-\frac{1}{2}}}.$$

Since the proofs of (Theorems 2 and 3) are similar to those in [4,6], respectively, we omit them. We set the Borel sums $\hat{W}_1^k = W_1^k + \hat{W}_{1,\leq 0}$ of $\hat{W}_1$ in the case where (α, β) is contained in $\tilde{\omega}_k$. Combining (Theorems 2 and 3), we obtain the following theorem:

Theorem 4. (I) *If (α, β) is contained in $\tilde{\omega}_3$, the Borel sum $\hat{W}_1^3 = W_1^3 + \hat{W}_{1,\leq 0}$ has the form*

$$\hat{W}_1^3 = \log \frac{\sqrt{2\pi}\Gamma((\alpha + \beta)\eta)((\alpha + \beta)\eta)^{\frac{1}{2}} e^{\frac{\alpha - \beta}{2}\eta\pi i}}{\Gamma(\frac{1}{2} + \alpha\eta)\Gamma(\frac{1}{2} + \beta\eta)}. \qquad (41)$$

(II) *If (α, β) is contained in $\tilde{\omega}_4$, the Borel sum $\hat{W}_1^4 = W_1^4 + \hat{W}_{1,\leq 0}$ has the form*

$$\hat{W}_1^4 = \log \frac{\Gamma((\alpha + \beta)\eta)\Gamma(\frac{1}{2} - \alpha\eta)((\alpha + \beta)\eta)^{\frac{1}{2}} e^{\frac{3\beta - \alpha}{2}\eta\pi i}}{\sqrt{2\pi}\Gamma(\frac{1}{2} + \beta\eta)}. \qquad (42)$$

4. Kummer's Solutions in the Neighborhood of 1 and the WKB Solutions

In this section, we consider the relation between u_2 and u_6 to which the large parameter η is introduced as (4) and Borel sums of the WKB solutions $\psi_{\tau,\pm}$ in the case where $(\alpha, \beta) \in \tilde{\omega}_k$ $(k = 3, 4)$. Let $\tilde{u}_2$ and $\tilde{u}_6$ denote u_2 and u_6 with η, respectively:

$$\tilde{u}_2 = {}_2F_1\left(\frac{1}{2} + \alpha\eta, \frac{1}{2} + \beta\eta, 1 + \alpha\eta + \beta\eta; 1 - x\right), \qquad (43)$$

$$\tilde{u}_6 = (1 - x)^{-(\alpha+\beta)\eta} {}_2F_1\left(\frac{1}{2} - \alpha\eta, \frac{1}{2} - \beta\eta, 1 - \alpha\eta - \beta\eta; 1 - x\right). \qquad (44)$$

We show the configuration of the Stokes curves in Fig. 2 in the case where $(\alpha, \beta) \in \tilde{\omega}_k$. Let $\mathcal{R}_{\mathrm{I}}$ (resp., $\mathcal{R}_{\mathrm{I}'}$; resp., $\mathcal{R}_{\mathrm{II}}$) be the Stokes region surrounded by $s_{\tau 1}$, s_{01} and the branch cut (resp., $s_{\tau 1'}$, s_{01} and the branch cut; resp., $s_{\tau 1}$, $s_{\tau 1'}$ and $s_{\tau\infty}$) in Fig. 2(a). Similarly, $\mathcal{R}_{\mathrm{I}}$ denotes the Stokes region surrounded by $s_{\tau 1}$, $s_{\tau\infty}$ and $s_{\tau\infty'}$ in Fig. 2(b). Let x_0 be a point in $\mathcal{R}_{\mathrm{I}}$ and C_1 be a closed path starting from x_0, going around 1 once counterclockwise and returning to x_0.

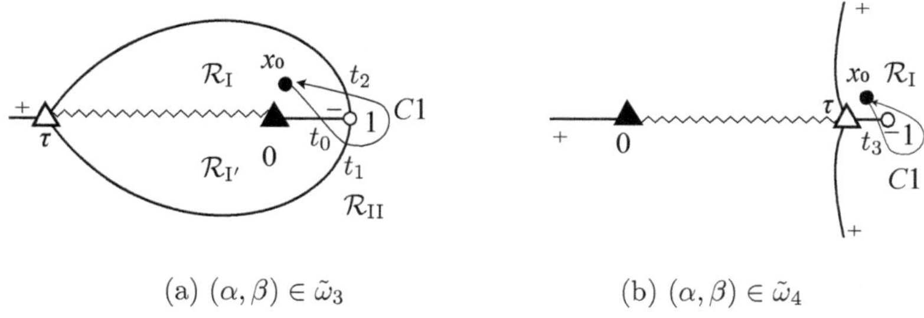

(a) $(\alpha, \beta) \in \tilde{\omega}_3$ (b) $(\alpha, \beta) \in \tilde{\omega}_4$

Fig. 2. Stokes curves and paths of continuation for $\tilde{\omega}_3$ and for $\tilde{\omega}_4$.

Here, $\mathcal{R}_{\mathrm{I}'}$ denotes a Stokes region near the segment 0 and 1 in $\{x \in \mathbb{C};$ Im $x < 0\}$. The WKB solutions $\psi_{\tau,\pm}$ and $\psi_{\pm}^{(1)}$ are Borel summable in $\mathcal{R}_h$ ($h = \mathrm{I}, \mathrm{I}', \mathrm{II}$). Let $\tilde{\Psi}_{\tau,\pm}^{h}$ and $\tilde{\Psi}_{\pm}^{(1),h}$ be the Borel sums of $\psi_{\tau,\pm}$ and $\psi_{\pm}^{(1)}$ in $\mathcal{R}_h$, respectively.

Firstly, we investigate the relation between Kummer's solution and the Borel sum of the recessive WKB solution $\psi_{+}^{(1)}$ in $\mathcal{R}_\mathrm{I}$. Since $\mathrm{Re}(\alpha + \beta) > 0$ in $\tilde{\omega}_k$ and $\psi_{+}^{(1)}$ is the recessive solution, we relate $\tilde{u}_2$ and the Borel sum $\tilde{\Psi}_{+}^{(1),\mathrm{I}}$ of $\psi_{+}^{(1)}$. The characteristic exponents of $x = 1$ of (1) are $(1 \pm (\alpha + \beta)\eta)/2$ and the recessive solution $\psi_{+}^{(1)}$ has the exponent $(1 + (\alpha + \beta)\eta)/2$. Thus, Frobenius function $(x - 1)^{-((\alpha+\beta)\eta+1)/2} \tilde{\Psi}_{+}^{(1),\mathrm{I}}$ has $x = 1$ as a removable singularity. Then $x^{-1/2}(1 - x)^{-((\alpha+\beta)\eta+1)/2} \tilde{\Psi}_{+}^{(1),\mathrm{I}}$ is a holomorphic solution to (10). We can evaluate this function at 1 and obtain the following theorem:

Theorem 5. *If* $\mathrm{Re}(\alpha + \beta) > 0$ *and the branch of* $\sqrt{Q_0/x}$ *is chosen as* (21), *then Kummer's solution* $\tilde{u}_2$ *and the Borel sum* $\tilde{\Psi}_{+}^{(1),k}$ *are related by*

$$\tilde{u}_2 = \sqrt{\frac{(\alpha + \beta)\eta}{2}} x^{-\frac{1}{2}} (1 - x)^{-\frac{1}{2}((\alpha+\beta)\eta+1)} e^{-\frac{(\alpha+\beta)\eta+1}{2}\pi i} \tilde{\Psi}_{+}^{(1),\mathrm{I}} \qquad (45)$$

in a neighborhood of 1.

Remark 4. The above relation holds under the assumption $\mathrm{Re}(\alpha + \beta) > 0$. The explicit form of the relation between the hypergeometric function (5) and the Borel sum of the WKB solution at the origin in the general case, i.e. the large parameter η is introduced as (6), are given in [5,6] and in the case where η is put as (7) are obtained in [3].

Next, we consider the relation between $(\tilde{u}_2, \tilde{u}_6)$ and $(\tilde{\Psi}^h_{\tau,+}, \tilde{\Psi}^h_{\tau,-})$ in the case where $(\alpha, \beta) \in \tilde{\omega}_k$ $(k = 3, 4)$.

Theorem 6. *If* $(\alpha, \beta) \in \tilde{\omega}_k$, *there exists a regular matrix* $\tilde{A}_k$ *for which*

$$(\tilde{u}_2, \tilde{u}_6) = \sqrt{\frac{(\alpha + \beta)\eta}{2}} x^{-\frac{1}{2}} (1 - x)^{-\frac{(\alpha+\beta)\eta+1}{2}} (\tilde{\Psi}^I_{\tau,+}, \tilde{\Psi}^I_{\tau,-}) \tilde{A}_k \qquad (46)$$

holds, where we set $(\tilde{u}_2, \tilde{u}_6)$ *as* (43) *and* (44), *respectively, and*

$$\tilde{A}_3 = \begin{pmatrix} e^{-\frac{(\alpha+\beta)\eta+1}{2}\pi i} e^{\hat{W}_1^3} & i\frac{\left(1+2e^{2\beta\eta\pi i}+e^{2(\beta-\alpha)\eta\pi i}\right)}{e^{2(\alpha+\beta)\eta\pi i}-1} e^{\frac{(\alpha+\beta)\eta+1}{2}\pi i} e^{-\hat{W}_1^3} \\[2mm] 0 & e^{\frac{(\alpha+\beta)\eta+1}{2}\pi i} e^{-\hat{W}_1^3} \end{pmatrix},$$

$$\qquad (47)$$

$$\tilde{A}_4 = \begin{pmatrix} e^{-\frac{(\alpha+\beta)\eta+1}{2}\pi i} e^{\hat{W}_1^4} & i\frac{e^{(2\pi i(\alpha+\beta)\eta)}}{e^{(2\pi i(\alpha+\beta)\eta)}-1} e^{\frac{(\alpha+\beta)\eta+1}{2}\pi i} e^{-\hat{W}_1^4} \\[2mm] 0 & e^{\frac{(\alpha+\beta)\eta+1}{2}\pi i} e^{-\hat{W}_1^4} \end{pmatrix}. \qquad (48)$$

Here, the Borel sums $\hat{W}_1^k$ *of* $\hat{W}_1$ *are given by Theorem 4.*

Proof. The relation between $\tilde{u}_2$ and $\tilde{\Psi}^I_{\tau,+}$ is obtained by using (Theorems 3 and **5**. To derive the relation between $\tilde{u}_6$ and $(\tilde{\Psi}^I_{\tau,+}, \tilde{\Psi}^I_{\tau,-})$, we compute monodromy matrix of $(\tilde{\Psi}^I_{\tau,+}, \tilde{\Psi}^I_{\tau,-})$ in the case where $(\alpha, \beta) \in \tilde{\omega}_k$ by using the method developed in [13]. We only prove the relation in the case where $(\alpha, \beta) \in \tilde{\omega}_3$ since the others can be similar. Let $\varphi_\pm$ be the WKB solutions normalized at x_0 given by

$$\varphi_\pm = \frac{1}{\sqrt{S_{\text{odd}}}} \exp\left(\pm \int_{x_0}^x S_{\text{odd}}\, dx\right) \qquad (49)$$

and their Borel sums $\Phi^h_\pm$ in $\mathcal{R}_h$. Since $\text{Re}(\alpha + \beta) > 0$ and the branch of $\sqrt{Q_0/x}$ at 1 is taken as (21), we have $\text{Re}\int_0^x \sqrt{Q_0/x}\, dx < 0$ on s_{01}, s_{11}, s_{12}. Thus, φ_- is the dominant solution on s_{01}, s_{11}, s_{12}. We calculate the monodromy matrix $\tilde{M}_1$ with respect to $(\tilde{\Psi}^I_{\tau,+}, \tilde{\Psi}^I_{\tau,-})$

along the closed curve C_1 depicted in Fig. 2(a). That is, C_1 crosses the Stokes curves in order $t_0 \to t_1 \to t_2$ and passes though the Stokes regions $\mathcal{R}_h$ in order $\mathcal{R}_\mathrm{I} \to \mathcal{R}_{\mathrm{I}'} \to \mathcal{R}_\mathrm{II} \to \mathcal{R}_\mathrm{I}$. We compute the connection matrices T_m at t_m ($m = 0, 1, 2$) for the Borel sums of (φ_+, φ_-). To describe the monodromy matrix $\tilde{M}_1$ along C_1, we define e_a, $\nu_j^{\pm}$, u_a, $u_{aa'}$ ($a = \tau, 0$; $a' = \tau, 0$) by

$$e_a = \exp\left(\int_{\delta_a} S_\mathrm{odd}\, dx\right), \quad \nu_j^{\pm} = \exp\left(\pi i\left(1 \pm \operatorname*{Res}_{x=b_j} S_\mathrm{odd}\, dx\right)\right), \tag{50}$$

$$u_a = \exp\left(2\int_{\delta_a} S_\mathrm{odd}\, dx\right), \quad u_{aa'} = u_a^{-1} u_{a'} \tag{51}$$

and set

$$P_a = \begin{pmatrix} e_a & 0 \\ 0 & e_a^{-1} \end{pmatrix}, \quad D_j = \begin{pmatrix} \nu_j^+ & 0 \\ 0 & \nu_j^- \end{pmatrix}, \quad U_{aa'} = \begin{pmatrix} u_{aa'} & 0 \\ 0 & u_{aa'}^{-1} \end{pmatrix}, \tag{52}$$

where $b_j = 0, 1, \infty$, ($j = 0, 1, \infty$) and δ_a is an oriented curve starting from x_0 and terminating at a. We investigate the Borel sums $\tilde{\Psi}_{\tau,\pm}^\mathrm{I}$ in the intersection of $\mathcal{R}_1$ and a small neighborhood of τ and take the analytic continuation of them along appropriate paths. Since C_1 crosses s_{01} emanating from the simple pole of Q_0/x in a clockwise manner as seen from $x = 0$, the connection matrix between $(\tilde{\Psi}_{\tau,+}^\mathrm{I}, \tilde{\Psi}_{\tau,-}^\mathrm{I})$ and $(\tilde{\Psi}_{\tau,+}^{\mathrm{I}'}, \tilde{\Psi}_{\tau,-}^{\mathrm{I}'})$ is given as follows:

$$(\tilde{\Psi}_{\tau,+}^\mathrm{I}, \tilde{\Psi}_{\tau,-}^\mathrm{I}) = (\tilde{\Psi}_{\tau,+}^{\mathrm{I}'}, \tilde{\Psi}_{\tau,-}^{\mathrm{I}'}) \begin{pmatrix} 1 & -2i \\ 0 & 1 \end{pmatrix} \tag{53}$$

by using the method in [12]. We consider two pairs of the WKB solutions (φ_+, φ_-) and $(\psi_{\tau,+}, \psi_{\tau,-})$ are related by

$$(\varphi_+, \varphi_-) = (\psi_{\tau,+}, \psi_{\tau,-}) P_0 \tag{54}$$

near $x = t_0$. Hence, we have

$$(\Phi_+^\mathrm{I}, \Phi_-^\mathrm{I}) = (\tilde{\Psi}_{\tau,+}^\mathrm{I}, \tilde{\Psi}_{\tau,-}^\mathrm{I}) P_0 \tag{55}$$

and

$$(\Phi_+^{\mathrm{I}'}, \Phi_-^{\mathrm{I}'}) = (\tilde{\Psi}_{\tau,+}^{\mathrm{I}'}, \tilde{\Psi}_{\tau,-}^{\mathrm{I}'}) P_0. \tag{56}$$

Combining (53), (55) and (56), we obtain

$$(\Phi_+^{\mathrm{I}}, \Phi_-^{\mathrm{I}}) = (\Phi_+^{\mathrm{I}'}, \Phi_-^{\mathrm{I}'}) P_0^{-1} \begin{pmatrix} 1 & -2i \\ 0 & 1 \end{pmatrix} P_0. \tag{57}$$

Hence, we have $T_0 = P_0^{-1} \begin{pmatrix} 1 & -2i \\ 0 & 1 \end{pmatrix} P_0$. Next, we calculate T_1. The closed path C_1 crosses s_{11} in a clockwise manner as seen from $x = \tau$. Therefore, we have

$$(\tilde{\Psi}_{\tau,+}^{\mathrm{I}'}, \tilde{\Psi}_{\tau,-}^{\mathrm{I}'}) = (\tilde{\Psi}_{\tau,+}^{\mathrm{II}}, \tilde{\Psi}_{\tau,-}^{\mathrm{II}}) \begin{pmatrix} 1 & -i \\ 0 & 1 \end{pmatrix} \tag{58}$$

by Voros connection formula (see [13,19]). Investigating a closed oriented curve consisting of the portion of C_1 from x_0 through t_1, the portion of the Stokes curve from t_1 through τ and from τ terminating at x_0, we get

$$(\Phi_+^{\mathrm{I}'}, \Phi_-^{\mathrm{I}'}) = (\tilde{\Psi}_{\tau,+}^{\mathrm{I}'}, \tilde{\Psi}_{\tau,-}^{\mathrm{I}'}) P_\tau D_0^{-1} U_{0\tau} \tag{59}$$

and

$$(\Phi_+^{\mathrm{II}}, \Phi_-^{\mathrm{II}}) = (\tilde{\Psi}_{\tau,+}^{\mathrm{II}}, \tilde{\Psi}_{\tau,-}^{\mathrm{II}}) P_\tau D_0^{-1} U_{0\tau}. \tag{60}$$

Thus, we obtain

$$(\Phi_+^{\mathrm{I}'}, \Phi_-^{\mathrm{I}'}) = (\Phi_+^{\mathrm{II}}, \Phi_-^{\mathrm{II}})(P_\tau D_0^{-1} U_{0\tau})^{-1} \begin{pmatrix} 1 & -i \\ 0 & 1 \end{pmatrix} P_\tau D_0^{-1} U_{0\tau}. \tag{61}$$

Then we have $T_1 = (P_\tau D_0^{-1} U_{0\tau})^{-1} \begin{pmatrix} 1 & -i \\ 0 & 1 \end{pmatrix} P_\tau D_0^{-1} U_{0\tau}$. In a similar way as (61), we obtain $T_2 = (P_\tau D_1)^{-1} \begin{pmatrix} 1 & -i \\ 0 & 1 \end{pmatrix} P_\tau D_1$, which satisfies

$$(\Phi_+^{\mathrm{II}}, \Phi_-^{\mathrm{II}}) = (\Phi_+^{\mathrm{I}''}, \Phi_-^{\mathrm{I}''}) T_2. \tag{62}$$

Here, $(\Phi_+^{\mathrm{I}''}, \Phi_-^{\mathrm{I}''})$ is the Borel sum of (φ_+, φ_-) in $\mathcal{R}_{\mathrm{I}}$ for the second pass. The Borel sum $(\Phi_+^{\mathrm{I}''}, \Phi_-^{\mathrm{I}''})$ of continuation of $(\Phi_+^{\mathrm{I}}, \Phi_-^{\mathrm{I}})$ along C_1 gains the local monodromy at $x = 1$:

$$(\Phi_+^{\mathrm{I}''}, \Phi_-^{\mathrm{I}''}) = (\Phi_+^{\mathrm{I}}, \Phi_-^{\mathrm{I}}) D_1. \tag{63}$$

Hence, we have

$$(\Phi^{\mathrm{I}}_{+}, \Phi^{\mathrm{I}}_{-})_{C_1} = (\Phi^{\mathrm{I}}_{+}, \Phi^{\mathrm{I}}_{-})D_1 T_2 T_1 T_0, \tag{64}$$

where $(\Phi^{\mathrm{I}}_{+}, \Phi^{\mathrm{I}}_{-})_{C_1}$ designates the analytic continuation of $(\Phi^{\mathrm{I}}_{+}, \Phi^{\mathrm{I}}_{-})$ along C_1. If $x_0 \to \tau$, we get

$$(\tilde{\Psi}^{\mathrm{I}}_{\tau,+}, \tilde{\Psi}^{\mathrm{I}}_{\tau,-})_{C_1} = (\tilde{\Psi}^{\mathrm{I}}_{\tau,+}, \tilde{\Psi}^{\mathrm{I}}_{\tau,-})\tilde{M}_1$$

$$= (\tilde{\Psi}^{\mathrm{I}}_{\tau,+}, \tilde{\Psi}^{\mathrm{I}}_{\tau,-})D_1 T_2 T_1 T_0 \tag{65}$$

with

$$e_0 = e^{\pi i \sum_{x=b_j} \mathrm{Res}_{x=b_j} S_{\mathrm{odd}}\, dx} = e^{\alpha \eta \pi i}, \quad e_\tau = 1, \quad u_{0\tau} = e^{2\alpha \eta \pi i}. \tag{66}$$

Here, $(\tilde{\Psi}^{\mathrm{I}}_{\tau,+}, \tilde{\Psi}^{\mathrm{I}}_{\tau,-})_{C_1}$ is the analytic continuation of $(\tilde{\Psi}^{\mathrm{I}}_{\tau,+}, \tilde{\Psi}^{\mathrm{I}}_{\tau,-})$ along C_1. The explicit form of monodromy matrix $\tilde{M}_1$ along C_1 is

$$\tilde{M}_1 = \begin{pmatrix} -e^{(\alpha+\beta)\eta\pi i} & i\left(e^{-(\alpha+\beta)\eta\pi i} + 2e^{(\beta-\alpha)\eta\pi i} + e^{(\beta-3\alpha)\eta\pi i}\right) \\ 0 & -e^{-(\alpha+\beta)\eta\pi i} \end{pmatrix}. \tag{67}$$

We can assume that the relation of the following form holds:

$$(\tilde{u}_2, \tilde{u}_6) = \tilde{p}(x)(\tilde{\Psi}^{\mathrm{I}}_{\tau,+}, \tilde{\Psi}^{\mathrm{I}}_{\tau,-})\begin{pmatrix} \tilde{a}_{11} & \tilde{a}_{12} \\ 0 & \tilde{a}_{22} \end{pmatrix}, \tag{68}$$

where

$$\tilde{p}(x) = x^{-\frac{1}{2}}(1-x)^{-\frac{(\alpha+\beta)\eta+1}{2}}, \quad \tilde{a}_{11} = \sqrt{\frac{(\alpha+\beta)\eta}{2}}\, e^{-\frac{(\alpha+\beta)\eta\pi i}{2}} e^{\hat{W}_1^3} \tag{69}$$

and $\tilde{a}_{12}$, $\tilde{a}_{22}$ are constants. We take analytic continuation of (68) along C_1. Then we have

$$(\tilde{u}_2, \tilde{u}_6)\begin{pmatrix} 1 & 0 \\ 0 & e^{-2\pi i(\alpha+\beta)\eta} \end{pmatrix}$$

$$= \tilde{p}(x)(\tilde{\Psi}^{\mathrm{I}}_{\tau,+}, \tilde{\Psi}^{\mathrm{I}}_{\tau,-})\begin{pmatrix} -e^{-(\alpha+\beta)\eta\pi i} & 0 \\ 0 & -e^{-(\alpha+\beta)\eta\pi i} \end{pmatrix} \tilde{M}_1 \begin{pmatrix} \tilde{a}_{11} & \tilde{a}_{12} \\ 0 & \tilde{a}_{22} \end{pmatrix}. \tag{70}$$

196   T. Takahashi & M. Tanda

We rewrite (70) as follows:

$$(\tilde{u}_2, \tilde{u}_6) = \tilde{p}(x)(\tilde{\Psi}^{\mathrm{I}}_{\tau,+}, \tilde{\Psi}^{\mathrm{I}}_{\tau,-})\tilde{A}_3. \tag{71}$$

Here, $\tilde{A}_3$ has the following form:

$$\begin{pmatrix} \tilde{a}_{11} & e^{2(\alpha+\beta)\eta\pi i}\tilde{a}_{12} - i\left(1 + 2e^{2\beta\eta\pi i} + e^{2(\beta-\alpha)\eta\pi i}\right)\tilde{a}_{22} \\ 0 & \tilde{a}_{22} \end{pmatrix}. \tag{72}$$

Comparing (68) and (72), we obtain

$$\tilde{a}_{12} = i\frac{\left(1 + 2e^{2\beta\eta\pi i} + e^{2(\beta-\alpha)\eta\pi i}\right)}{e^{2(\alpha+\beta)\eta\pi i} - 1}\tilde{a}_{22}. \tag{73}$$

To evaluate $\tilde{a}_{22}$, we use the following Lemma:

Lemma 1. *We set $s(x) = \int_\tau^x S_{-1}\, dx$. Under the assumptions, notation as above and a suitable choice of $d_0 > 0$, we have*

$$\lim_{\substack{x\to 1, x\in\mathcal{R}_1 \\ \mathrm{Im}s(x)=d_0}} (1-x)^{(\alpha+\beta)\eta}\tilde{p}(x)\tilde{\Psi}^{\mathrm{I}}_{\tau,-} = \sqrt{\frac{2}{(\alpha+\beta)\eta}}e^{-\frac{(\alpha+\beta)\eta+1}{2}\pi i}e^{\hat{W}_1^3}. \tag{74}$$

The proof is essentially the same as the proof of Lemma 4.8 in [6]. The relation (68) yields

$$(1-x)^{(\alpha+\beta)\eta}\tilde{a}_{22}\tilde{p}(x)\tilde{\Psi}^{\mathrm{I}}_{\tau,-} = (1-x)^{(\alpha+\beta)\eta}\tilde{u}_6 - \tilde{a}_{12}\tilde{a}_{11}^{-1}(1-x)^{(\alpha+\beta)\eta}\tilde{u}_2. \tag{75}$$

Since the definition of $\tilde{u}_2$ and $\tilde{u}_6$ and $(\alpha+\beta) > 0$, we have

$$\lim_{x\to 1}(1-x)^{(\alpha+\beta)\eta}\tilde{u}_2 = 0, \quad \lim_{x\to 1}(1-x)^{(\alpha+\beta)\eta}\tilde{u}_6 = 1. \tag{76}$$

Hence, we obtain

$$\tilde{a}_{22}\sqrt{\frac{2}{(\alpha+\beta)\eta}}e^{-\frac{(\alpha+\beta)\eta+1}{2}\pi i}e^{\hat{W}_1^3} = 1. \tag{77}$$

Thus, we have (47). $\square$

5. The Gauss Equation with $a = 1/2 + \alpha\eta$, $b = 1/2 + \beta\eta$ and $c = 1 + \gamma\eta$

Since each method to derive the following theorems is similar to that used in the previous section, we show only the results here.

We consider the following differential equation:

$$x(1 - x)\frac{d^2 w}{dx^2} + (1 + \gamma\eta - ((\alpha + \beta)\eta + 2)x)\frac{dw}{dx}$$
$$-\left(\frac{1}{2} + \alpha\eta\right)\left(\frac{1}{2} + \beta\eta\right)w = 0. \tag{78}$$

This equation is equation (3) with (7). By the guage transformation

$$\psi = x^{(1+\gamma\eta)/2}(1 - x)^{(1+(\alpha+\beta-\gamma)\eta)/2}w,$$

we have the normal form of equation (78):

$$\left(-\frac{d^2}{dx^2} + \eta^2 Q\right)\psi = 0, \tag{79}$$

where $Q = Q_0/x^2 + \eta^{-2}Q_1/x^2$ with (9). Note that if we let $\gamma = 0$, then Q_0 coincides with Q_0/x in (1). We let ω_k ($k = 3, 4$) be the regions in the space of parameters $(\alpha, \beta, \gamma) \in \mathbb{C}^3$ defined by

$$\omega_3 = \{(\alpha, \beta, \gamma) \in \mathbb{C}^3 \mid 0 < \operatorname{Re}\gamma < \operatorname{Re}\alpha < \operatorname{Re}\beta\},$$
$$\omega_4 = \{(\alpha, \beta, \gamma) \in \mathbb{C}^3 \mid 0 < \operatorname{Re}\gamma < \operatorname{Re}\alpha + \operatorname{Re}\beta < \operatorname{Re}\beta\}. \tag{80}$$

As in the case treated in the previous section, we construct the WKB solutions normalized at a turning point to equation (79):

$$\psi_{\tau_0,\pm} = \frac{1}{\sqrt{S_{\text{odd}}}}\exp\left(\pm\int_{\tau_0}^{x} S_{\text{odd}}dx\right).$$

We choose τ_0 as follows:

(i) If $(\alpha, \beta, \gamma) \in \omega_3$,

$$\tau_0 = \frac{\beta\gamma + \gamma\alpha - 2\alpha\beta - \sqrt{\alpha\beta(\alpha - \gamma)(\beta - \gamma)}}{(\alpha - \beta)^2}.$$

(ii) If $(\alpha, \beta, \gamma) \in \omega_4$,

$$\tau_0 = \frac{\beta\gamma + \gamma\alpha - 2\alpha\beta + \sqrt{\alpha\beta(\alpha - \gamma)(\beta - \gamma)}}{(\alpha - \beta)^2}.$$

Definition 1. We set $\hat{V}_{\pm,1}$ as

$$\hat{V}_{\pm,1} = \int_1^{\tau_0} \left(S_{\text{odd}} \mp \frac{(\alpha + \beta - \gamma)\eta}{2(x - 1)} \right) dx \mp \frac{1}{2}(\alpha + \beta - \gamma)\eta \log(\tau_0 - 1).$$

$$(81)$$

This $\hat{V}_{\pm,1}$ is a sum of two quantities $V_{\pm,1}$ and $V_{1,\leq 0}$:

$$\hat{V}_{\pm,1} = V_{\pm,1} \pm V_{1,\leq 0}.$$

We call $V_{\pm,1}$ the Voros coefficient at $x = 1$. The explicit form of $V_{\pm,1}$ (cf. [4, Theorem 2.3]) is the following:

Theorem 7. *The explicit form of the Voros coefficient is given as follows:*

$$V_{\pm,1} = \pm \frac{1}{2} \sum_{n=2}^{\infty} \frac{B_n \eta^{1-n}}{n(n-1)} \left\{ (1 - 2^{1-n}) \left(\frac{1}{\alpha^{n-1}} + \frac{1}{\beta^{n-1}} \right. \right.$$

$$\left. \left. - \frac{1}{(\gamma - \alpha)^{n-1}} - \frac{1}{(\gamma - \beta)^{n-1}} \right) + \frac{1}{(\alpha + \beta - \gamma)^{n-1}} \right\}. \quad (82)$$

Here, B_n is the Bernoulli number given by (38).

The residue of $S_{\text{odd}}dx$ at each singularity can be computed and given in Table 1.

Table 1.   The residue of $S_{\text{odd}}dx$.

	0	1	∞
ω_3	$\dfrac{-\gamma\eta}{2}$	$\dfrac{\alpha + \beta - \gamma}{2}\eta$	$\dfrac{\alpha - \beta}{2}\eta$
ω_4	$\dfrac{\gamma\eta}{2}$	$\dfrac{\alpha + \beta - \gamma}{2}\eta$	$\dfrac{\alpha - \beta}{2}\eta$

We obtain the following theorem.

Theorem 8. *If* $(\alpha, \beta, \gamma) \in \omega_3$, *then* $V_{1,\leq 0}$ *is*

$$
V_{1,\leq 0} = -\frac{\alpha\eta}{2}\log\alpha - \frac{\beta\eta}{2}\log\beta - \frac{(\alpha-\gamma)\eta}{2}\log(\gamma-\alpha)
$$

$$
- \frac{(\beta-\gamma)\eta}{2}\log(\gamma-\beta) - (\gamma-\alpha-\beta)\eta\log(\alpha+\beta-\gamma)
$$

$$
+ \frac{(\alpha-\beta)\eta\pi i}{2}. \tag{83}
$$

If $(\alpha, \beta, \gamma) \in \omega_4$, *then* $V_{1,\leq 0}$ *is*

$$
V_{1,\leq 0} = -\frac{\alpha\eta}{2}\log(-\alpha) - \frac{\beta\eta}{2}\log\beta + \frac{(\gamma-\alpha)\eta}{2}\log(\gamma-\alpha)
$$

$$
- \frac{(\beta-\gamma)\eta}{2}\log(\beta-\gamma) + (\alpha+\beta-\gamma)\eta\log(\alpha+\beta-\gamma)
$$

$$
+ \frac{(3\beta-\alpha-\gamma)\eta\pi i}{2}. \tag{84}
$$

The formal series $\hat{V}_1$ is the Borel summable in ω_k $(k = 3, 4)$. Then we have

Theorem 9. *Let* $\hat{V}_1^k$ *be the Borel sums of* $\hat{V}_1$. *Then* $\hat{V}_1^k(k = 3, 4)$ *is given by the following forms:*

If $(\alpha, \beta, \gamma) \in \omega_3$, *then* $\hat{V}_1^3$ *is*

$$
\hat{V}_1^3 = \frac{1}{2}\log\frac{2\pi\Gamma((\alpha+\beta-\gamma)\eta)\Gamma((\alpha+\beta-\gamma)\eta+1)\eta e^{(\alpha-\beta)\eta\pi i}}{\Gamma(1/2+\alpha\eta)\Gamma(1/2+\beta\eta)\Gamma(1/2+(\alpha-\gamma)\eta)\Gamma(1/2+(\beta-\gamma)\eta)}. \tag{85}
$$

If $(\alpha, \beta, \gamma) \in \omega_4$, *then* $\hat{V}_1^4$ *is*

$$
\hat{V}_1^4 = \frac{1}{2}\log\frac{\Gamma(1/2-\alpha\eta)\Gamma(1/2+(\gamma-\alpha)\eta)\Gamma((\alpha+\beta-\gamma)\eta)^2(\alpha+\beta-\gamma)\eta}{\Gamma(1/2+\beta\eta)\Gamma(1/2+(\beta-\gamma)\eta)2\pi e^{(\alpha-\gamma-3\beta)\eta\pi i}}. \tag{86}
$$

We let $(\Psi^{\mathrm{I}}_{\tau_0,+}, \Psi^{\mathrm{I}}_{\tau_0,-})$ denote the Borel sums of the WKB solutions defined in the region R_{I} in Fig. 3. Note that, in both cases, $\psi_{\tau_0,+}$ is recessive on the Stokes curves flowing into $x = 1$. Computing the monodromy matrices of (u_2, u_6) and $(\Psi^{\mathrm{I}}_{\tau_0,+}, \Psi^{\mathrm{I}}_{\tau_0,-})$ along the paths

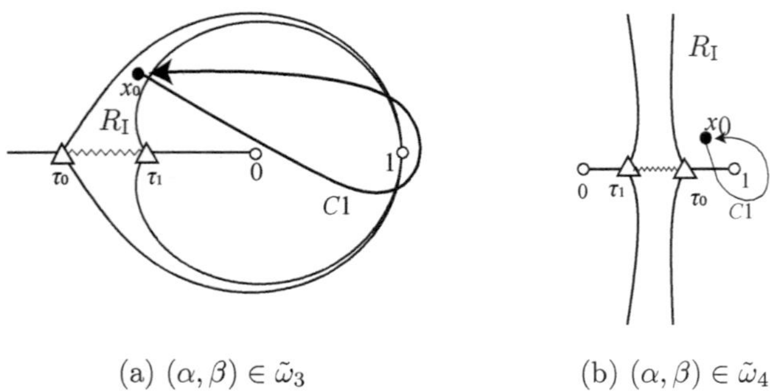

(a) $(\alpha, \beta) \in \tilde{\omega}_3$ (b) $(\alpha, \beta) \in \tilde{\omega}_4$

Fig. 3.   Stokes curves and paths of continuation.

C_1 depicted in Fig. 3, using the quantities we obtained so far, we get the relations between Kummer's solutions and the Borel sums of the WKB solutions.

Theorem 10. *The relations between* (u_2, u_6) *and* $(\Psi^I_{\tau_0,+}, \Psi^I_{\tau_0,-})$ *for the cases* $(\alpha, \beta, \gamma) \in \omega_3, \omega_4$ *are given as follows:*

(i) *If* $(\alpha, \beta, \gamma) \in \omega_3$, *then*

$$(u_2, u_6) = \sqrt{\frac{(\alpha + \beta - \gamma)\eta}{2}} \, x^{-\frac{\gamma\eta+1}{2}} (1 - x)^{-\frac{(\alpha+\beta-\gamma)\eta+1}{2}} (\Psi^I_{\tau_0,+}, \Psi^I_{\tau_0,-}) A,$$

$$(87)$$

where

$$A = \begin{pmatrix} a_{11} & a_{12} \\ 0 & a_{22} \end{pmatrix}$$

with

$$a_{11} = e^{\hat{V}_1^3} e^{-(((\alpha+\beta-\gamma)\eta+1)\pi i)/2},$$

$$a_{12} = i \frac{e^{((\alpha+\beta-\gamma)\eta-1))\pi i)/2} \left(1 + e^{(2\pi i(\beta+\gamma)\eta)} + e^{(2\beta\pi i)} + e^{2(\beta-\alpha)\eta\pi i}\right)}{e^{(2\pi i(\alpha+\beta-\gamma)\eta)} - 1}$$

$$\times e^{(-\hat{V}_1^3)},$$

$$a_{22} = e^{-\hat{V}_1^3} e^{((\alpha+\beta-\gamma)\eta-1))\pi i)/2}. \tag{88}$$

(ii) *If $(\alpha, \beta, \gamma) \in \omega_4$, then*

$$(u_2, u_6) = \sqrt{\frac{(\alpha + \beta - \gamma)\eta}{2}}\, x^{-\frac{\gamma\eta+1}{2}}(1-x)^{-\frac{(\alpha+\beta-\gamma)\eta+1}{2}}(\Psi^{\mathrm{I}}_{\tau_0,+}, \Psi^{\mathrm{I}}_{\tau_0,-})A,$$

(89)

where

$$A = \begin{pmatrix} a_{11} & a_{12} \\ 0 & a_{22} \end{pmatrix}$$

with

$$a_{11} = e^{\hat{V}^4_1}e^{-(((\alpha+\beta-\gamma)\eta+1)\pi i)/2},$$

$$a_{12} = i\frac{e^{((\alpha+\beta-\gamma)\eta-1))\pi i)/2}e^{(2\pi i(\alpha+\beta-\gamma)\eta)}}{e^{(2\pi i(\alpha+\beta-\gamma)\eta)}-1}e^{-\hat{V}^4_1},$$

(90)

$$a_{22} = e^{-\hat{V}^4_1}e^{((\alpha+\beta-\gamma)\eta-1))\pi i)/2}.$$

Remark 5. The explicit forms of the monodromy matrices at $x = 1$ of the Borel sums of the WKB solutions are given in [18], however, the starting point x_0 is in the different Stokes region from R_{I} in Fig. 3 there. Therefore, the monodromy matrices obtained here are different from those in [18].

It immediately follows that the relations in Theorem 7 in the previous section can be obtained by taking the limit $\gamma \to 0$ in the relations in Theorem 10.

Acknowledgment

The authors are very grateful to Professors Takashi Aoki for helpful discussions and advice. This work was partially supported by JSPS KAKENHI Grant Number 18K13433.

References

1. T. Aoki, T. Kawai, and Y. Takei, The Bender-Wu analysis and the Voros theory. *Special Functions, ICM-90 Satellite Conference Proceedings*, Springer-Verlag, Tokyo, 1991, pp. 1–29.

2. T. Aoki and M. Tanda, Some concrete shapes of degenerate Stokes curves of hypergeometric differential equations with a large parameter. *J. School Sci. Eng. Kinki Univ.*, **47**(2011), 5–8.

3. T. Aoki, T. Takahashi, and M. Tanda, The hypergeometric function and WKB solutions. *RIMS Kôkyûroku Bessatsu*, **B57**(2016), 61–68.

4. T. Aoki and M. Tanda, Parametric Stokes phenomena of the Gauss hypergeometric differential equation with a large parameter. *J. Math. Soc. Japan*, **68**(2016), 1099–1132.

5. T. Aoki, T. Takahashi, and M. Tanda, Relation between the hypergeometric function and WKB solutions. *RIMS Kôkyûroku Bessatsu* **B61**(2017), 1–7.

6. T. Aoki, T. Takahashi, and M. Tanda, The hypergeometric function. the confluent hypergeometric function and WKB solutions. *J. Math. Soc. Japan*, **73(4)**(2021), 1019–1062.

7. T. Aoki, T. Takahashi, and M. Tanda, Voros coefficients of the Gauss hypergeometric differential equation with a large parameter. *Integral Transforms and Special Functions*, **32, OPSFA15**(2021), 336–345.

8. T. Aoki and M. Tanda, Borel sums of Voros coefficients of hypergeometric differential equations with a large parameter, to appear in *RIMS Kôkyûroku*.

9. A. Erdélyi *et al.*, *Higher Transcendental Functions*, Bateman Manuscript Project, Vol. I. California Institute of Technology, McGraw-Hill, 1953.

10. E. Delabaere, H. Dillinger, and F. Pham, Résurgence de Voros et périodes des courbes hyperelliptiques. *Ann. Inst. Fourier, Grenoble*, **43**(1993), 153–199.

11. K. Iwaki, T. Koike, and Y.-M. Takei, Voros coefficients for the hypergeometric differential equations and Eynard-Orantin's Topological recursion, Part II: For the confluent family of hypergeometric equations, arXiv:1810.02946.

12. T. Koike, On the exact WKB analysis of second order linear ordinary differential equations with simple poles. *Publ. RIMS, Kyoto Univ.*, **36**(2000), 297–319.

13. T. Kawai and Y. Takei, *Algebraic Analysis of Singular Perturbation Theory*, Translation of Mathematical Monographs, Vol. 227. AMS, Providence, Rhode Island, 2005.

14. T. Koike and R. Schäfke, On the Borel summability of WKB solutions of Schrödinger equations with polynomial potential and its application. In: *Private Communication and [in preparation for the publication in RIMS Kôkyûroku Bessatsu.]*

15. H. Shen and H. J. Silverstone, Observations on the JWKB treatment of the quadratic barrier. *Algebr. Anal. Diff. Eqn.* Springer-Verlag, 2008, 307–319.

16. Y. Takei, Sato's conjecture for the Weber equation and transformation theory for Schrödinger equations with a merging pair of turning points. *RIMS Kôkyûroku Bessatsu*, **B 10**(2008), 205–224.

17. Y. Takei, WKB analysis and Stokes geometry of differential equations. *Analytic, Algebraic and Geometric Aspects of Differential Equations*, Birkhäuser, Będlewo, Poland, 2017, pp. 263–304.

18. M. Tanda, Exact WKB analysis of hypergeometric differential equations, *to appear in RIMS Kôkyûroku Bessatsu*.

19. A. Voros, The return of the quartic oscillator, The complex WKB method, *Ann. Inst. Henri Poincaré*, **39**(1983), 211–338.

© 2022 World Scientific Publishing Europe Ltd.
https://doi.org/10.1142/9781800611368_0009

Chapter 9

Unified Value Sharing of Meromorphic Functions

Kuldeep Singh Charak[*,‡], Risto Korhonen[†,§]
and Gaurav Kumar[*,¶]

*Department of Mathematics
University of Jammu
Jammu 180 006, India
†Department of Physics and Mathematics
University of Eastern Finland
P.O. Box 111, FI 80101, Joensuu
Finland
‡kscharak7@rediffmail.com
§risto.korhonen@uef.fi
¶kgaurav866@gmail.com

Let f be a meromorphic function in the complex plane. In this chapter, we introduce a general class $\mathcal{B}_f$ of functions which includes as a special case the derivative functions $f^{(n)}$, $n \in \mathbb{N}$, the differences $\Delta^n f(z)$ and shifts $f(z+c)$, $c \in \mathbb{C} \setminus \{0\}$, of f if the hyper-order of f is less than one, and the q-shifts and q-differences of f in the case f is of zero order. Using $\mathcal{B}_f$, we present a unified way to consider uniqueness problems concerning f and its derivatives, shifts, differences and their combinations, as well as many other functions.

1. Introduction

According to Nevanlinna's five values theorem [13], if two meromorphic functions share five distinct values in the extended complex plane,

205

then these two functions must be identically the same. Nevanlinna also showed that if the functions share only four values instead of five, but with the same multiplicities, then the functions are unique up to a Möbius transformation. Further reduction in the number of shared values can only lead to uniqueness if something more is known about the functions. Rubel and Yang [14] proved that if a non-constant entire function f and its derivative f' share two distinct finite values, taking multiplicities into account, then $f' \equiv f$. Mues and Steinmetz [12] and Gundersen [6] extended this result for meromorphic functions. Heittokangas *et al.* [8,9] initiated the investigations of the uniqueness of meromorphic functions sharing values (or small functions) with their shifts. In this chapter, we present a method for simultaneous study of uniqueness problems of f sharing values with its derivatives, shifts, differences and their linear combinations.

Two non-constant meromorphic functions f and g are said to share a meromorphic function a, counting multiplicity (CM) if $E(a, f) = E(a, g)$, where $E(a, f)$ is the set of zeros of $f(z) - a(z)$ in which each zero is counted according to its multiplicity. Here, and throughout the chapter, by a meromorphic function we mean a non-constant meromorphic function on $\mathbb{C}$. Further, we say that f and g shares a, ignoring multiplicity (IM) if $\overline{E}(a, f) = \overline{E}(a, g)$, where $\overline{E}(a, f)$ is the set of zeros of $f(z) - a(z)$ in which each zero is counted only once. Also, we say that f shares a partially [4] with g if $\overline{E}(a, f) \subseteq \overline{E}(a, g)$. We denote by $\rho(f)$ and $\rho_2(f)$ the order and hyper-order of f defined as

$$\rho(f) = \limsup_{r \to \infty} \frac{\log T(r, f)}{\log r}$$

and

$$\rho_2(f) = \limsup_{r \to \infty} \frac{\log \log T(r, f)}{\log r}.$$

For a meromorphic function f, $a \in \mathbb{C}$ and k a positive integer, we use the symbol $\overline{n}_{k)}(r, 1/(f - a))$ or $\overline{n}_{k)}(r, a)$ to denote the counting function of zeros of $f - a$ in $|z| \leq r$, whose multiplicities are not greater than k and are counted only once. Similarly, we use $\overline{n}_{(k+1}(r, 1/(f - a))$ or $\overline{n}_{(k+1}(r, a)$ to denote the counting function of those zeros of $f - a$ in $|z| \leq r$ whose multiplicities are greater than k and are counted only once. The corresponding integrated counting functions are denoted by $\overline{N}_{k)}(r, 1/(f-a))$ or $\overline{N}_{k)}(r, a)$ and $\overline{N}_{(k+1}(r, 1/(f-a))$ or $\overline{N}_{(k+1}(r, a)$. When the zeros are counted according to their multiplicities, we use the notations $n_{k)}(r, 1/(f-a))$, $n_{(k+1}(r, 1/(f-a))$, $N_{k)}(r, 1/(f-a))$ and $N_{(k+1}(r, 1/(f-a))$.

For a meromorphic function f, a meromorphic function a such that $T(r, a) = o(T(r, f))$, as $r \to \infty$ outside a set of finite logarithmic measure, is called a small function of f. The family of all such small functions of f is denoted by $\mathcal{S}(f)$ and $\hat{\mathcal{S}}(f) = \mathcal{S}(f) \cup \{\infty\}$.

As we mentioned above, Heittokangas *et al.* were the first to study the uniqueness of meromorphic functions sharing values with their shifts. They proved, for instance, the following result.

Theorem A ([9]). *Let f be a meromorphic function of finite order and let $c \in \mathbb{C} \setminus \{0\}$, and let $a_1, a_2, a_3 \in \hat{\mathcal{S}}(f)$ be three distinct periodic functions with period c. If $f(z)$ and $f(z+c)$ share a_1, a_2 CM and a_3 IM, then $f(z) \equiv f(z+c)$.*

In recent years, several studies concerning uniqueness of meromorphic functions sharing values or small functions with their difference operators have appeared. For instance, Chen and Yi [5] proved that if $f(z)$ is a transcendental meromorphic function such that $f(z+c) \not\equiv f(z)$, and $\Delta_c f(z)$ and $f(z)$ share three distinct values a, b, ∞ CM, then $f(z+c) \equiv 2f(z)$. Li and Gao considered the nth order difference operator, and proved the following result.

Theorem B ([11]). *Let $f(z)$ be a entire function of finite order, $c \in \mathbb{C}$, and n be a positive integer. Suppose that $f(z)$ and $\Delta_c^n f(z)$ share two distinct finite values a, b CM and one of the following two conditions is satisfied:*

(i) *$ab=0$;*
(ii) *$ab \neq 0$ and $\rho(f) \notin \mathbb{N}$.*

Then $f(z) = \Delta_c^n f(z)$.

Li and Chen [10] generalized the result above and obtained a uniqueness result for a meromorphic function f of hyper-order $\rho_2(f) < 1$ sharing three values with a linear difference polynomial $L(z, f)$.

Note that each preceding result involves sharing of small functions by a given function and its shift that guarantees their uniqueness. Now, one can naturally think of such a problem from a broader perspective. Let f be a given meromorphic function and consider the following class of meromorphic functions:

$$\mathcal{B}_f = \{g : g \text{ is meromorphic in } \mathbb{C} \text{ and } m\left(r, \frac{g}{f}\right) = S(r, f)\},$$

where $S(r, f)$ is a quantity of the form $o(T(r, f))$ as $r \to \infty$ outside of a small exceptional set. The size of the exceptional set can be chosen to be, for instance, of finite linear or logarithmic measure, or of zero logarithmic density, depending on context.

Clearly, $\mathcal{B}_f$ contains at least all derivatives of f; all shifts of f, if f is of hyper-order less than one; all q-shifts of f, if f is of zero order, and of course all linear sums of all such functions, but in general the elements of $\mathcal{B}_f$ do not necessarily need to be related to f in terms of linear operators. This type of approach has been taken by Halburd and the second author in the context of general value distribution theory in [7], where a generalization of the second fundamental theorem for special subfields of meromorphic functions was obtained with the derivative in the ramification term replaced by a more general linear operator.

The aim of this chapter is to consider the sharing of small functions by f and $g \in \mathcal{B}_f$ that ensures their uniqueness. In particular, our results generalize the results of Chen and Yi [5], Li and Gao [11] and Li and Chen [10]. Here, by a small function we mean a small function of both f and g. We start with the following two results on entire functions.

Theorem 9.1. *Let f be an entire function and $g \in \mathcal{B}_f$ be an entire function such that f and g share 0 CM and 1 IM. Then $f \equiv g$.*

Theorem 9.2. *Let f be an entire function and $g \in \mathcal{B}_f$ be an entire function. Suppose f shares the small functions a_1 and a_2 partially with g such that*

$$T(r, f) \neq \sum_{k=1}^{2} \overline{N}\left(r, \frac{1}{f - a_k}\right) + S(r, f).$$

Then $f \equiv g$.

The condition in Theorem 9.2 cannot be dropped. Consider meromorphic functions $f(z) = \sin z$ and $g(z) = \sin z + \cos z$. Then $g \in \mathcal{B}_f$. Here, f share $-1, 1$ partially with g but $f \not\equiv g$.

The following result gives an extension of Theorem 9.1 to the meromorphic case:

Theorem 9.3. *Let f be a meromorphic function and $g \in \mathcal{B}_f$ be such that f and g share $0, 1, \infty$ CM with $\overline{N}_{1)}(r, f) = S(r, f)$. Then $f \equiv g$.*

The condition stated in Theorem 9.3 cannot be dropped. Consider meromorphic functions $f(z) = i \tan z$ and $g(z) = -i \tan z$. Then $g \in \mathcal{B}_f$, and f and g share $0, 1, \infty$ CM but $f(z) \not\equiv g(z)$.

We turn next to the case of meromorphic functions sharing four values.

Theorem 9.4. *Let f be a non-constant meromorphic function and $g \in \mathcal{B}_f$ such that f and g share $0, 1, \infty, c$ CM, where $c \neq 0, 1, \infty$. Then either $f(z) \equiv g(z)$ or $f(z) \equiv -g(z)$. The latter case holds if $c = -1$ and $1, -1$ are Picard exceptional values of f and g.*

The second case in Theorem 9.4 is not superfluous. Let $f(z) = i \tan z$, $g(z) = -i \tan z$. Then $g \in \mathcal{B}_f$, f and g share $0, 1, -1, \infty$ and $f(z) = -g(z)$. Also $1, -1$ are Picard exceptional values of f and g.

By restricting to a subclass

$$\mathcal{C}_f := \{ g : g \in \mathcal{B}_f \text{ and } m\left(r, \frac{g}{f-1}\right) = S(r, f) \}$$

of $\mathcal{B}_f$, the assertions Theorems 9.3 and 9.4 can be substantially improved.

Theorem 9.5. *Let f and $g \in \mathcal{C}_f$ be meromorphic functions such that f and g share $0, 1, \infty$ CM. Then $f \equiv g$.*

The class $\mathcal{C}_f \subset \mathcal{B}_f$ still contains all derivative functions of f, and all exact differences of f, if f is of hyper-order less than one, as well as all exact q-differences of f, if f is of zero order. For instance, if f is meromorphic in $\mathbb{C}$ and $g = f^{(n)}$, we have

$$m\left(r, \frac{f^{(n)}}{f}\right) = S(r, f)$$

and

$$m\left(r, \frac{f^{(n)}}{f-1}\right) = m\left(r, \frac{(f-1)^{(n)}}{f-1}\right) = S(r, f)$$

by the lemma on the logarithmic derivatives. Corresponding identities for $g = \Delta_c^n f(z)$ and the analogous q-difference operator follow by the difference and q-difference analogues of the lemma on the logarithmic derivatives, respectively.

Let k be a non-negative integer or infinity. For $a \in \overline{\mathbb{C}}$, we denote by $E_k(a, f)$ the set of all $a-$points of f, where an $a-$point of multiplicity m is counted m times if $m \le k$ and $k+1$ times if $m > k$. If for two meromorphic functions f and g, $E_k(a, f) = E_k(a, g)$, we say that f and g share a with weight k. Moreover, for $S \subset \overline{\mathbb{C}}$, we define $E_f(S, k)$ as $E_f(S, k) = \bigcup_{a \in S} E_k(a, f)$, where k is a non-negative integer or infinity. If $E_f(S, k) = E_g(S, k)$, we say that f and g share the set S with weight k. Clearly, if the set S is a singleton, say $S = \{a\}$, then the concept of sharing of set S reduces to sharing of value a.

In the present discussion, we assume that $S_1 = \{1, \omega, ..., \omega^{n-1}\}$, where ω is the nth root of unity and $S_2 = \{\infty\}$ where $n \in \mathbb{N}$.

Recently, many papers (e.g. [2,3]) on uniqueness problems of meromorphic functions sharing sets with their differences or shifts have appeared. In this context, we prove the following result:

Theorem 9.6. *Let f be a non-constant meromorphic function and $g \in \mathcal{B}_f$ such that $E_f(S_1, 2) = E_g(S_1, 2)$ and $E_f(S_2, \infty) = E_g(S_2, \infty)$. If $n \ge 7$, then $f = tg$, where $t^n = 1$.*

2.   The Proofs of Main Results

First, we state a few results required to prove the main results of this chapter. The first one is [15, Theorem 5.5].

Theorem C ([15]). *Let $f(z)$ and $g(z)$ be non-constant meromorphic functions sharing $0, 1$ CM and ∞ IM. If $N^*(r, \infty) \ne S(r, f)$, where $N^*(r, \infty)$ is defined to be the counting function of multiple poles of f and g counted according to the smaller multiplicity, then $f(z) \equiv g(z)$.*

We also need [15, Theorem 3.29].

Theorem D **([15])**. *Let $f(z), g(z) \in \mathcal{A} := \{h : \overline{N}(r,h) + \overline{N}(r,1/h) = S(r,h)\}$ and a, b be finite non-zero values. Further, let k be a positive integer.*

(i) *If $\overline{E}_{1)}(a, f) = \overline{E}_{1)}(a, g)$, then $f(z) \equiv g(z)$ or $f(z) \cdot g(z) = a^2$.*
(ii) *If $\overline{E}_{1)}(b, f^{(k)}) = \overline{E}_{1)}(b, g^{(k)})$, then $f(z) \equiv g(z)$ or $f(z) \cdot g(z) = b^2$.*

The following theorem is a refined version of the Nevanlinna four values theorem [15, Theorem 4.1].

Theorem E **([15])**. *Let $f(z)$ and $g(z)$ be non-constant meromorphic functions. If $f(z)$ and $g(z)$ share $0, 1, \infty, c$ CM, where $c \neq 0, 1, \infty$, then f and g have one of the following relations:*

(i) *$f(z) \equiv g(z)$;*
(ii) *$f(z) \equiv -g(z)$ holds if $c = -1$ and $-1, 1$ are Picard exceptional values of f and g;*
(iii) *$f(z) \equiv 2 - g(z)$ holds if $c = 2$ and $0, 2$ are Picard exceptional values of f and g;*
(iv) *$f(z) \equiv 1 - g(z)$ holds if $c = \frac{1}{2}$ and $0, 1$ are Picard exceptional values of f and g;*
(v) *$f(z) \equiv \frac{g(z)}{2g(z)-1}$ holds if $c = \frac{1}{2}$ and $\frac{1}{2}, \infty$ are Picard exceptional values of f and g;*
(vi) *$f(z) \equiv \frac{g(z)}{g(z)-1}$ holds if $c = 2$ and $1, \infty$ are Picard exceptional values of f and g;*
(vii) *$f(z) \equiv \frac{1}{g(z)}$ holds if $c = -1$ and $0, \infty$ are Picard exceptional values of f and g.*

The final auxiliary result we need is a Lemma due to Banerjee [1].

Lemma F **([1])**. *Let F and G be two non-constant meromorphic functions. If $E_2(1, F) = E_2(1, G)$ and $E_k(\infty, F) = E_k(\infty, G)$, where $0 \leq k \leq \infty$, then one of the following cases occurs:*

(1) *$T(r, F) + T(r, G) \leq 2\{N_2\left(r, \frac{1}{F}\right) + N_2\left(r, \frac{1}{G}\right) + \overline{N}(r, F) + \overline{N}(r, G) + \overline{N}_*(r, \infty; F, G)\} + S(r, F) + S(r, G)$;*
(2) *$F \equiv G$;*
(3) *$FG \equiv 1$,*

where

$$N_2\left(r, \frac{1}{F}\right) = \overline{N}(r,0) + \overline{N}_{(2}(r,0)$$

and $\overline{N}_*(r, \infty; F, G)$ denotes the reduced counting function of those poles of F whose multiplicities differ from the multiplicities of corresponding poles of G.

2.1. *Proof of Theorem 9.1*

Suppose on the contrary, that $f \not\equiv g$. Since f and g are entire functions sharing two finite values 0 and 1, we have

(1) $S(r,f) = S(r,g)$,

(2) $\frac{g}{f} = e^{Q(z)}$, for some entire function $Q(z)$,

(3) $\overline{N}\left(r, \frac{1}{f-1}\right) = \overline{N}\left(r, \frac{1}{g-1}\right)$,

(4) $\overline{N}\left(r, \frac{1}{f}\right) + \overline{N}\left(r, \frac{1}{f-1}\right) \leq N\left(r, \frac{1}{f-g}\right)$.

Here, the first identity follows by the Nevanlinna three values theorem [13] (see also, e.g. [15, Theorem 5.1]).

Now by the second fundamental theorem of Nevanlinna, we have

$$T(r,f) + S(r,f) \leq \overline{N}\left(r, \frac{1}{f}\right) + \overline{N}\left(r, \frac{1}{f-1}\right) + S(r,f)$$

$$\leq \overline{N}\left(r, \frac{1}{g-f}\right) + S(r,f)$$

$$\leq T(r, g-f) + S(r,f)$$

$$= m(r, g-f) + S(r,f)$$

$$\leq m(r,f) + m\left(r, \frac{g}{f} - 1\right) + S(r,f)$$

$$= m(r,f) + S(r,f)$$

$$= T(r,f) + S(r,f),$$

where we have used the fact that $g \in \mathcal{B}_f$ to obtain the second to last identity. Thus,

$$T(r, f) = \overline{N}\left(r, \frac{1}{g-f}\right) + S(r, f) = \overline{N}\left(r, \frac{1}{f}\right) + \overline{N}\left(r, \frac{1}{f-1}\right) + S(r, f).$$

Also,

$$\overline{N}\left(r, \frac{1}{g-f}\right) + S(r, f) \leq \overline{N}\left(r, \frac{1}{f}\right) + \overline{N}\left(r, \frac{1}{\frac{g}{f} - 1}\right) + S(r, f)$$

$$\leq \overline{N}\left(r, \frac{1}{f}\right) + T\left(r, \frac{g}{f}\right) + S(r, f)$$

$$\leq \overline{N}\left(r, \frac{1}{f}\right) + T(r, e^{Q(z)}) + S(r, f)$$

$$= \overline{N}\left(r, \frac{1}{f}\right) + m(r, e^{Q(z)}) + S(r, f)$$

$$= \overline{N}\left(r, \frac{1}{f}\right) + m\left(r, \frac{g}{f}\right) + S(r, f)$$

$$= \overline{N}\left(r, \frac{1}{f}\right) + S(r, f)$$

$$\leq \overline{N}\left(r, \frac{1}{g-f}\right) + S(r, f),$$

and so

$$T(r, f) = \overline{N}\left(r, \frac{1}{g-f}\right) + S(r, f)$$

$$= \overline{N}\left(r, \frac{1}{f}\right) + \overline{N}\left(r, \frac{1}{f-1}\right) + S(r, f)$$

$$= \overline{N}\left(r, \frac{1}{f}\right) + S(r, f).$$

Thus, $\overline{N}\left(r, \frac{1}{f-1}\right) = S(r, f)$ and $\overline{N}\left(r, \frac{1}{f}\right) = T(r, f) + S(r, f)$. Therefore, by (2) and (3) it follows that $\overline{N}\left(r, \frac{1}{g-1}\right) = S(r, g)$. This, and the fact that g is entire, implies by the second fundamental theorem that $\overline{N}\left(r, \frac{1}{g}\right) = T(r, g) + S(r, g)$.

Define $F = f - 1$ and $G = g - 1$. Then $F, G \in \mathcal{A}$ and share -1 CM, so by Theorem D, we find that either $F \equiv G$ and so the assertion follows, or $FG = 1$ and thus 0 is an omitted value of F and G. Consequently, we must have

$$F = e^h \text{ and } G = e^{-h}, \text{ for some entire function } h.$$

Equivalently,

$$f = e^h + 1, \ g = e^{-h} + 1.$$

Thus,

$$e^Q = \frac{g}{f} = \frac{1 + e^{-h}}{1 + e^h} = e^{-h}$$

and so

$$f = e^{-Q} + 1, g = e^Q + 1,$$

which implies that

$$T(r, f) = T(r, e^Q) = S(r, f),$$

and this is a contradiction. Hence, $f \equiv g$. $\square$

2.2. *Proof of Theorem* 9.2

Suppose $f \not\equiv g$. Since $\overline{E}(a_k, f) \subseteq \overline{E}(a_k, g)$, $k = 1, 2$, by the second fundamental theorem of Nevanlinna, we have

$$T(r, f) + S(r, f) \leq \overline{N}\left(r, \frac{1}{f - a_1}\right) + \overline{N}\left(r, \frac{1}{f - a_2}\right) + S(r, f)$$

$$\leq \overline{N}\left(r, \frac{1}{f - g}\right) + S(r, f)$$

$$\leq T(r, g - f) + S(r, f)$$

$$= m(r, g - f) + S(r, f)$$

$$\leq m(r, f) + m\left(r, \frac{g}{f} - 1\right) + S(r, f)$$

$$= m(r, f) + S(r, f)$$

$$= T(r, f) + S(r, f).$$

Thus, $T(r, f) = \overline{N}\left(r, \frac{1}{f-a_1}\right) + \overline{N}\left(r, \frac{1}{f-a_2}\right) + S(r, f)$, which is a contradiction. Hence, $f \equiv g$. $\square$

2.3. *Proof of Theorem 9.3*

Suppose $f \not\equiv g$. Then since f and g share $0, 1, \infty$ CM, we have

$$\frac{g}{f} = e^{\phi_1(z)} \quad \text{and} \quad \frac{g-1}{f-1} = e^{\phi_2(z)}$$

for some entire functions $\phi_1(z)$ and $\phi_2(z)$. On simplification of the last equalities, we get

$$f = \frac{1 - e^{\phi_2}}{e^{\phi_1} - e^{\phi_2}}, \quad g = \frac{1 - e^{\phi_2}}{1 - e^{\phi_2 - \phi_1}}.$$

Then

$$T(r, e^{\phi_1}) = m(r, e^{\phi_1}) = m\left(r, \frac{g}{f}\right) = S(r, f).$$

Also,

$$T(r, e^{\phi_2}) + S(r, f) = T(r, e^{\phi_2} - 1) + S(r, f)$$

$$= T\left(r, \frac{g-f}{f-1}\right) + S(r, f)$$

$$= T\left(r, (e^{\phi_1} - 1)\frac{f}{f-1}\right) + S(r, f)$$

$$\leq T\left(r, \frac{f}{f-1}\right) + S(r, f)$$

$$= T(r, f) + S(r, f).$$

Again,

$$T(r, f) + S(r, f) = T(r, f - 1) + S(r, f)$$

$$= T\left(r, \frac{1 - e^{\phi_1}}{e^{\phi_1} - e^{\phi_2}}\right) + S(r, f)$$

$$\leq T(r, e^{\phi_2}) + S(r, f).$$

Thus, $T(r, f) = T(r, e^{\phi_2}) + S(r, f)$.

Note that

$$N\left(r,\frac{1}{f-1}\right) \le N\left(r,\frac{1}{e^{\phi_1}-1}\right) \le T(r,e^{\phi_1}) = S(r,f).$$

Now if $N^*(r,\infty) \ne S(r,f)$, then by Theorem C, $f \equiv g$, which is a contradiction.

Thus, we must have $N^*(r,\infty) = S(r,f)$. Also, since $\overline{N}_{1)}(r,f) = S(r,f)$, we have that $\delta(1,f) = 1$ and $\delta(\infty,f) = 1$.

Define $F = f-1$ and $G = g-1$. Then $F, G \in \mathcal{A}$ and hence either $F \equiv G$, in which case the assertion follows, or $FG \equiv 1$ implying that 0 and ∞ are omitted values of F and G. Therefore, $F(z) = e^{h(z)}$ and $G(z) = e^{-h(z)}$, for some entire function $h(z)$ and thus

$$f(z) = 1 + e^{h(z)}, \quad g(z) = 1 + e^{-h(z)}.$$

Since $e^{\phi_1} = \frac{g}{f} = e^{-h}$, we have

$$T(r,e^h) = T(r,e^{-h}) = T(r,e^{\phi_1}) = S(r,f),$$

and so

$$T(r,f) = T(r,e^h) + O(1) = S(r,f),$$

which is not possible. Thus, $f \equiv g$. $\qquad\qquad\square$

2.4.   *Proof of Theorem* **9.4**

We shall accomplish the proof of this theorem by showing that the cases (iii) to (vii) in Theorem E do not occur for the class $\mathcal{B}_f$. Suppose case (iii) holds, that is $f(z) = 2 - g(z)$, $c = 2$ and $0, 2$ are Picard exceptional values of f and g. Then,

$$\frac{1}{f} = \frac{1}{2}\left(1+\frac{g}{f}\right)$$

and so

$$m\left(r,\frac{1}{f}\right) = m\left(r,\frac{1}{2}\left(1+\frac{g}{f}\right)\right)$$

$$\le m\left(r,\frac{g}{f}\right) + O(1)$$

$$\le S(r,f).$$

Thus, $m\left(r,\frac{1}{f}\right) = S(r,f)$. Also, since 0 is Picard exceptional value of f, we have $N\left(r,\frac{1}{f}\right) = 0$ implying that $T\left(r,\frac{1}{f}\right) = S(r,f)$, which is a contradiction.

Suppose (iv) holds, that is $f(z) \equiv 1 - g(z), c = \frac{1}{2}$ and $0,1$ are Picard exceptional values of f and g, then following the way in (iii), we get a contradiction.

Again, if (v) holds, that is $f(z) = \frac{g(z)}{2g(z)-1}, c = \frac{1}{2}$ and $\frac{1}{2},\infty$ are Picard exceptional values of f and g. On simplification, we have $g(z) = \frac{1}{2}\left(\frac{g(z)}{f(z)} + 1\right)$. So,

$$
\begin{aligned}
m(r,g) &= m\left(r, \frac{1}{2}\left(\frac{g(z)}{f(z)} + 1\right)\right) \\
&\leq m\left(r, \frac{g}{f}\right) + O(1) \\
&= S(r,f) \\
&= S(r,g).
\end{aligned}
$$

Thus, $m(r,g) = S(r,g)$. Also, since ∞ is Picard exceptional value of g, we have $N(r,g) = 0$ implying that $T(r,g) = S(r,g)$, which is a contradiction.

Similarly, we can show that the case (vi) cannot hold.

Finally, if case (vii) holds, then $f(z) = \frac{1}{g(z)}, c = -1$ and $1,\infty$ are Picard exceptional values of f and g and so $\frac{1}{f^2} = \frac{g}{f}$, which implies that $m\left(r,\frac{1}{f}\right) = S(r,f)$. Also, since 0 is a Picard exceptional value of f, $N\left(r,\frac{1}{f}\right) = 0$ implying that $T\left(r,\frac{1}{f}\right) = S(r,f)$, and this is a contradiction. $\qquad\square$

2.5. *Proof of Theorem 9.5*

Suppose contrary to the assertion that $f \not\equiv g$, since f and g share $0,1,\infty$ CM, we have

$$
\frac{g}{f} = e^{\phi_1} \tag{1}
$$

and

$$\frac{g-1}{f-1} = e^{\phi_2} \tag{2}$$

for some entire functions ϕ_1 and ϕ_2. Also, since $T(r,f) \sim T(r,g)$ by the Nevanlinna three values theorem, it follows that $S(r,f) = S(r,g)$. Solving (1) and (2), we can easily obtain

$$f = \frac{1 - e^{\phi_2}}{e^{\phi_1} - e^{\phi_2}}.$$

Now,

$$T(r, e^{\phi_1}) = m(r, e^{\phi_1}) = m\left(r, \frac{g}{f}\right) = S(r,f)$$

since $g \in \mathcal{C}_f$. Also,

$$N\left(r, \frac{1}{f-1}\right) \leq N\left(r, \frac{1}{e^{\phi_1} - 1}\right)$$
$$\leq T(r, e^{\phi_1}) + O(1) \tag{3}$$
$$= S(r,f).$$

Now define

$$\phi = \frac{g}{f(f-1)}.$$

Then,

$$m(r, \phi) = m\left(r, \frac{g}{f(f-1)}\right)$$
$$= m\left(r, \frac{g}{f-1} - \frac{g}{f}\right)$$
$$\leq m\left(r, \frac{g}{f-1}\right) + m\left(r, \frac{g}{f}\right) + O(1)$$
$$= S(r,f),$$

by the assumption $g \in \mathcal{C}_f$. Moreover,

$$N(r, \phi) = N\left(r, \frac{g}{f(f-1)}\right)$$

$$\leq N\left(r, \frac{g}{f}\right) + N\left(r, \frac{1}{f-1}\right)$$

$$= N\left(r, \frac{1}{f-1}\right) + S(r, f),$$

and hence by (3), we have

$$T(r, \phi) = S(r, f). \tag{4}$$

Since $f(f-1) = \frac{g}{\phi}$, using (4), we find that

$$2T(r, f) + O(1) = T(r, f(f-1))$$

$$= T\left(r, \frac{g}{\phi}\right)$$

$$\leq T(r, \phi) + T(r, g) + S(r, f)$$

$$\leq T\left(r, \frac{g}{f}\right) + T(r, f) + S(r, f)$$

$$= m\left(r, \frac{g}{f}\right) + N\left(r, \frac{g}{f}\right) + T(r, f) + S(r, f)$$

$$= T(r, f) + S(r, f)$$

and this leads to a contradiction. $\qquad \square$

2.6. *Proof of Theorem* 9.6

Let $F = f^n$ and $G = g^n$. Then by the given conditions, F and G share 1 with weight 2 and ∞ CM. Clearly, $N_*(r, \infty, F, G) = 0$.

Also,

$$N_2\left(r, \frac{1}{F}\right) \leq 2\overline{N}\left(r, \frac{1}{F}\right) = 2\overline{N}\left(r, \frac{1}{f^n}\right) = 2\overline{N}\left(r, \frac{1}{f}\right).$$

Similarly,

$$N_2\left(r, \frac{1}{G}\right) \leq 2\overline{N}\left(r, \frac{1}{g}\right),$$

$$\overline{N}\left(r, F\right) = \overline{N}\left(r, f^n\right) = \overline{N}\left(r, f\right)$$

and

$$\overline{N}\left(r, G\right) = \overline{N}\left(r, g^n\right) = \overline{N}\left(r, g\right).$$

Suppose now that (1) in Lemma F holds. Then

$$T(r, F) + T(r, G) \leq 4\overline{N}\left(r, \frac{1}{f}\right) + 4\overline{N}\left(r, \frac{1}{g}\right) + 2\overline{N}\left(r, f\right) + 2\overline{N}\left(r, g\right)$$

$$+ \; S(r, f) + S(r, g)$$
$$\leq 6T(r, f) + 6T(r, g) + S(r, f) + S(r, g),$$

which implies that $nT(r, f) + nT(r, g) \leq 6T(r, f) + 6T(r, g) + S(r, f) + S(r, g)$. That is, $(n - 6)(T(r, f) + T(r, g)) \leq S(r, f) + S(r, g)$, which is absurd since $n \geq 7$.

Therefore, by Lemma F, either $FG \equiv 1$ or $F \equiv G$. If $FG \equiv 1$, we have $f^n g^n = 1$. Now since f and g share ∞ CM, we must have $N\left(r, \frac{g}{f}\right) \leq N\left(r, \frac{1}{f}\right) \leq T(r, f)$. Now,

$$2nT(r, f) = T\left(r, \frac{1}{f^{2n}}\right) = T\left(r, \frac{g^n}{f^n}\right)$$

$$= nT\left(r, \frac{g}{f}\right)$$

$$= nm\left(r, \frac{g}{f}\right) + nN\left(r, \frac{g}{f}\right)$$

$$\leq nN\left(r, \frac{1}{f}\right) + S(r, f) \leq nT(r, f) + S(r, f),$$

and so $2nT(r, f) \leq nT(r, f) + S(r, f)$, which is not possible. Thus, we must have $F \equiv G$ and so $f \equiv tg$, where $t^n = 1$. $\square$

Acknowledgments

The work of the first author is partially supported by Mathematical Research Impact Centric Support (MATRICS) grant, File No. MTR/2018/000446, by the Science and Engineering Research Board (SERB), Department of Science and Technology (DST), Government of India. The second author was partially supported by the Academy of Finland grant (#286877). The work of the third author is supported by University Grants Commission (UGC), India.

References

1. A. Banerjee, Meromorphic functions sharing two sets. *Czechoslovak Math. J.* **57(132)** (2007), 1199–1214, MR 2357586.
2. A. Banerjee and G. Haldar, Uniqueness of meromorphic functions sharing two finite sets in $\mathbb{C}$ with finite weight. *Konuralp J. Math.*, **2**(2) (2014), 42–52.
3. S. Bhoosnurmath and S. Kabbur, Value distribution and uniqueness theorems for difference of entire and meromorphic functions. *Intl. J. Anal. Appl.*, **2**(2) (2013), 124–136.
4. K. S. Charak, R. J. Korhonen, and G. Kumar, A note on partial sharing of values of meromorphic functions with their shifts. *J. Math. Anal. Appl.*, **435**(2) (2016), 1241–1248, MR 3429639.
5. Z.-X. Chen and H.-X. Yi, On sharing values of meromorphic functions and their differences. *Results Math.*, **63**(1–2) (2013), 557–565, MR 3009705.
6. G. G. Gundersen, Meromorphic functions that share two finite values with their derivative. *Pacific J. Math.*, **105**(2) (1983), 299–309, MR 691606.
7. R. G. Halburd and R. J. Korhonen, Value distribution and linear operators. *Proc. Edinb. Math. Soc.*, (2) **57**(2) (2014), 493–504, MR 3200320.
8. J. Heittokangas, R. Korhonen, I. Laine, and J. Rieppo, Uniqueness of meromorphic functions sharing values with their shifts. *Complex Var. Elliptic Eqn.*, **56**(1–4) (2011), 81–92, MR 2774582.
9. J. Heittokangas, R. Korhonen, I. Laine, J. Rieppo, and J. Zhang, Value sharing results for shifts of meromorphic functions, and sufficient conditions for periodicity. *J. Math. Anal. Appl.*, **355**(1) (2009), 352–363, MR 2514472.
10. S. Li and B. Q. Chen, Meromorphic functions sharing small functions with their linear difference polynomials. *Adv. Diff. Eqn.*, **2013**(6), (2013), 58, MR 3042707.

11. S. Li and Z. S. Gao, Entire functions sharing one or two finite values CM with their shifts or difference operators. *Arch. Math. (Basel)*, **97**(5) (2011), 475–483, MR 2860575.

12. E. Mues and N. Steinmetz, Meromorphe Funktionen, die mit ihrer Ableitung zwei Werte teilen. *Results Math.*, **6**(1) (1983), 48–55, MR 714657.

13. R. Nevanlinna, Le théorème de Picard-Borel et la théorie des fonctions méromorphes, VII + 174 p. Paris, Gauthier-Villars (Collections de monographies sur la théorie des fonctions), 1929 (French).

14. L. A. Rubel and C. C. Yang, *Values shared by an entire function and its derivative.* In: *Complex Analysis.* (Proc. Conf., Univ. Kentucky, Lexington, Ky., 1976), Springer, Berlin, 1977, pp. 101–103. *Lecture Notes in Math.*, Vol. 599.

15. C.-C. Yang and H.-X. Yi, *Uniqueness Theory of Meromorphic Functions*, Mathematics and its Applications, Vol. 557. Kluwer Academic Publishers Group, Dordrecht, 2003.

© 2022 World Scientific Publishing Europe Ltd.
https://doi.org/10.1142/9781800611368_0010

Chapter 10

Global Borel Summability of Some Partial Differential Equations

Masafumi Yoshino

Hiroshima University
Higashi-Hiroshima, Japan
yoshinom@hiroshima-u.ac.jp

We study the global Borel summability of certain partial differential equations which appear in the study of a movable singularity of a Hamiltonian system or a blowup phenomenon of equations in mathematical physics.

1. Introduction and Results

In this chapter, we study the global Borel summability of formal solutions of partial differential equations. Borel summation of formal solutions of PDEs is a problem of significant interest in mathematics and there is a vast literature devoted to this subject, see the recent works [2–5], and references therein.

In order to state our results we introduce some notations. Let $(t, r) \in \mathbb{C}^2$ and let $\mathcal{R}(z, u)$ be analytic in $z \in C_0$ for some open set $C_0 \subset \mathbb{C}$ containing the origin, and polynomial in u. Let $u_0 = -t + t^{(0)} - \frac{2}{\lambda \nu} \log(1 - \lambda c_0 r^\nu)$, where c_0, $t^{(0)}$, $\lambda \neq 0$ and $\nu \geq 1$ are constants. We consider the Borel summability for the partial differential equation for the unknown function $u = u(t, r)$

$$h\left(r^{\nu-1} c_0 \frac{\partial u}{\partial t} + (1 - \lambda c_0 r^\nu) \frac{\partial u}{\partial r} \right) - r^{\nu-1} (\lambda u + \mathcal{R}(u_0, u)) = 0, \quad (1)$$

where h is a constant. Equation (1) is the model equation of the normalizing equation of a Hamiltonian vector field, which is crucial in the study of the movable singular point of a Hamiltonian system. The argument in this note still works with minor modifications in the case when $u_0(t,r)$ is a certain analytic function of t and r. We take the special form in this chapter for the sake of simplicity of the presentation and in view of the application to the concrete Hamiltonian system.

The main object in this chapter is to discuss the Borel summability of the formal series solution given in (2) below. We note that the summability yields the existence of the transformation stated in the above. As for the construction of a singular solution, we do not go into the detail and refer to [7]. Consider the formal power series solution of (1)

$$w(t,r,h) = \sum_{n=0}^{\infty} w_n(t,r)h^n, \tag{2}$$

where $w_n(t,r)$'s are holomorphic when $(t,r) \in U_1 \times \Omega_1$ for some open sets U_1 and Ω_1. If $\nu = 1$, then the summability of (2) holds for some neighborhoods of the origin, U_1 and Ω_1 (cf. [6]). Suppose $\nu \geq 1$. We are interested in the Borel summability on the Riemannian surface of r^ν, namely the Borel summability when r moves on a bounded open set of the Riemannian surface of r^ν, and t moves on a bounded domain. Note that ν may not be an integer and a phenomenon similar to the so-called turning point appears at $r = 0$ when $\nu > 1$.

We give some definitions concerning the Borel summability (cf. [1]). The formal Borel transform of $\tilde{w} := w - w_0$ is defined by

$$\mathcal{B}(\tilde{w})(t,r,y) := \sum_{n=1}^{\infty} \frac{w_n(t,r)}{(n-1)!} y^{n-1}, \tag{3}$$

where y is a dual variable of h. For $\theta > 0$ and the direction ξ, let

$$E(\xi,\theta) := \left\{ z \in \mathbb{C} \,\middle|\, \mathrm{dist}(z, \mathbb{R}_+ e^{i\xi}) < \frac{\theta}{2} \right\}, \tag{4}$$

where $\mathbb{R}_+$ is the set of nonnegative real numbers, and $\mathrm{dist}(z, \mathbb{R}_+ e^{i\xi})$ denotes the distance from z to the set $\mathbb{R}_+ e^{i\xi}$.

We say that the formal power series $w(t,r,h)$ is Borel summable in the direction ξ with respect to h if there exists $\theta > 0$ such that

$B(\tilde{w})(t,r,y)$ converges when $(t,r) \in U_1 \times \Omega_1$ and y is in some neighborhood of $y = 0$, can be analytically continued to the set $\{(t,r,y) \in U_1 \times \Omega_1 \times E(\xi,\theta)\}$, and is of exponential type of order 1 in $y \in E(\xi,\theta)$ for every $(t,r) \in U_1 \times \Omega_1$. That is, there exist $K_0 > 0$ and $K_2 > 0$ such that

$$|B(\tilde{w})(t,r,y)| \leq K_0 e^{K_2|y|}, \quad y \in E(\xi,\theta)$$

holds for $(t,r) \in U_1 \times \Omega_1$. Then the Borel sum of the formal series $w(t,r,h)$, $W(t,r,h)$ is defined by the Laplace transform

$$W(t,r,h) := w_0 + \int_0^{\infty e^{i\xi}} e^{-yh^{-1}} B(\tilde{w})(t,r,y)dy. \tag{5}$$

For $0 < \theta < \pi$, set $\Omega := E(0,\theta)$. Let $H(\Omega)$ be the set of holomorphic functions in Ω. For $c > 0$, define $\mathcal{H}_c(\Omega)$ as those $h \in H(\Omega)$ such that there exists $K \geq 0$ for which

$$|h(z)| \leq K e^{c|Re\,z|}(1 + |z|)^{-2} \quad \text{for all } z \in \Omega. \tag{6}$$

Clearly, $\mathcal{H}_c(\Omega)$ is a Banach space with the norm

$$\|h\|_{\Omega,c} := \sup_{z\in\Omega} |h(z)|(1 + |z|)^2 e^{-c|Re\,z|}. \tag{7}$$

Construction of the formal solution. Substitute (2) into (1) and compare the power of h. By comparing the power of $h^0 = 1$, w_0 satisfies

$$\lambda w_0 + \mathcal{R}(u_0, w_0) = 0. \tag{8}$$

We assume that w_0 is determined as a holomorphic function of u_0. For example, this follows from the implicit function theorem if we assume $\mathcal{R}(u_0, u) = O(|u|^2)$, $|u| \to 0$. If $\nu = 1$, we assume that $w_0 = O(u_0^2)$ when $u_0 \to 0$. We can show that w_n's $(n \geq 1)$ are holomorphic at $r = 0$. As for the estimate of w_n, there exist constants $K_1 > 0$ and $\rho > 0$ independent of n such that $|w_n| \leq K_1 \rho^n n!$ $(n = 0, 1, 2, \ldots)$ by Nagumo's lemma. In the following, we assume that every coefficient of the formal series (2) is holomorphic in $(t.r) \in U_1 \times \Omega_1$ for some open sets U_1 and Ω_1.

When $\nu \geq 1$ is an integer, let $\Omega_0 \subset \Omega_1$ be a bounded domain containing the origin $r = 0$ such that the closure $\overline{\Omega_0}$ is a bounded set in the universal covering space of $\log(1 - \lambda c_0 r^\nu)$. When $\nu > 1$ is not an integer, let $\Omega_0 \subset \Omega_1$ be a bounded domain contained in the sector $\{r|\,|\arg r| < \alpha\}$ in the universal covering space of $\mathbb{C}\backslash 0$ for some $\alpha > 0$ such that the closure $\overline{\Omega_0}$ is a bounded set in the universal covering space of $\log\,(1 - \lambda c_0 r^\nu)$. Then we have

Theorem 1. *Assume that $t^{(0)} \in U_1$. Then there exists a neighborhood U_0 of $t = t^{(0)}$ and $r_0 > 0$ such that (1) is Borel summable in the direction 0 if $(t, r) \in U_0 \times (\Omega_0 \cap \{r|\,|r| < r_0\})$.*

Next, let $U_0 \subset U_1$ be a bounded domain and let $\epsilon_1 > 0$. Then we have

Theorem 2. *Assume that there exists $\ell_1 > 0$ such that $\mathcal{R}(z, u)$ is analytic with respect to z in $\{z|\,|Im\,(\lambda z)| \geq \ell_1\}$. Moreover, assume that $\mathcal{R}(z, u)$ has the growth order $O((1 + |z|)^{-1})$ when $z \to \infty$ in $\{z|\,|Im\,(\lambda z)| \geq \ell_1\}$ and for every $u \in \mathbb{C}$. Then there exists an $\ell_0 > 0$ such that, if $U_0 \subset \{z|\,|Im\,(\lambda z)| > \ell_0\}$, then (1) is Borel summable in the direction 0 when $(t, r) \in U_0 \times \Omega_0$.*

By Theorem 2, we have the solvability of (1).

Corollary 1. *Suppose that there exists $\ell_1 > 0$ such that $\mathcal{R}(z, u)$ is analytic with respect to z in $\{z|\,|Im\,(\lambda z)| \geq \ell_1\}$. Moreover, assume that $\mathcal{R}(z, u)$ has the growth order $O((1 + |z|)^{-1})$ when $z \to \infty$ in $\{z|\,|Im\,(\lambda z)| \geq \ell_1\}$ and $u \in \mathbb{C}$ is fixed. Let $h_0 > 0$ be given. Then there exists an $\ell_0 > 0$ and a sector S_0 in h-variable with vertex at the origin such that if $|h| < h_0$, $h \in S_0$ and $U_0 \subset \{z|\,|Im\,(\lambda\nu z)| > \ell_0\}$, then (1) has an analytic solution u when $(t, r) \in U_0 \times \Omega_0$.*

2. Reduction and Lemmas

In order to apply the formal Borel transform to (1) we eliminate the constant term. Set $u := w_0 + v$, where $v = O(h)$. Let $\hat{v} := \mathcal{B}(v)$. Substitute the relations $u := w_0 + v$ into (1) and cancel the terms of $h^0 = 1$. Divide the equation with h and apply the formal Borel transform to both sides of the equation, to obtain

$$\left(r^{\nu-1}c_0\frac{\partial\hat{v}}{\partial t} + (1 - \lambda c_0 r^\nu)\frac{\partial\hat{v}}{\partial r}\right) = r^{\nu-1}\frac{\partial}{\partial y}\left(\lambda\hat{v} + \mathcal{R}^*(u_0, \hat{u})\right), \quad (9)$$

where $\mathcal{R}^*$ is defined by regarding the product in $\mathcal{R}$ as the convolution product. Let D be a domain in $(t, r) \in \mathbb{C}^2$ and define $\Omega := E(0, \theta)$ for $\theta > 0$. Let $H(D, \Omega)$ be the set of all holomorphic functions of $(t, r) \in D$ and $y \in \Omega$. For $c > 0$, let $\mathcal{H}_c(D, \Omega)$ be the set of $f \equiv f(t, r, y) \in H(D, \Omega)$ such that there exists $K \geq 0$ satisfying

$$\sup_{(t,r) \in D} |f(t, r, y)| \leq K e^{c|\mathrm{Re}\, y|}(1 + |y|)^{-2} \quad \text{for all } y \in \Omega. \tag{10}$$

In order to prove Theorem 1, it is sufficient to show

Theorem 3. *Assume that $t^{(0)} \in U_1$. Let $c > 0$ be given. Then, there exist an $E(0, \theta) = \Omega$, a neighborhood U_0 of $t = t^{(0)}$ and $r_0 > 0$ such that (9) has a solution $\hat{u}$ in $\mathcal{H}_c(D, \Omega)$, where $D = U_0 \times (\Omega_0 \cap \{r|\, |r| < r_0\})$.*

We note that there exists a neighborhood $y = 0$, W_0 such that the formal Borel transform $\mathcal{B}(v)$ is analytic in $(t, r, y) \in D \times W_0$ and is the unique analytic solution in W of (9).

In order to prove Theorem 3, we prepare some lemmas. Define $\lambda(z) := \lambda + \mathcal{R}'_u(z, w_0(0))$. In order to solve (9), we consider

$$\mathcal{L}\hat{w} \equiv r^{\nu-1} c_0 \frac{\partial \hat{w}}{\partial t} + (1 - \lambda c_0 r^\nu) \frac{\partial \hat{w}}{\partial r} - r^{\nu-1} \lambda(u_0) \frac{\partial \hat{w}}{\partial y} = r^{\nu-1} \hat{g}, \tag{11}$$

where $\hat{g} \in \mathcal{H}_c(D, \Omega)$. In order to solve (11), we use the method of characteristics. Consider

$$\frac{dt}{r^{\nu-1} c_0} = \frac{dr}{1 - \lambda c_0 r^\nu} = \frac{dy}{-r^{\nu-1} \lambda(u_0)}. \tag{12}$$

By taking r as an independent variable, t is given by

$$t = t^{(0)} - (\lambda\nu)^{-1} \log\left(1 - \lambda c_0 r^\nu\right). \tag{13}$$

Substituting (13) into (12) and solving (12) with respect to y, we obtain

$$y = y_0 - \Phi(r), \quad \Phi(r) := c_0^{-1} \int_0^\sigma \lambda(s)\, ds, \tag{14}$$

$$\sigma = -(\lambda\nu)^{-1} \log(1 - \lambda c_0 r^\nu), \tag{15}$$

where $y_0 = y(0) \in \Omega$ is an initial value of $y = y(r)$ at $r = 0$. By (13) and (15), we have

$$\sigma = -t + t^{(0)} - (2/\lambda\nu)\log(1 - \lambda c_0 r^\nu) = u_0(r).$$

In the following lemmas we assume $1 - \lambda c_0 r_0^\nu \neq 0$.

Lemma 1. *There exists a curve γ_{r_0} on $\mathbb{C}$ passing through r_0 such that $\operatorname{Im} \Phi$ is constant on γ_{r_0}, where Φ is given by (14).*

Proof. First we consider the case where $\lambda(s)$ is a constant function. Assume that it is equal to $\lambda(\sigma_0) \equiv -\lambda\nu c_0 \mu$. Set $\zeta = 1 - c_0\lambda r^\nu$ and $\zeta_0 = 1 - c_0\lambda r_0^\nu$. Then $\operatorname{Im} \Phi$ is constant on γ_{r_0} if $\operatorname{Im}(\mu \log \zeta) = \operatorname{Im}(\mu \log \zeta_0)$ on γ_{r_0}. Consider the change of variable $\zeta \mapsto \tilde{\zeta}$, $\tilde{\zeta} = \mu \log \zeta$. Set $\tilde{\zeta} = \tilde{x} + i\tilde{y}$ and $\tilde{\zeta}_0 = \tilde{x}_0 + i\tilde{y}_0 := \mu \log \zeta_0$. Define the curve $\tilde{\gamma}_{r_0}$ by $\tilde{y} = \tilde{y}(\tilde{x}) = \tilde{y}_0$. Define γ_{r_0} as the image of $\tilde{\gamma}_{r_0}$ by the map, $\tilde{\zeta} \mapsto \zeta$.

Next, we consider the case $\lambda(\sigma)$ is not a constant function. The condition $\operatorname{Im} \Phi(r) = \operatorname{Im} \Phi(r_0)$ can be written in $\operatorname{Im}\left(c_0^{-1} \int_{\sigma_0}^{\sigma} \lambda(s)ds\right) = 0$ with $\sigma_0 = -(\lambda\nu)^{-1}\log(1 - c_0\lambda r_0^\nu)$. By the Taylor expansion we have

$$\lambda(s) = \lambda(\sigma_0) + (s - \sigma_0)\tilde{\lambda}(s), \tag{16}$$

where $\tilde{\lambda}(z)$ is holomorphic at $z = 0$. Therefore, we have

$$\operatorname{Im}\left(c_0^{-1}\lambda(\sigma_0)(\sigma - \sigma_0)\right) + \operatorname{Im}\left(c_0^{-1}(\sigma - \sigma_0)^2 \tilde{R}\right) = 0, \tag{17}$$

where $\tilde{R}$ is holomorphic. The first term in the right-hand side of (17) is equal to $\tilde{y} - \tilde{y}_0$. The second term is sufficiently small. Hence, by the implicit function theorem we have the assertion. $\square$

Lemma 2. *For $r \in \gamma_{r_0}$, define*

$$P\hat{g} := \int_{r_0}^{r} \hat{g}(u_0(s), s, y_0 - \Phi(s)) \frac{s^{\nu-1}}{1 - \lambda c_0 s^\nu} ds, \quad \hat{g} \in \mathcal{H}_c(D, \Omega), \tag{18}$$

where the integral is taken along γ_{r_0}. Then $P\hat{g}$ satisfies (11). Especially, $P\hat{g}$ is analytic in r in some neighborhood of the origin $r = 0$.

Proof. We show that P is well defined. By the definition of Φ, we have

$$y_0 - \Phi(s) = y + \Phi(r) - \Phi(s). \tag{19}$$

By Lemma 1 and the definition of γ_{r_0} we have $\mathrm{Im}\,(\Phi(r) - \Phi(s)) = 0$ on γ_{r_0}. Because $\hat{g}$ is holomorphic in $y \in E(0,\theta)$, the integral in P converges.

Next, we show that $P\hat{g}$ satisfies (11). By (12) and (18), we have

$$\hat{g}(u_0(r), r, y_0 - \Phi(r)) \frac{r^{\nu-1}}{1 - \lambda c_0 r^\nu} = \frac{d\hat{w}}{dr}$$

$$= \frac{\partial q_1}{\partial r} \frac{\partial \hat{w}}{\partial q_1} + \frac{\partial \hat{w}}{\partial r} + \frac{\partial y}{\partial r} \frac{\partial \hat{w}}{\partial y}$$

$$= \frac{r^{\nu-1} c_0}{1 - \lambda c_0 r^\nu} \frac{\partial \hat{w}}{\partial q_1} + \frac{\partial \hat{w}}{\partial r} - \frac{r^{\nu-1} \lambda(u_0)}{1 - \lambda c_0 r^\nu} \frac{\partial \hat{w}}{\partial y}. \tag{20}$$

By multiplying $1 - \lambda c_0 r^\nu$ to both sides, we obtain (11). $\qquad\square$

Lemma 3. *Let r_1 satisfy $1 - \lambda c_0 r_1^\nu = 0$. Assume that there exists a neighborhood W_{r_1} of r_1 and $M_1 > 0$ such that, if $\gamma_{r_0} \subset W_{r_1}$, then $M_1^{-1} \leq |r_1 - s||r_1 - t|^{-1} \leq M_1$ for every $s, t \in \gamma_{r_0}$. Then, there exists a constant $c_1 \geq 0$ such that, for every $\hat{g} \in \mathcal{H}_c(D, \Omega)$ we have*

$$\|P\hat{g}\|_c \leq c_1 \|\hat{g}\|_c, \quad \left\|\frac{\partial}{\partial y}(P\hat{g})\right\|_c \leq c_1 \|\hat{g}\|_c. \tag{21}$$

Proof. We show the first inequality of (21). In view of (19), we estimate $\Phi(r) - \Phi(s)$. By the continuity of $\lambda(z)$, there exists $C_1 > 0$ such that

$$|\Phi(r) - \Phi(s)| \leq C_1 |\lambda \nu c_0|^{-1} \left| \log \left(\frac{1 - \lambda c_0 s^\nu}{1 - \lambda c_0 r^\nu} \right) \right| =: C_1 |\lambda \nu c_0|^{-1} A(s,t). \tag{22}$$

If the path γ_{r_0} lies outside a neighborhood of r_1 with $1 - \lambda c_0 r_1^\nu = 0$, then $A(s,t)$ is estimated by some constant. If otherwise, then, by noting $1 - \lambda c_0 s^\nu = \lambda c_0 (r_1^\nu - s^\nu)$ and $1 - \lambda c_0 r^\nu = \lambda c_0 (r_1^\nu - r^\nu)$, $A(s,t)$ is estimated by $|\log(r_1^\nu - s^\nu)(r_1^\nu - r^\nu)^{-1}|$. The last term is bounded by some constant by assumption.

Therefore, there exists $K_1 > 0$ independent of c_0 and r such that

$$\exp(c|\mathrm{Re}\,(y_0 - \Phi(s))|) = \exp(c|\mathrm{Re}\,(y + \Phi(r) - \Phi(s))|)$$

$$\leq \exp(c|\mathrm{Re}\,y| + c|\Phi(r) - \Phi(s))|) \leq K_1 e^{c|\mathrm{Re}\,y|}. \tag{23}$$

Similarly there exists $K_2 > 0$ such that

$$(1 + |y_0 - \Phi(s)|)^{-2} \leq K_2(1 + |y|)^{-2}, \quad y \in \Omega. \tag{24}$$

Therefore, there exist $C_2 > 0$ and $C_3 > 0$ such that

$$\|P\hat{g}\|_c \leq \sup\left((1 + |y|)^2 e^{-c|\mathrm{Re}\, y|} \int \|\hat{g}\|_c \frac{e^{|\mathrm{Re}\,(y_0 - \Phi(s))|}}{(1 + |y_0 - \Phi(s)|)^2}|ds|\right)$$

$$\leq C_2\|\hat{g}\|_c \int |ds| \leq C_3\|\hat{g}\|_c. \tag{25}$$

Next we show the second estimate of (21). Since $\lambda(u_0) \neq 0$, we have

$$\frac{\partial \hat{w}}{\partial y} = -\frac{\hat{g}}{\lambda(u_0)} + \frac{c_0}{\lambda(u_0)} \frac{\partial \hat{w}}{\partial q_1} + \frac{1 - \lambda c_0 r^\nu}{r^{\nu-1}\lambda(u_0)} \frac{\partial \hat{w}}{\partial r}. \tag{26}$$

The first term of the right-hand side can be estimated by $\|\hat{g}\|_c$. As for the third term, by differentiating $P\hat{g} = \hat{w}$ we obtain

$$\frac{1 - \lambda c_0 r^\nu}{r^{\nu-1}\lambda(u_0)} \frac{\partial \hat{w}}{\partial r} = \frac{1 - \lambda c_0 r^\nu}{r^{\nu-1}\lambda(u_0)} \hat{g}(u_0(r), r, y_0 - \Phi(r)) \frac{r^{\nu-1}}{1 - \lambda c_0 r^\nu}$$

$$= \frac{\hat{g}(u_0(r), r, y_0 - \Phi(r))}{\lambda(u_0)}. \tag{27}$$

The right-hand side is estimated by $\|\hat{g}\|_c$. We consider the second term in (26). Since $\hat{w}$ is analytic in q_1 we take the directional derivative of q_1 given by (12). Then we have

$$\frac{\partial \hat{w}}{\partial q_1} = \frac{dr}{dq_1} \frac{\partial \hat{w}}{\partial r} = \frac{1 - \lambda c_0 r^\nu}{r^{\nu-1}c_0} \frac{\partial \hat{w}}{\partial r}. \tag{28}$$

The right-hand side is estimated by $\|\hat{g}\|_c$. $\square$

3.   Proof of Theorems

We give the proof of Theorem 3.

Proof of Theorem 3. Set $v := u - w_0$ and expand in the Taylor series

$$\mathcal{R}(u_0, u) = \mathcal{R}(u_0, w_0) + \mathcal{R}'_u(u_0, w_0)v + \sum_{\beta \geq 2} r_\beta(u_0, w_0)v^\beta. \tag{29}$$

Hence, we have

$$\frac{\partial}{\partial y}\left(\mathcal{R}^*(u_0,\hat{u}) - \mathcal{R}(u_0,w_0)\right) = \mathcal{R}'_u(u_0,w_0)\hat{v}_y + \sum_{\beta\geq 2} r_\beta(u_0,w_0)(\hat{v}_y)^{*\beta},$$

(30)

where $\hat{v}_y = (\partial/\partial y)\hat{v}$ and $(\hat{v}_y)^{*\beta}$ is β-times the convolution product of $\hat{v}_y$. Hence, it is sufficient to solve the following equation:

$$\mathcal{L}\hat{v} + r^{\nu-1}(\lambda(u_0) - \lambda)\hat{v}_y = r^{\nu-1}\mathcal{R}'_u(u_0,w_0)\hat{v}_y + \sum_{\beta\geq 2} r_\beta(u_0,w_0)(\hat{v}_y)^{*\beta}.$$

(31)

By comparing the terms of h in (9), we have $\mathcal{L}(\hat{v}_1 + w_0) = 0$. Hence, we have

$$\hat{v}_1 = -P\mathcal{L}w_0. \tag{32}$$

Define inductively, for $k = 1, 2, \ldots$

$$\hat{v}_{k+1} = P\left(r^{\nu-1}\left(\mathcal{R}'_u(u_0,w_0) - (\lambda(u_0) - \lambda)\right)(\hat{v}_k)_y\right)$$

$$+ P\left(\sum_\beta r^{\nu-1} r_\beta(\hat{v}_k^{*\beta})_y\right). \tag{33}$$

If the limit $\hat{v} := \lim_{k\to\infty}\hat{v}_k$ exists, then $v + w_0 =: u$ is the unique analytic solution of (9). In order to show the convergence, it is sufficient to show the following *a priori* estimate. $\qquad\square$

Lemma 4. *There exist an $\epsilon_1 > 0$ and $K_1 > 0$ independent of k such that for every ϵ, $0 < \epsilon < \epsilon_1$ there exist a neighborhood U_0 of $t^{(0)}$ and $r_0 > 0$ such that if $(t,r) \in U_0 \times (\Omega_0 \cap \{r\,|\,|r| < r_0\})$, then we have*

$$\|\hat{v}_k\|_c \leq \epsilon K_1, \quad \|(\hat{v}_k)_y\|_c \leq \epsilon K_1, \quad k = 1, 2, \ldots \tag{34}$$

Proof. We estimate the sequence inductively. The smallness of $\hat{v}_1$ is assured by taking $r_0 > 0$ sufficiently small, i.e. by taking r_0 sufficiently small since P is a bounded operator by Lemma 3. For the estimate of general $\hat{v}_k$, we use (33) inductively. Indeed, the second term of the right-hand side can be made small by taking ϵ sufficiently small since $\beta \geq 2$. On the other hand, the first term in the right-hand side can be estimated if $\nu > 1$ by taking r_0 sufficiently small. If otherwise, the term can be made sufficiently small in terms of the definition of $\lambda(z)$. This proves the *a priori* estimate. $\qquad\square$

In order to prove Theorem 2, it is sufficient to show

Theorem 4. *Suppose that there exists $\ell_1 > 0$ such that $\mathcal{R}(z, u)$ is analytic with respect to z in $\{z \mid |Im\,(\lambda z)| \geq \ell_1\}$. Moreover, assume that $\mathcal{R}(z, u)$ has the growth order $O((1 + |z|)^{-1})$ when $z \to \infty$ in $\{z \mid |Im\,(\lambda z)| \geq \ell_1\}$ and u being fixed. Then there exists an $\ell_0 > 0$ such that, if $U_0 \subset \{\zeta \mid |Im\,(\lambda \nu \zeta)| > \ell_0\}$, then (9) has a solution $\hat{u}$ in $\mathcal{H}_c(D, \Omega)$, where $D = U_0 \times \Omega_0$.*

Proof. The argument is almost similar to the proof of Theorem 3. We define the approximate sequence by the same formula as in the proof of Theorem 3. We need to show the *a priori* estimate as in Lemma 4. We note that r may not be small since r moves on some bounded set such that the formal solution and u_0 are analytic. We make w_0 sufficiently small by taking ℓ large since the coefficients decay by assumption. Then we have the *a priori* estimate. $\qquad \square$

Acknowledgments

This chapter is partially supported by Grant-in-Aid for Scientific Research (No. 20K03683), JSPS, Japan.

References

1. W. Balser, *Formal Power Series and Linear Systems of Meromorphic Ordinary Differential Equations*, Universitext, Springer-Verlag, New York, 2000.
2. W. Balser and M. Loday-Richaud, Summability of solutions of the heat equation with inhomogeneous thermal conductivity in two variables. *Adv. Dyn. Syst. Appl.*, **4**(2) (2009), 159–177.
3. K. Ichinobe and M. Miyake, On k-summability of formal solutions for certain partial differential operators with polynomial coefficients. *Opuscula Mathematica*, **35**(5) (2015), 625–653.
4. A. Lastra, S. Michalik, and M. Suwinska, Summability of formal solutions for some generalized moment partial differential equations. *Results Math.*, **76**(2021), 22.
5. H. Tahara and H. Yamazawa, Multisummability of formal solutions to the Cauchy problem for some linear partial differential equations. *J. Diff. Eqn.*, **255**(2013), 3592–3637.

6. H. Yamazawa and M. Yoshino, Borel summability of some semilinear system of partial differential equations. *Opuscula Mathematica*, **35**(6) (2015), 825–845.

7. M. Yoshino, Movable singularity of semi linear Heun equation and application to Blowup phenomenon. *Nonlin. Differ. Eqn. Appl.*, **26**(2019), 8.

© 2022 World Scientific Publishing Europe Ltd.
https://doi.org/10.1142/9781800611368_0011

Chapter 11

A Remark of k-Summability of Divergent Solutions to Some Linear q-Difference-Differential Equations with Entire Cauchy Data

Kunio Ichinobe

Aichi University of Education
Hirosawa, Igaya, Kariya City
Aichi Pref. 448-8542, Japan
ichinobe@auecc.aichi-edu.ac.jp

We consider the Cauchy problem (CP) to some linear q-difference-differential equations with entire Cauchy data. We see that the summability condition is equivalent with the convergence for the formal solution in some case.

1. Introduction

We shall consider the following Cauchy problem (CP) for q-difference-differential equations:

$$\begin{cases} D_{q,t}^{\kappa} u(t,x) = \partial_x^{\nu} u(t,x), \\ u(0,x) = \varphi(x) \in \mathcal{O}_x, \quad D_{q,t}^{i} u(0,x) = 0 \quad (1 \le i \le \kappa - 1), \end{cases} \tag{CP}$$

where $(t,x) \in \mathbb{C}^2$, κ and ν are natural numbers, $\partial_x = \partial/\partial x$ and $\mathcal{O}_x$ denotes the set of all holomorphic functions in a neighborhood of the origin $x = 0$. Here, setting $0 < q < 1$, we define the q-difference

235

operator $D_{q,t}$ as

$$D_{q,t}u(t,x) := \frac{u(qt,x) - u(t,x)}{qt - t}.$$

The CP has a unique formal solution of the form

$$\hat{u}(t,x) = \sum_{n \geq 0} \frac{\varphi^{(\nu n)}(x)}{[\kappa n]_q!} t^{\kappa n} \overset{\text{put}}{=} \sum_{n \geq 0} u_n(x) t^n \in \mathcal{O}_x[[t]]_{1/k} \quad \left(k = \frac{\kappa}{\nu}\right), \tag{1}$$

where $[n]_q = \frac{q^n - 1}{q - 1}$ for $n \geq 0$ and

$$[n]_q! = \begin{cases} 1 & (n = 0), \\ [n]_q[n-1]_q \cdots [1]_q & (n \geq 1). \end{cases}$$

Here, the notation $\mathcal{O}_x[[t]]_{1/k}$ denotes the set of formal power series in t with the coefficients $u_n(x)$ which are holomorphic in a common closed disc $B(r) := \{x \in \mathbb{C}; |x| \leq r\}$ for some $r > 0$ and satisfy the following Gevrey type estimates:

$$\max_{|x| \leq r} |u_n(x)| \leq CK^n\Gamma(1 + n/k) \tag{2}$$

with some positive constants C and K for any nonnegative integer n. In this case, we say that the Gevrey order of $\hat{u}$ is (at most) $1/k$.

In order to explain our problem, we define a class $\mathrm{Exp}(\gamma; \mathbb{C})$ or $\mathrm{Exp}_x(\gamma; \mathbb{C})$ of entire functions for $\gamma > 0$ by

$$\mathrm{Exp}(\gamma; \mathbb{C}) := \{f(x) \in \mathcal{O}(\mathbb{C}); |f(x)| \leq C\exp(\delta|x|^\gamma) \text{ for some } C, \delta > 0\},$$

where $\mathcal{O}(\mathbb{C})$ denotes the set of entire functions. By an easy calculation, we see that

$$f(x) \in \mathrm{Exp}(\gamma; \mathbb{C}) \iff |f^{(n)}(0)| \leq AB^n(n!)^{1-1/\gamma}$$

for all n by some positive constants A and B. From this fact, a characterization of the convergence of the formal solution (1) is stated as follows [4,5].

Theorem 1. *The formal solution $\hat{u}(t,x)$ is convergent if and only if $\varphi(x) \in \mathrm{Exp}(1, \mathbb{C})$ for the Cauchy data.*

We assume that the Cauchy data $\varphi(x)$ belongs to a class of entire functions as follows:

$$\varphi(x) \in \mathrm{Exp}\,(\nu/\ell; \mathbb{C}), \quad 0 \le \ell \le \nu - 1 \ (\ell \in \mathbb{N}). \tag{3}$$

When $\ell = 0$, we understand that $\mathrm{Exp}(\nu/0; \mathbb{C}) = \mathcal{O}_x$ which does not make any contradiction in the results. In this case, we have for the Gevrey order of the formal solution

$$\hat{u}(t, x) \in \mathcal{O}_x[[t]]_{1/k(\ell)}, \quad k(\ell) = \kappa/(\nu - \ell) \ge k(0) = \kappa/\nu. \tag{4}$$

Our problem is to give a characterization of $k(\ell)$-summability of the formal solution (1) under the assumption (3) for the Cauchy data.

The organization of the chapter is as follows. In Section 2, we review a result of $k(0)$-summability of $\hat{u}(t, x)$, and we give some definitions and propositions which are needed for proofs of a main theorem. In Section 3, we give the main theorem (Theorem 4) and its corollary. For proving the main theorem, we give a result of $k(\ell)$-summability of $\hat{u}(t, x)$ and prove it in Section 4. In Section 5, we give a proof of the main theorem.

Remark 1. The equation (CP) is simple, but the generalization is possible for an equation in two variables which has a quasi-homogeneity in the operator $D_{q,t}$ and ∂_x.

2. Review of a Result of $k(0)$-Summability

We first give the definition of Gevrey asymptotic expansion and k-summability [1].

For $d \in \mathbb{R}, \alpha > 0$ and $\rho \ (0 < \rho \le \infty)$, we define a sector $S = S(d, \alpha, \rho)$ by

$$S(d, \alpha, \rho) := \{t \in \mathbb{C}; |d - \arg t| < \alpha/2, \ 0 < |t| < \rho\},$$

where d, α and ρ are called the direction, the opening angle and the radius of S, respectively. We write $S(d, \alpha, \infty) = S(d, \alpha)$ for short.

Let $k > 0, \hat{v}(t, x) = \sum_{n=0}^{\infty} v_n(x) t^n \in \mathcal{O}_x[[t]]_{1/k}$ and $v(t, x)$ be an analytic function on $S(d, \alpha, \rho) \times B(r)$. Then we say that $v(t, x)$ has a Gevrey asymptotic expansion $\hat{v}(t, x)$ of order k in $S(d, \alpha, \rho)$, which is denoted by

$$v(t, x) \cong_k \hat{v}(t, x) \quad \text{in } S(d, \alpha, \rho),$$

if for any closed subsector S' of $S(d, \alpha, \rho)$, there exist some positive constants C and K such that for any N, we have

$$\max_{|x| \leq r} \left| v(t, x) - \sum_{n=0}^{N-1} v_n(x) t^n \right| \leq C K^N |t|^N \Gamma(1 + N/k), \quad t \in S'. \quad (5)$$

For $k > 0, d \in \mathbb{R}$ and $\hat{v}(t, x) \in \mathcal{O}_x[[t]]_{1/k}$, we say that $\hat{v}(t, x)$ is k-summable in d direction, which is denoted by $\hat{v}(t, x) \in \mathcal{O}_x\{t\}_{k,d}$, if there exists a sector $S = S(d, \alpha, \rho)$ with $\alpha > \pi/k$ and an analytic function $v(t, x)$ on $S \times B(r)$ such that $v(t, x) \cong_k \hat{v}(t, x)$ in S.

We remark that the function $v(t, x)$ above for a k-summable $\hat{v}(t, x)$ is unique if it exists. Therefore, such a function $v(t, x)$ is called the k-sum of $\hat{v}(t, x)$ in d direction.

Now, we give results of the existence of Gevrey asymptotic solutions and $k(0)$-summability for the formal solution (1) of (CP).

Theorem 2 ([4]). *For any α with $0 < \alpha < \pi/k(0)$ and $d \in \mathbb{R}$, we can find positive constants ρ and r such that there exists an analytic solution $u(t, x)$ of (CP) in $S(d, \alpha, \rho) \times B(r)$ with $u(t, x) \cong_{k(0)} \hat{u}(t, x)$ in $S(d, \alpha, \rho)$.*

Actually, there are infinitely many such solutions on the sector.

Theorem 3 ([5]). *For $d \in \mathbb{R}$ and $\varepsilon > 0$, we put*

$$\Omega_\nu^\kappa(d, \varepsilon) := \bigcup_{j=0}^{\nu-1} S\left(\frac{\kappa d + 2\pi j}{\nu}, \varepsilon \right). \quad (6)$$

Then $\hat{u}(t, x) \in \mathcal{O}_x\{t\}_{k(0),d}$ if and only if $\varphi(x) \in \mathrm{Exp}(1; \Omega_\nu^\kappa(d, \varepsilon))$ for the Cauchy data.

We next give definitions and propositions which are needed for proving the main theorem.

For $n = 0, 1, 2, \ldots$, the q-shift factorial is defined by $(a; q)_0 = 1$ and

$$(a; q)_n := \prod_{j=0}^{n-1} (1 - aq^j) \ (n \geq 1), \quad (a; q)_\infty := \prod_{j=0}^{\infty} (1 - aq^j).$$

For any $a \in \mathbb{C}$, the infinite product $(a;q)_\infty$ is convergent. We set $(a_1, a_2, \ldots, a_r; q)_n := \prod_{j=1}^{r}(a_j;q)_n$ for $n = 0, 1, 2, \ldots$ or $n = \infty$.

The basic hypergeometric series with the base q is defined as

$$_{\kappa+1}\phi_\kappa \left(\begin{matrix} a_1, a_2, \ldots, a_{\kappa+1} \\ b_1, \ldots, b_\kappa \end{matrix} ; q, x \right) := \sum_{n \geq 0} \frac{(a_1, a_2, \ldots, a_{\kappa+1}; q)_n}{(b_1, \ldots, b_\kappa; q)_n (q; q)_n} x^n,$$

which is convergent in $|x| < 1$.

An analytic continuation of basic hypergeometric series is given by the following proposition [5].

Proposition 1. *We assume* $0 < q < 1$ *and* $\kappa \geq 0$. *Let* $\boldsymbol{a}_\kappa = (a_1, a_2, \ldots, a_\kappa)$, $\boldsymbol{c}_\kappa = (c_1, c_2, \ldots, c_\kappa) \in \mathbb{C}^\kappa$ *and* $b \in \mathbb{C}$, *where* $|a_j| < 1$ *for* $j = 1, 2, \ldots, \kappa$. *Then the following equality holds:*

$$_{\kappa+1}\phi_\kappa \left(\begin{matrix} \boldsymbol{a}_\kappa, b \\ \boldsymbol{c}_\kappa \end{matrix} ; q, x \right) = \frac{(\boldsymbol{a}_\kappa, bx; q)_\infty}{(\boldsymbol{c}_\kappa, x; q)_\infty} \prod_{i=1}^{\kappa} \sum_{m_i \geq 0} \frac{(c_i/a_i, q^{M_i}x; q)_{m_i}}{(bq^{M_i}x, q; q)_{m_i}} a_i^{m_i},$$

$$(7)$$

where $M_i := \sum_{j=0}^{i-1} m_j$ *and* $m_0 = 0$. *Here, when* $\kappa = 0$, *we regard* $\boldsymbol{a}_0$ *and the product* $\prod_{i=1}^{0} \cdots$ *as the empty and 1, respectively.*

Remark 2. When $\kappa = 0$ and $\kappa = 1$ in the above proposition, the equalities (7) are nothing but the q-binomial theorem and Heine's transformation formula, respectively (cf. [3, §1.3, 1.4]). Therefore, our proposition gives a generalization of their formulas. (for the proof, see [5, §6].)

We give a result that guarantees the uniqueness of the solution for q-difference-differential equations.

We consider the following Cauchy problem for q-difference-differential equations:

$$\begin{cases} \partial_t^m u(t, x) = f(t, x, \{D_{q,x}^\ell \partial_x^\alpha \partial_t^{m-j} u(t, x)\}_\Lambda), \\ \partial_t^k u(0, x) = \varphi_k(x) \in \mathcal{O}_x \quad (0 \leq k \leq m - 1), \end{cases} \quad (\text{K})$$

where $t, x \in \mathbb{C}$, m is a natural number, $\Lambda = \{(\ell, \alpha, j); 0 \leq \ell \leq L, 0 \leq \alpha \leq j, 1 \leq j \leq m\}$ for $L \geq 0$. We assume that $f(t, x, \xi)$ ($\xi = \{\xi_{\ell,\alpha,j}\}_\Lambda$) is holomorphic at $(t, x, \xi) = (0, 0, 0)$ and $f(t, x, \xi)$ is an entire function in ξ for any fixed (t, x). Then we have the following proposition [13].

Proposition 2. *The Cauchy problem* (K) *has a unique holomorphic solution at* $(t, x) = (0, 0)$.

Remark 3. When $L = 0$, this proposition is nothing but the Cauchy–Kowalevski theorem for PDEs. Therefore, we call the equations (K) "of Kowalevski-type" for q-difference-differential equations. Proposition 2 claims that for any given local holomorphic Cauchy data, the formal solutions are convergent although the equations (K) are, in general, reduced to non-Kowalevski-type for partial differential equations when $q \to 1$.

3.   Main Theorem

In the following, we assume that

$$\varphi(x) \in \mathrm{Exp}\,(\nu/\ell; \mathbb{C})\,, \quad 0 \le \ell \le \nu - 1 \ (\ell \in \mathbb{N}). \tag{8}$$

In this case, the Gevrey order of the formal solution is given by

$$\hat{u}(t, x) \in \mathcal{O}_x[[t]]_{1/k(\ell)}, \quad k(\ell) = \kappa/(\nu - \ell) \ge k(0) = \kappa/\nu. \tag{9}$$

We remark that since Theorem 2 holds without any conditions for the Cauchy data $\varphi(x)$, we can immediately obtain the corresponding result to Theorem 2 without any conditions under the assumption (8). Therefore, our problem is to obtain the corresponding result with $\ell \ge 1$ to Theorem 3 with $\ell = 0$ under the assumption (8). Before giving our results, we give the definition of k-summability on an interval for formal series.

Let $I \subset \mathbb{R}$ be an interval. For $k > 0$ and $\hat{v}(s, x) \in \mathcal{O}_x[[t]]_{1/k}$, we say that $\hat{v}(s, x)$ is k-summable on I, which is denoted by $\hat{v}(s, x) \in \mathcal{O}_x\{t\}_{k,I}$, if for any $\tilde{d} \in I$, $\hat{v}(s, x) \in \mathcal{O}_x\{t\}_{k,\tilde{d}}$.

Now, our main theorem is stated as follows.

Theorem 4. *We assume that* $\varphi(x) \in \mathrm{Exp}\,(\nu/\ell; \mathbb{C})$.

- *When* $\ell = 1$, *for given* $d \in \mathbb{R}$ *and* $\varepsilon > 0$, *let* $I_d(\kappa, \varepsilon)$ *be an interval* $(d - \pi/(2\kappa) - \varepsilon, d + \pi/(2\kappa) + \varepsilon)$. *Then we have*

$$\hat{u}(t, x) \in \mathcal{O}_x\{t\}_{k(1), I_d(\kappa, \varepsilon)} \iff \varphi(x) \in \mathrm{Exp}(1; \Omega_\nu^\kappa(d, \varepsilon'))$$

for a sufficiently small $\varepsilon' > 0$, *where* $\Omega_\nu^\kappa(d, \varepsilon')$ *is given by* (6).

- *When $\ell \geq 2$, we have*

$$\hat{u}(t,x) \in \mathcal{O}_{t,x} \iff \varphi(x) \in \mathrm{Exp}(1; \Omega_\nu^\kappa(d,\varepsilon)).$$

Remark 4. This result can be applied to partial differential equations of non-Kowalevski type which are obtained as the limit $q \to 1$ when $\kappa < \nu$.

From Theorems 1 and 4, we have the following corollary.

Corollary 1. *We assume that $\varphi(x) \in \mathrm{Exp}(\nu/\ell; \mathbb{C})$. When $\ell \geq 2$, we have*

$$\varphi(x) \in \mathrm{Exp}(1; \Omega_\nu^\kappa(d,\varepsilon)) \iff \varphi(x) \in \mathrm{Exp}(1; \mathbb{C}).$$

Corollary 1 can be also proved directly by using a theorem of Phragmén [12, p. 342]. We omit the detail.

We give an important lemma of the summability for proving Theorem 5. [1,6,7].

Lemma 1. *Let $k > 0$, $d \in \mathbb{R}$ and $\hat{f}(t,x) \in \mathcal{O}_x[[t]]_{1/k}$. Then the following two statements are equivalent:*

(i) $\hat{f}(t,x) \in \mathcal{O}_x\{t\}_{k,d}$.
(ii) *Let $\{k_j\}_{j=1}^J$ $(k_j > 0, J \geq 1)$ satisfy*

$$\frac{1}{k} = \frac{1}{k_1} + \cdots + \frac{1}{k_J},$$

and we define $g(s,x)$ by iterated formal Borel transforms of $\hat{f}(t,x)$

$$g(s,x) := (\hat{\mathcal{B}}_{k_J} \circ \cdots \circ \hat{\mathcal{B}}_{k_1} \hat{f})(s,x), \tag{10}$$

where the formal k-Borel transform $\hat{\mathcal{B}}_k$ is defined as follows: for $\hat{f}(t,x) = \sum_{n \geq 0} f_n(x) t^n$, we define

$$(\hat{\mathcal{B}}_k \hat{f})(s,x) = \sum_{n \geq 0} f_n(x) \frac{s^n}{\Gamma(1 + n/k)} \in \mathcal{O}_{s,x}.$$

Then $g(s,x) \in \mathrm{Exp}(k; S(d,\varepsilon) \times B(r))$ for some $\varepsilon > 0$ and $r > 0$.

Moreover, under the condition (1), the k-sum $f(t,x)$ of $\hat{f}(t,x)$ is given by the iterated Laplace transforms of $g(s,x)$

$$f(t,x) = (\mathcal{L}_{k_1,d} \circ \cdots \circ \mathcal{L}_{k_J,d}g)(t,x), \tag{11}$$

where

$$(\mathcal{L}_{k,d}g)(t,x) := \frac{1}{t^k} \int_0^{\infty(d)} \exp\left(-\left(\frac{s}{t}\right)^k\right) g(s,x)d(s^k)$$

with the path taken from 0 to ∞ along $\arg s = d$. We write $\mathcal{L}_{k,d}g = \mathcal{L}_k g$ for short.

At the end of this section, we give the definition of Borel transform [1]. Let $S = S(d,\alpha,\rho)$ be a sector with $\alpha > \pi/k$. If $f(s)$ is analytic in S and is bounded at the origin, we define the k-Borel transform of f by

$$(\mathcal{B}_k f)(\tau) = \frac{-k}{2\pi i} \int_{\gamma_k(d)} f(s)e^{(\tau/s)^k} ds/s,$$

where $\gamma_k(d)$ denotes the path from the origin along $\arg s = d + (\varepsilon + \pi)/(2k)$ to some point s_1 with a positive ε, then along the circle $|s| = |s_1|$ to the ray $\arg s = d - (\varepsilon + \pi)/(2k)$, and back to the origin along this ray such that $\gamma_k(d) \subset S$.

4.   A Result of $k(\ell)$-Summability

Before giving a proof of Theorem 4, we give a result of $k(\ell)$-summability for the formal solution $\hat{u}(t,x)$ of (CP) and its proof.

Theorem 5. *Let $\varphi(x) \in \mathrm{Exp}(\nu/\ell;\mathbb{C})$ $(1 \le \ell \le \nu-1)$ for the Cauchy data and $d \in \mathbb{R}$. Then $\hat{u}(t,x) \in \mathcal{O}_x\{t\}_{k(\ell),d}$ $(k(\ell) = \kappa/(\nu-\ell))$ if $\varphi(x) \in \mathrm{Exp}(1;\Omega_\nu^\kappa(d,\varepsilon))$, where $\Omega_\nu^\kappa(d,\varepsilon)$ is given by (6) in Theorem 3.*

Remark 5. The results for partial differential equations corresponding to Theorem 4 for q-difference-differential equations have already been obtained [8,9]. In their results, the integral representation of the Borel sum was also given by using a special function as the kernel function.

Remark 6. We can find a necessary and sufficient condition for the $k(\ell)$-summability in different form by the following Balser argument [2,10,11].

Let

$$\hat{u}(t,x) = \sum_{n\geq 0} \varphi^{(\nu n)}(x) \frac{t^{\kappa n}}{[\kappa n]_q!} \in \mathcal{O}_x[[t]]_{1/k} \quad (k > 0)$$

be a formal solution of the Cauchy problem (CP). (The restriction for k such as $k(\ell)$ is not necessary.) We define

$$\hat{\psi}_j(x)(= (\partial_x^j \hat{u})(x,0)) = \sum_{n\geq 0} \varphi^{(\nu n+j)}(0) \frac{x^{\kappa n}}{[\kappa n]_q!} \quad (0 \leq j \leq \nu - 1).$$

Then we can prove that

$$\hat{u}(t,x) \in \mathcal{O}_x\{t\}_{k,d} \iff \hat{\psi}_j(x) \in \mathbb{C}\{x\}_{k,d} \quad (0 \leq j \leq \nu - 1). \quad (12)$$

In fact, the sufficiency of the condition is easily proved. For the proof of the necessity, we take the k-sum $\psi_j(x)$ of $\hat{\psi}_j(x)$. Then we can prove

$$u(t,x) = \sum_{j=0}^{\nu-1}\sum_{n\geq 0} D_{q,t}^{\kappa n}\psi_j(t) \frac{x^{\nu n+j}}{(\nu n + j)!}$$

is the k-sum of $\hat{u}(t,x)$. Here we have to mention that it is not an easy task to realize the condition (12) as those for the Cauchy data $\varphi(x)$ itself as in Theorem 5.

Proof of Theorem 5. We recall $k(\ell) = \kappa/(\nu - \ell)$ and we assume that

$$\varphi(x) \in \mathrm{Exp}\,(\nu/\ell; \mathbb{C})\,.$$

Let $v(s,x)$ be $(\nu - \ell)$ times iterated formal κ-Borel transforms of $\hat{u}$

$$v(s,x) := (\hat{\mathcal{B}}_\kappa^{\nu-\ell}\hat{u})(s,x) = \sum_{n\geq 0} \frac{\varphi^{(\nu n)}(x)}{[\kappa n]_q!} \frac{s^{\kappa n}}{n!^{\nu-\ell}}, \quad (13)$$

which is convergent in a neighborhood of $(s,x) = (0,0)$. Then for the proof of $k(\ell)$-summability of $\hat{u}(t,x)$ in a direction d, it is enough to prove

$$v(s,x) \in \mathrm{Exp}_s(k(\ell); S(d,\varepsilon_1) \times B(r)) \tag{14}$$

for some positive constants ε_1 and r under the assumption

$$\varphi(x) \in \mathrm{Exp}(1; \Omega_\nu^\kappa(d,\varepsilon)), \quad \Omega_\nu^\kappa(d,\varepsilon) = \bigcup_{j=0}^{\nu-1} S\left(\frac{\kappa d + 2\pi j}{\nu}, \varepsilon\right).$$

For that purpose, we further take ℓ times iterated κ-Borel transforms of $v(s,x)$ and put

$$w(\tau, x) := (\mathcal{B}_\kappa^\ell v)(\tau, x) = (\hat{\mathcal{B}}_\kappa^\nu \hat{u})(\tau, x) = \sum_{n \geq 0} \frac{\varphi^{(\nu n)}(x)}{[\kappa n]_q!} \frac{\tau^{\kappa n}}{n!^\nu}. \tag{15}$$

In this case, from the fact $v(s,x) \in \mathcal{O}_{s,x}$ or $\varphi(x) \in \mathrm{Exp}(\nu/\ell; \mathbb{C})$, we see that $w(\tau, x) \in \mathrm{Exp}_\tau(\kappa/\ell; \mathbb{C}^2)$. Moreover, we can prove the following lemma.

Lemma 2.

$$w(\tau, x) \in \mathrm{Exp}_\tau\left(\kappa/\nu; S(d,\varepsilon_1') \times B(r)\right) \tag{16}$$

for a sufficiently small $\varepsilon_1' > 0$.

By admitting Lemma 2, we immediately get the desired property (14) for $v(s,x)$. In fact, since $v(s,x)$ is given by

$$v(s,x) = (\mathcal{L}_\kappa^\ell w)(s,x),$$

we get the desired estimate for $v(s,x)$ by repeating the following lemma (without the proof [1]).

Lemma 3. *We assume $f(\tau) \in \mathrm{Exp}_\tau\left(p/q; S(d,\varepsilon)\right)$, where $p, q \in \mathbb{N}$. We put $F(s) := (\mathcal{L}_p f)(s)$. Then we have $F(s) \in \mathrm{Exp}_s(p/(q-1); S(d, \pi/p + \varepsilon'))$ with $\varepsilon' < \varepsilon$.*

Now, in order to complete the proof, we give a proof of Lemma 2, whose outlines are as follows. In the expression (15), after using the Cauchy integral formula for $\varphi(x)$, we represent the integral kernel function as the basic hypergeometric series. By using the representation of an analytic continuation of the kernel function, and after some calculations of residue, we obtain some representation of $w(\tau, x)$ which derives the desired estimate for w (see [5, §4] for the details).

We recall that $w(\tau, x)$ is given by

$$w(\tau, x) = \sum_{n \geq 0} \frac{\varphi^{(\nu n)}(x)}{[\kappa n]_q!} \frac{\tau^{\kappa n}}{n!^\nu}.$$

By using Cauchy integral formula, we get for some σ

$$w(\tau, x) = \frac{1}{2\pi i} \oint_{|\xi|=\sigma} \frac{\varphi(x + \xi)}{\xi} \sum_n \frac{(\nu n)!}{n!^\nu} \frac{1}{[\kappa n]_q!} \left(\frac{\tau^\kappa}{\xi^\nu}\right)^n d\xi.$$

Here since we have

$$\frac{(\nu n)!}{n!^\nu} = \frac{\nu^{\nu n}}{\boldsymbol{\Gamma}(\boldsymbol{\nu}/\nu)^2} \prod_{j=1}^{\nu-1} B\left(\frac{j}{\nu} + n, 1 - \frac{j}{\nu}\right)$$

$$= \frac{\nu^{\nu n}}{\boldsymbol{\Gamma}(\boldsymbol{\nu}/\nu)^2} \prod_{j=1}^{\nu-1} \int_0^1 y_j^{n+\frac{j}{\nu}-1}(1 - y_j)^{-\frac{j}{\nu}} dy_j,$$

where $\boldsymbol{\nu} = (1, 2, \ldots, \nu) \in \mathbb{N}^\nu$ and $\boldsymbol{\Gamma}(\boldsymbol{\nu}/\nu) = \prod_{j=1}^\nu \Gamma(j/\nu)$ $(= \prod_{j=1}^{\nu-1} \Gamma(j/\nu))$, we get

$$w(\tau, x) = \frac{1}{2\pi i \boldsymbol{\Gamma}(\boldsymbol{\nu}/\nu)^2} \prod_{j=1}^{\nu-1} \int_0^1 y_j^{\frac{j}{\nu}-1}(1 - y_j)^{-\frac{j}{\nu}} dy_j$$

$$\times \oint_{|\xi|=\sigma} \frac{\varphi(x + \xi)}{\xi} \sum_n \frac{1}{[\kappa n]_q!} \left(\frac{\nu^\nu \boldsymbol{y} \tau^\kappa}{\xi^\nu}\right)^n d\xi$$

$$= \frac{1}{2\pi i \boldsymbol{\Gamma}(\boldsymbol{\nu}/\nu)^2} \prod_{j=1}^{\nu-1} \int_0^1 y_j^{\frac{j}{\nu}-1}(1 - y_j)^{-\frac{j}{\nu}} dy_j$$

$$\times \oint_{|\xi|=\sigma} \frac{\varphi(x + \xi)}{\xi} {}_\kappa \phi_{\kappa-1}$$

$$\times \left(\begin{matrix} \mathbf{0}_\kappa \\ p^{1/\kappa}, p^{2/\kappa}, \ldots, p^{(\kappa-1)/\kappa} \end{matrix}; p, \frac{(1-q)^\kappa \nu^\nu \boldsymbol{y} \tau^\kappa}{\xi^\nu}\right) d\xi,$$

where $\boldsymbol{y} = y_1 y_2 \cdots y_{\nu-1}$, $\mathbf{0}_\kappa = (0, 0, \ldots, 0) \in \mathbb{C}^\kappa$, $p = q^\kappa$ and $(1-q)^\kappa \nu^\nu |\tau|^\kappa < \sigma^\nu$.

246   *K. Ichinobe*

From Proposition 1 with $\boldsymbol{a}_\kappa = \boldsymbol{0}_\kappa$ and $b = 0$, and the fact that $(c/a; p)_m a^m \to (-c)^m p^{m(m-1)/2}$ as $a \to 0$, we have

$$w(\tau, x) = \frac{(p; p)_\infty}{2\pi i \Gamma(\boldsymbol{\nu}/\nu)^2 (\boldsymbol{p}_\kappa; p)_\infty} \prod_{j=1}^{\nu-1} \int_0^1 y_j^{\frac{i}{\nu}-1} (1 - y_j)^{-\frac{i}{\nu}} dy_j$$

$$\times \oint_{|\xi|=\sigma} \frac{\varphi(x + \xi)}{\xi} \frac{1}{\left(\frac{(1-q)^\kappa \nu^\nu \boldsymbol{y} \tau^\kappa}{\xi^\nu}; p \right)_\infty}$$

$$\times \prod_{l=1}^{\kappa-1} \sum_{m_l \geq 0} \frac{\left(p^{M_l} \frac{(1-q)^\kappa \nu^\nu \boldsymbol{y} \tau^\kappa}{\xi^\nu}; p \right)_{m_l}}{(p; p)_{m_l}} (-1)^{m_l} p^{\frac{m_l(m_l-1)}{2} + l\frac{m_l}{\kappa}} d\xi,$$

where $\boldsymbol{p}_\kappa = (p^{1/\kappa}, p^{2/\kappa}, \ldots, p^{\kappa/\kappa})$ and $M_i = \sum_{j=0}^{i-1} m_j$ with $m_0 := 0$.

We remark that the integrand has poles at

$$\xi = \xi_{j,n}(\tau) := \omega_\nu^j (1 - q)^{\kappa/\nu} \nu \boldsymbol{y}^{1/\nu} \tau^{\kappa/\nu} p^{n/\nu} \quad (0 \leq j \leq \nu - 1, \ n \geq 0),$$

where $\omega_\nu = e^{2\pi i/\nu}$, and $\xi = 0$ is a removable singularity of the integrand. Therefore, after the residue calculations, we have

$$w(\tau, x) = \frac{1}{\nu \Gamma(\boldsymbol{\nu}/\nu)^2 (\boldsymbol{p}_\kappa; p)_\infty} \prod_{j=1}^{\nu-1} \int_0^1 y_j^{\frac{i}{\nu}-1} (1 - y_j)^{-\frac{i}{\nu}} dy_j$$

$$\times \sum_{j=1}^{\nu-1} \sum_{n \geq 0} \varphi(x + \omega_\nu^j (1 - q)^{\kappa/\nu} \nu \boldsymbol{y}^{1/\nu} \tau^{\kappa/\nu} p^{n/\nu})(-1)^n$$

$$\times \sum_{m_1=0}^{n-M_1} \cdots \sum_{m_{\kappa-1}=0}^{n-M_{\kappa-1}} \frac{p^{F_n}}{(p; p)_{m_1} \cdots (p; p)_{m_{\kappa-1}} (p; p)_{n-M_\kappa}}$$

$$=: W(\tau, x; \varphi),$$

where $F_n = F_n(m_1, \ldots, m_{\kappa-1})$ is given by

$$F_n := \sum_{l=1}^{\kappa-1} \left(\frac{m_l(m_l - 1)}{2} + \frac{lm_l}{\kappa} \right) + \frac{(n - M_\kappa)(n - M_\kappa + 1)}{2}. \quad (17)$$

Since $\max_{|x|\le r}|W(\tau,x;1)|$ is bounded (see [5, §4] for the detail), it is enough to check $\Phi_j(\tau) := \varphi(\omega_\nu^j \tau^{\kappa/\nu}) \in \mathrm{Exp}_\tau(\kappa/\nu; S(d,\varepsilon_1))$ ($j = 0,1,\ldots,\nu-1$), and we immediately see it from the assumption that $\varphi(x) \in \mathrm{Exp}_x(1;\Omega_\nu^\kappa(d,\varepsilon))$ with $\Omega_\nu^\kappa(d,\varepsilon) = \cup_{j=0}^{\nu-1}S((\kappa d + 2\pi j)/\nu,\varepsilon)$. $\square$

Remark 7. We see that for any j, $\Phi_j(\tau) \in \mathrm{Exp}(\kappa/\nu; S(d + 2\pi i/\kappa, \varepsilon_1))$ for $0 \le i \le \kappa - 1$. This means that under the assumption $\varphi(x) \in \mathrm{Exp}(1;\Omega_\nu^\kappa(d,\varepsilon))$, we have $\hat{u}(t,x) \in \mathcal{O}_x\{t\}_{k(\ell),d+2\pi i/\kappa}$ for $0 \le i \le \kappa - 1$.

5. Proof of Theorem 4

By Remark 7 in the previous section, we see that Lemma 2 is rewritten as follows:

$$w(\tau,x) \in \mathrm{Exp}_\tau\left(\kappa/\nu; S_\kappa(d,\varepsilon_1') \times B(r)\right),$$

where $S_\kappa(d,\varepsilon_1') = \cup_{i=0}^{\kappa-1}S(d + 2\pi i/\kappa, \varepsilon_1')$. Moreover, by Lemma 3, we have

$$v(s,x) = (\mathcal{L}_\kappa^\ell w)(s,x) \in \mathrm{Exp}_s(k(\ell); S_\kappa(d,\varepsilon_2 + \ell\pi/\kappa) \times B(r)), \quad (18)$$

where $\varepsilon_2 < \varepsilon_1'$.

5.1. *Proof of Theorem 4 when $\ell \ge 2$*

When $\ell \ge 2$, we see from (18) that $v(s,x) \in \mathrm{Exp}_s(k(\ell); \mathbb{C} \times B(r))$, which means that $\hat{u}(t,x) \in \mathcal{O}_{t,x}$. Inversely, if $\hat{u}(t,x) \in \mathcal{O}_{t,x}$, it is obvious that $\varphi(x) \in \mathrm{Exp}(1;\mathbb{C})$ from Theorem 1.

5.2. *Proof of the necessity of Theorem 4 when $\ell = 1$*

When $\ell = 1$, we have $v(s,x) \in \mathrm{Exp}_s(k(1); S_\kappa(d,\varepsilon_2 + \pi/\kappa) \times B(r))$. Therefore, by putting the interval $I_d(\kappa,\varepsilon')$ for a sufficiently small $\varepsilon' > 0$ by

$$I_d(\kappa,\varepsilon') = \left(d - \pi/(2\kappa) - \varepsilon', d + \pi/(2\kappa) + \varepsilon'\right),$$

we see that for any $\tilde{d} \in I_d(\kappa,\varepsilon')$, we have $\hat{u}(t,x) \in \mathcal{O}_x\{t\}_{k(1),\tilde{d}}$.

5.3.   *Proof of the sufficiency of Theorem 4 when $\ell = 1$*

We prove that $\varphi(x) \in \mathrm{Exp}(1; \Omega_\nu^\kappa(d, \varepsilon'))$ under the assumption $\hat{u}(t, x) \in \mathcal{O}_x\{t\}_{k(1), I_d(\kappa, \varepsilon)}$ for some positive ε' and ε, where $k(1) = \kappa/(\nu - 1)$ and

$$\Omega_\nu^\kappa(d, \varepsilon') = \bigcup_{j=0}^{\nu-1} S\left(\frac{\kappa d + 2\pi j}{\nu}, \varepsilon'\right)$$

$$I_d(\kappa, \varepsilon) = \left(d - \frac{\pi}{2\kappa} - \varepsilon, d + \frac{\pi}{2\kappa} + \varepsilon\right).$$

We note the following fact.

Proposition 3. *Let $\hat{u}(t, x)$ be the formal solution of* (CP). *Then the following statements are equivalent.*

(i) $\hat{u}(t, x) \in \mathcal{O}_x\{t\}_{k,d}.$
(ii) $\hat{u}(t, x) \in \mathcal{O}_x\{t\}_{k,d(i)}$ *with $d(i) = d + 2\pi i/\kappa$ for $0 \le i \le \kappa - 1$.*

In fact, it is easily seen from the fact that if we define σ_q by $\sigma_q f(t) = f(qt)$ for a function $f(t)$, then by induction we have

$$D_{q,t}^\kappa = \frac{\prod_{i=0}^{\kappa-1}(\sigma_q - q^i)}{(q-1)^\kappa q^{\frac{1}{2}\kappa(\kappa-1)} t^\kappa}.$$

Let $v(s, x) = (\hat{\mathcal{B}}_\kappa^{\nu-1}\hat{u})(s, x)$. Then $v(s, x)$ satisfies the following Cauchy problem which is obtained by the iterated formal Borel transforms of (CP)

$$\begin{cases} D_{q,s}^\kappa \widetilde{\delta_s}^{\,\nu-1} v(s, x) = \partial_x^\nu v(s, x), \\ v(0, x) = \varphi(x) \in \mathcal{O}_x(\mathbb{C}), \\ D_{q,s}^i v(0, x) = 0 \quad (1 \le i \le \kappa - 1), \end{cases} \qquad (19)$$

where $\widetilde{\delta_s} = (1/\kappa)s\partial_s$. In fact, it is deduced from the following commutative diagram:

$$
\begin{array}{ccc}
t^{\kappa n} & \xrightarrow{\ \hat{\mathcal{B}}_\kappa^{\nu-1}\ } & \dfrac{s^{\kappa n}}{n!^{\nu-1}} \\[2em]
{\scriptstyle D_{q,t}^\kappa}\Big\downarrow & & \Big\downarrow{\scriptstyle D_{q,s}^\kappa \widetilde{\delta_s}^{\,\nu-1}} \\[2em]
\dfrac{[\kappa n]_q!}{[\kappa(n-1)]_q!}t^{\kappa(n-1)} & \xrightarrow{\ \hat{\mathcal{B}}_\kappa^{\nu-1}\ } & \dfrac{[\kappa n]_q!}{[\kappa(n-1)]_q!}\dfrac{s^{\kappa(n-1)}}{(n-1)!^{\nu-1}}
\end{array}
$$

Then we remark that $v(s,x) \in \mathrm{Exp}_s(k(1); S_{\kappa,R}(d,\varepsilon_1 + \pi/\kappa) \times \mathbb{C})$ for some positive R and ε_1 since $\hat{u}(t,x) \in \mathcal{O}_x\{t\}_{k(1),I_{d(i)}(\kappa,\varepsilon)}$ for $0 \le i \le \kappa - 1$, where

$$S_{\kappa,R}(d,\alpha) := B(R) \cup S_\kappa(d,\alpha), \quad S_\kappa(d,\alpha) = \bigcup_{i=0}^{\kappa-1} S\left(d + \frac{2\pi i}{\kappa}, \alpha\right)$$

for $\alpha > 0$.

Now, we can regard $v(s,x)$ as a unique solution in a neighborhood at $(s,x) = (0,0)$ of the following Cauchy problem with respect to x direction:

$$\begin{cases} D_{q,s}^\kappa \widetilde{\delta}_s^{\,\nu-1} v(s,x) = \partial_x^\nu v(s,x), \\ \partial_x^j v(s,0) = \phi_j(s) \in \mathcal{O}_s \quad (0 \le j \le \nu - 1), \end{cases} \tag{20}$$

for some functions $\phi_j(s)$. Here, we remark that the equation (20) is of Kowalevski-type for q-difference-differential equations.

In this case, we can assume that $\phi_j(s) \in \mathrm{Exp}(k(1); S_{\kappa,R}(d,\varepsilon_1 + \pi/\kappa))$ for all j.

We put

$$w(\tau,x) := (\mathcal{B}_\kappa v)(\tau,x) = (\hat{\mathcal{B}}_\kappa^\nu \hat{u})(\tau,x) = \sum_{n\ge 0} \frac{\varphi^{(\nu n)}(x)}{[\kappa n]_q!} \frac{\tau^{\kappa n}}{n!^\nu}. \tag{21}$$

Then $w(\tau,x)$ satisfies the following CP:

$$\begin{cases} D_{q,\tau}^\kappa \widetilde{\delta}_\tau^{\,\nu-1} \circ (\widetilde{\delta}_\tau) w(\tau,x) = \partial_x^\nu w(\tau,x), \\ w(0,x) = \varphi(x) \in \mathcal{O}_x(\mathbb{C}), \\ D_{q,\tau}^i w(0,x) = 0 \quad (1 \le i \le \kappa - 1). \end{cases} \tag{22}$$

For this $w(\tau,x)$, we can regard $w(\tau,x)$ as a unique solution in a neighborhood at $(\tau,x) = (0,0)$ of the following CP with respect to x direction:

$$\begin{cases} D_{q,\tau}^\kappa \widetilde{\delta}_\tau^{\,\nu} w(\tau,x) = \partial_x^\nu w(\tau,x), \\ \partial_x^j w(\tau,0) = \psi_j(\tau) \quad (0 \le j \le \nu - 1). \end{cases} \tag{23}$$

Here, for $0 \le j \le \nu - 1$, $\psi_j(\tau)$ are given by κ-Borel transform of $\phi_j(s)$

$$\psi_j(\tau) = (\mathcal{B}_\kappa \phi_j)(\tau) = \frac{-\kappa}{2\pi i} \int_{\gamma_\kappa(\tilde{d})} \phi_j(s) e^{\left(\frac{\tau}{s}\right)^\kappa} \frac{ds}{s}, \tag{24}$$

where $\tilde{d} \in \mathbb{R}$ is arbitrary. In this case, since $\phi_j(s) \in \mathcal{O}_s$, we see that $\psi_j(\tau) \in \operatorname{Exp}(\kappa; \mathbb{C})$ for all j. Moreover, since $\phi_j(s) \in \operatorname{Exp}(k(1); S_{\kappa,R}(d, \varepsilon_1 + \pi/\kappa))$ $(0 \le j \le \nu - 1)$, we have the following lemma [1], which is immediately obtained from the integral representation (24).

Lemma 4.

$$\psi_j(\tau) \in \operatorname{Exp}(\kappa/\nu; S_\kappa(d, \varepsilon_1')) \quad (0 \le j \le \nu - 1) \tag{25}$$

for a sufficiently small ε_1'.

Let us prove $\varphi(x) = w(0, x) \in \operatorname{Exp}(1, \Omega_\nu^\kappa(d, \varepsilon'))$ under the assumptions (25).

In the following, we write $\psi_0(\tau)$ by $\psi(\tau)$ and we assume that $\psi_j(\tau) = 0$ $(1 \le j \le \nu - 1)$ without loss of generality.

Since $w(\tau, x)$ satisfies the CP (23), by putting $w(\tau, x) = \sum_{n \ge 0} w_n(\tau) x^n / n!$, we have

$$w(\tau, x) = \sum_{n \ge 0} (D_{q,\tau}^\kappa \widetilde{\delta_\tau}^\nu)^n \psi(\tau) \frac{x^{\nu n}}{(\nu n)!}.$$

Therefore, since $(D_{q,\tau}^\kappa \widetilde{\delta_\tau}^\nu)^n \psi(0) = \psi^{(\kappa n)}(0) n!^\nu [\kappa n]_q! / (\kappa n)!$ after some calculations, we have

$$\varphi(x) = w(0, x) = \sum_{n \ge 0} \frac{\psi^{(\kappa n)}(0)}{(\kappa n)!} \frac{n!^\nu}{(\nu n)!} [\kappa n]_q! x^{\nu n}. \tag{26}$$

Here, since we have

$$\frac{n!^\nu}{(\nu n + \nu - 1)!} = \frac{1}{\nu^{\nu-1+\nu n}} \prod_{j=1}^{\nu-1} B\left(n+1, \frac{j}{\nu}\right)$$

$$= \frac{1}{\nu^{\nu-1+\nu n}} \prod_{j=1}^{\nu-1} \eta_j^n (1 - \eta_j)^{1-\frac{i}{\nu}} d\eta_j,$$

we obtain the following expression for $\widetilde{\Phi}(x) := D_x^{-(\nu-1)}\varphi(x)$, where $D_x^{-1} = \int_0^x$:

$$\widetilde{\Phi}(x) = D_x^{-(\nu-1)}\varphi(x) = x^{\nu-1}\sum_{n\geq 0}\frac{\psi^{(\kappa n)}(0)}{(\kappa n)!}\frac{n!^\nu}{(\nu n + \nu - 1)!}[\kappa n]_q! x^{\nu n}$$

$$= \frac{x^{\nu-1}}{\nu^{\nu-1}}\prod_{j=1}^{\nu-1}\int_0^1 (1-\eta_j)^{1-\frac{i}{\nu}}d\eta_j \sum_{n\geq 0}\frac{\psi^{(\kappa n)}(0)}{(\kappa n)!}[\kappa n]_q!\left(\frac{\boldsymbol{\eta}x^\nu}{\nu^\nu}\right)^n,$$

where $\boldsymbol{\eta} = \eta_1\eta_2\cdots\eta_{\nu-1}$.

We put

$$f(x; Z, \psi) := \sum_{n\geq 0}\frac{\psi^{(\kappa n)}(0)}{(\kappa n)!}[\kappa n]_q!(Zx^\nu)^n. \tag{27}$$

By using Cauchy integral formula and the formula

$$[\kappa n]_q! = \frac{(1-q^{\kappa n})(1-q^{\kappa n-1})\cdots(1-q)}{(1-q)^{\kappa n}} = \frac{(q^\kappa, q^{\kappa-1},\ldots,q;q^\kappa)_n}{(1-q)^{\kappa n}},$$

we get

$$f(x; Z, \psi) = \frac{1}{2\pi i}\oint_{|\zeta|=\rho}\frac{\psi(\zeta)}{\zeta}\sum_{n\geq 0}(q^\kappa, q^{\kappa-1},\ldots,q;q^\kappa)_n\left(\frac{Zx^\nu}{(1-q)^\kappa\zeta^\kappa}\right)^n d\zeta$$

$$= \frac{1}{2\pi i}\oint_{|\zeta|=\rho}\frac{\psi(\zeta)}{\zeta}{}_{\kappa+1}\phi_\kappa\left(\begin{matrix}\boldsymbol{p}_\kappa, p\\ \boldsymbol{0}_\kappa\end{matrix}; p, \frac{Zx^\nu}{(1-q)^\kappa\zeta^\kappa}\right)d\zeta,$$

where $\rho > |Zx^\nu|^{1/\kappa}/(1-q)$ and we put $q^\kappa = p$ and $\boldsymbol{p}_\kappa = (p^{1/\kappa}, p^{2/\kappa},\ldots,p)$.

From Proposition 1 with $\mathbf{c}_\kappa = \boldsymbol{0}_\kappa$, we have

$$f(x; Z, \psi) = \frac{(\boldsymbol{p}_\kappa; p)_\infty}{2\pi i}\oint_{|\zeta|=\rho}\frac{\psi(\zeta)}{\zeta}\frac{\left(p\cdot\frac{Zx^\nu}{(1-q)^\kappa\zeta^\kappa}; p\right)_\infty}{\left(\frac{Zx^\nu}{(1-q)^\kappa\zeta^\kappa}; p\right)_\infty}$$

$$\times \prod_{i=1}^\kappa\sum_{m_i\geq 0}\frac{\left(p^{M_i}\cdot\frac{Zx^\nu}{(1-q)^\kappa\zeta^\kappa}; p\right)_{m_i}}{\left(p^{1+M_i}\cdot\frac{Zx^\nu}{(1-q)^\kappa\zeta^\kappa}, p; p\right)_{m_i}}p^{\frac{i}{\kappa}m_i}d\zeta.$$

252   K. Ichinobe

Since the integrand has simple poles at

$$\zeta = \zeta_{i,n}(x) := \omega_\kappa^i Z^{1/\kappa} x^{\nu/\kappa} p^{n/\kappa}/(1-q) \quad 0 \le i \le \kappa - 1, \quad n \ge 0,$$

$$\tag{28}$$

we have after some residue calculations

$$f(x; Z, \psi) = \frac{(\boldsymbol{p}_\kappa; p)_\infty}{\kappa} \sum_{i=0}^{\kappa-1} \sum_{n \ge 0} \psi \left(\frac{\omega_\kappa^i Z^{1/\kappa} p^{n/\kappa}}{1-q} x^{\nu/\kappa} \right)$$

$$\times \sum_{M_{\kappa+1}=n} \frac{\prod_{i=1}^\kappa p^{\frac{i}{\kappa} m_i}}{\prod_{i=1}^\kappa (p; p)_{m_i}},$$

where $M_{\kappa+1} = m_1 + m_2 + \cdots + m_\kappa$. Since $|f(x; Z, 1)|$ is bounded
(see [5, §5] for the detail), we obtain $f(x; 1, \psi) \in \mathrm{Exp}_x(1; \Omega_\nu^\kappa(d, \varepsilon'))$
under the assumption $\psi(\tau) \in \mathrm{Exp}(\kappa/\nu; S_\kappa(d, \varepsilon_1'))$ for some positive
ε'. Therefore, we have $\widetilde{\Phi}(x) \in \mathrm{Exp}(1; \Omega_\nu^\kappa(d, \varepsilon'))$ and then $\varphi(x)$ also
has the same property as $\widetilde{\Phi}(x)$.

References

1. W. Balser, *From Divergent Power Series to Analytic Functions.*
   Springer Lecture Notes, No. 1582, 1994.
2. W. Balser, Divergent solutions of the heat equations: On the article of
   Lutz, Miyake and Schäfke. *Pacific J. Math.*, **188**(1999), 53–63.
3. G. Gasper and M. Rahman, *Basic Hypergeometric Series,* 2nd edn.
   Cambridge University Press, Cambridge, 2004.
4. K. Ichinobe, Gevrey asymptotic solutions to the Cauchy problem
   of some linear q-difference-differential equations, to appear in *RIMS
   Kôkyûroku.*
5. K. Ichinobe and S. Adachi, On k-summability of formal solutions to
   the Cauchy problems of some linear q-difference-differential equations.
   Complex Diff. Diff. Equn, De Gruyter, Proc. Math., (2020), 447–464.
6. D. A. Lutz, M. Miyake and R. Schäfke, On the Borel summability of
   divergent solutions of the heat equation. *Nagoya Math. J.*, **154**(1999),
   1–29.
7. M. Miyake, Borel summability of divergent solutions of the Cauchy
   problem to non-Kowalevski equations, *Partial Differential Equations
   and Their Applications.* World Scientific Publishing, Wuhan, 1999,
   pp. 225–239.

8. M. Miyake and K. Ichinobe, Hierarchy of partial differential equations and fundamental solutions associated with summable formal solutions of a partial differential equation of non Kowalevski type. *Differential equations and Asymptotic Theory in Mathematical Physics. World Scientific Publishing, Hackensack; NJ, Ser. Anal.*, **2**(2004), 330–342.

9. M. Miyake and K. Ichinobe, A Remark on k-summability of divergent solutions of a non-Kowalevski type equation with Cauchy data of entire functions. *Global and Asymptotic Analysis of Differential Equations in the Complex Domain, Sūrikaisekikenkyūsho Kōkyūroku*, No. 1367, (2004), 59–72.

10. P. Remy, Gevrey order and summability of formal series solutions of some classes of inhomogeneous linear partial differential equations with variable coefficients. *J. Dyn Control Syst.*, **22**(4) (2016), 693–711.

11. P. Remy, Gevrey order and summability of formal series solutions of certain classes of inhomogeneous linear integro-differential equations with variable coefficients. *J. Dyn Control Syst.*, **23**(4) (2017), 853–878.

12. G. Sansone and J. Gerretsen, *Lectures on the Theory of Functions of a Complex Variable*. P. Noordhoff, Groninger, 1960.

13. A. Shirai and K. Ichinobe, Maillet type theorem for nonlinear q-difference-differential equations of Kowalevski type. *J. School Educ. Sugiyama Jogakuen Univ.*, **13**(2020), 189–198.

© 2022 World Scientific Publishing Europe Ltd.
https://doi.org/10.1142/9781800611368_0012

Chapter 12

Multi-specializations and Minimal Asymptotics

Naofumi Honda[*,§], Shingo Kamimoto[†,¶] and Luca Prelli[‡,‖]

Faculty of Science
Department of Mathematics
Hokkaido University
Sapporo 060-0810, Japan
[†]*Graduate School of Sciences*
Hiroshima University
1-3-1 Kagamiyama, Higashi-Hiroshima
Hiroshima 739-8526, Japan
[‡]*Department of Mathematics*
Padova University
Via Trieste 63, 35121 Padova, Italy
[§]*honda@math.sci.hokudai.ac.jp*
[¶]*kamimoto@hiroshima-u.ac.jp*
[‖]*lprelli@math.unipd.it*

The purpose of this chapter is to continue the algebraic study of asymptotics. In [2], the functorial nature of Majima strong asymptotic was established applying multi-specialization to the subanalytic sheaf of Whitney holomorphic functions. The functorial nature of flat asymptotics and asymptotics data proven as well. In [5], the author introduced another kind of asymptotically developable function, whose data have an holomorphic extension to the origin (minimal asymptotics). In order to build this object we need to modify the functorial construction of the asymptotic data. This is possible thanks to a natural functor relating classical and subanalytic sheaves.

1. Sheaves on Subanalytic Sites

The results of this section are extracted from [4] (see also [6] for a more detailed study).

Let X be a real analytic manifold and let k be a field. Denote by $\mathrm{Op}(X_{sa})$ the category of open subanalytic subsets of X. One endows $\mathrm{Op}(X_{sa})$ with the following topology: $S \subset \mathrm{Op}(X_{sa})$ is a covering of $U \in \mathrm{Op}(X_{sa})$ if for any compact K of X there exists a finite subset $S_0 \subset S$ such that $K \cap \bigcup_{V \in S_0} V = K \cap U$. We will call X_{sa} the subanalytic site.

Let $\mathrm{Mod}(k_{X_{sa}})$ denote the category of sheaves on X_{sa}. We denote by $\rho : X \to X_{sa}$ the natural morphism of sites. We have functors

$$\mathrm{Mod}(k_X) \underset{\rho^{-1}}{\overset{\rho_*}{\rightleftarrows}} \mathrm{Mod}(k_{X_{sa}}).$$

The functor ρ^{-1} admits a left adjoint, denoted by $\rho_!$ which is fully faithful and exact. The sheaf $\rho_! F$ is the sheaf associated to the presheaf

$$\mathrm{Op}(X_{sa}) \ni U \mapsto F(\overline{U}).$$

Let X, Y be two real analytic manifolds, and let $f : X \to Y$ be a real analytic map. We get the internal operations $\mathcal{H}om$, $\otimes$ and the external operations f^{-1} and f_*, which are always defined for sheaves on Grothendieck topologies. For subanalytic sheaves, we can also define the functor of proper direct image $f_{!!}$. In the derived category it is possible to define an exceptional inverse image $f^!$, right adjoint to $Rf_{!!}$ and we get the usual isomorphisms like projection formula, base change formula, Künneth formula.

2. Multi-normal Deformation and Multi-specialization

The results of this section are extracted from [2]. For simplicity, we assume $X = \mathbb{C}^n$, with coordinates $z = (z_1, \ldots, z_n)$. Set $M_i = \{z_i = 0\}$, $i = 1, \ldots, n$. We can define a multi-normal deformation $\widetilde{X} = \mathbb{C}^n \times \mathbb{R}^n$ with the map $p : \widetilde{X} \to X$ defined by

$$p(z, t) = (t_1 z_1, \ldots, t_n z_n).$$

We are interested in the zero section S of $\widetilde{X}$ defined by $\{t_i = 0,\ i = 1,\ldots,n\}$. We have $S \simeq T_{M_1}X \times_X \cdots \times_X T_{M_n}X$. Let $s : S \hookrightarrow \widetilde{X}$ be the inclusion, $\Omega = \{t_1,\ldots,t_\ell > 0\}$, $M = \bigcap_{i=1}^n M_i = \{0\}$. We get a commutative diagram

$$
\begin{array}{ccc}
S & \overset{s}{\longrightarrow} \widetilde{X} \overset{i_\Omega}{\longleftarrow} \Omega \\
\downarrow{\scriptstyle \tau} & \downarrow{\scriptstyle p} \,\,\swarrow{\scriptstyle \widetilde{p}} \\
M & \overset{i}{\longrightarrow} X.
\end{array}
$$

The multi-specialization along $M_1,\ldots,M_n$ is the functor

$$
\nu : D^b(k_{X_{sa}}) \to D^b(k_S)
$$
$$
F \mapsto \rho^{-1}s^{-1}R\Gamma_\Omega p^{-1}F.
$$

Here, $D^b(\cdot)$ denotes the bounded derived category of sheaves.

Thanks to the functor $\rho^{-1} : D^b(k_{S_{sa}}) \to D^b(k_S)$ we can calculate the fibers at $\xi = (\xi_1,\ldots,\xi_n) \in T_{M_1}X \times_X \cdots \times_X T_{M_n}X$ which are given by

$$
(H^j\nu F)_\xi \simeq \varinjlim_W H^j(W; F),
$$

where $W = W_1 \times \cdots \times W_n$ ranges through the family of polysectors such that $\xi_i \in W_i$, $i = 1,\ldots,n$.

3. Strong Asymptotics

In this section, we recall the definition of strong asymptotics, refer to [5] for more details. Let $\mathcal{P}_n$ be the set of non-empty subsets of $\{1,\ldots,n\}$. Let $J \in \mathcal{P}_n$ and assume the following notations:

- $M_J = \bigcap_{j \in J} M_j$,
- $z_J = (z_i)_{i \in J}$, $z_J^C = (z_i)_{i \notin J}$,
- $\mathbb{N}_0^J = \{(\alpha_1,\ldots,\alpha_n) \in \mathbb{N}_0^n,\ \alpha_i = 0,\ i \notin J\}$,
- $\pi_J : X \to M_J$ the projection,
- given $W \subset X$, $W_J = \pi_J(W)$.

Let $W := W_1 \times \cdots \times W_n$ be a polysector. We say that $F = \{F_J\}_{J \in \mathcal{P}_n}$ is a total family of coefficients of multi-asymptotic expansion on W if each F_J consists of a family $\{f_{J,\alpha}\}_{\alpha \in \mathbb{N}_0^J}$ of holomorphic functions on W_J. Given a total family of coefficients $F = \{F_J\}_{J \in \mathcal{P}_n}$ and $N = (n_1, \ldots, n_n) \in \mathbb{N}_0^n$, the approximate function of degree N of F is

$$\mathrm{App}^{<N}(F; z) = \sum_{J \in \mathcal{P}_n} (-1)^{\sharp J + 1} \sum_{\alpha \in A_J(N)} \frac{f_{J,\alpha}(z_J^C)}{\alpha!} z^\alpha,$$

where

$$A_J(N) = \left\{ \alpha \in \mathbb{N}_0^J;\ \alpha_j < n_j \text{ for any } j \in J \right\}.$$

A holomorphic function f on X is strongly multi-asymptotically developable to $F = \{F_J\}$ on a polysector W if and only if for any polysector W' properly contained in S and for any $N = (n_1, \ldots, n_n) \in \mathbb{N}_0^n$ there exists a constant C such that

$$\left| f(z) - \mathrm{App}^{<N}(F; z) \right| \leq C \prod_{1 \leq j \leq n} (|z_j|)^{n_j} \quad (z \in W').$$

4. Multi-specialization and Strong-Asymptotics

Let $U \subset X$ be a relatively compact subanalytic open subset, a $\mathcal{C}^\infty$-function f on U is Whitney if it is bounded with bounded derivatives. Such functions do not form a sheaf on X with the usual topology, because they do not satisfy gluing conditions. They define a sheaf once we replace the topological space X with the subanalytic site X_{sa}: roughly speaking, we consider only subanalytic open subsets and locally finite covers. We denote by $\mathcal{C}_X^{\infty,w}$ the subanalytic sheaf of Whitney functions. The sheaf $\mathcal{O}_X^w$ of Whitney holomorphic functions on a complex manifold is the Dolbeaut complex with coefficients in $\mathcal{C}_X^{\infty,w}$.

Set $D := \cup_{j=1}^n M_j$. Strongly asymptotically developable functions on a polysector W are Whitney [1] on each polysector W' such that $\overline{W'} \setminus D \subset W$. Moreover, the geometrical properties of a polysector imply the vanishing of the cohomology of multi-specialization, see [2] for more details. Hence, the sheaf $\nu\mathcal{O}_X^w$ is concentrated in degree zero and, given a multi-direction $\xi = (\xi_1, \ldots, \xi_n) \in T_{M_1}X \times_X \cdots \times_X T_{M_n}X$, we have

$$(\nu\mathcal{O}_X^w)_\xi = \varinjlim_W \{\text{Strongly asymptotically developable functions on } W\},$$

where $W = W_1 \times \cdots \times W_n$ ranges through the family of polysectors such that $\xi_i \in W_i$, $i = 1, \ldots, n$.

Let $\mathcal{O}_X^w$, $\mathcal{O}_{X|X\backslash D}^w$, $\mathcal{O}_{X|D}^w$ denote the sheaves on the subanalytic site X_{sa} of Whitney holomorphic functions, flat Whitney holomorphic functions and Whitney holomorphic functions on D, respectively. Thanks to the vanishing of the cohomology of flat asymptotics ([2, Proposition 8.4]), the sequence

$$0 \to \nu\mathcal{O}_{X|X\backslash D}^w \to \nu\mathcal{O}_X^w \to \nu\mathcal{O}_{X|D}^w \to 0 \tag{1}$$

is exact. In the case of Majima's asymptotics, we have the isomorphisms (outside the zero section)

$$\mathcal{A}_X \xrightarrow{\sim} \nu\mathcal{O}_X^w,$$

$$\mathcal{A}_X^{\leq 0} \xrightarrow{\sim} \nu\mathcal{O}_{X|X\backslash D}^w,$$

$$\mathcal{A}_X^{CF} \xrightarrow{\sim} \nu\mathcal{O}_{X|D}^w,$$

where as usual, we denote by $\mathcal{A}_X$, $\mathcal{A}_X^{\leq 0}$, $\mathcal{A}_X^{CF}$ the sheaves of strongly asymptotically developable functions, flat asymptotics and consistent families of coefficients. So (1) is the Borel–Ritt exact sequence for strong asymptotics.

5. Minimal Asymptotics

Now we would like to restrict the family of coefficients and consider the convergent ones, namely, we would like to restrict $\mathcal{A}_X^{CF}$ to the sheaf $\mathcal{O}_{X|\widehat{D}} = \varprojlim_k \mathcal{O}_X/\mathcal{I}_D^k\mathcal{O}_X$ (formal completion of $\mathcal{O}_X$ along D). From a sheaf theoretical point of view, this corresponds to replacing $\mathcal{O}_{X|D}^w$ with a subsheaf.

To perform the construction, we need first to consider Whitney $\mathcal{C}^\infty$-functions. Let $\mathcal{C}_X^{\infty,w}$, $\mathcal{C}_{X|X\backslash D}^{\infty,w}$, $\mathcal{C}_{X|D}^{\infty,w}$ denote the sheaves on the subanalytic site X_{sa} of Whitney $\mathcal{C}^\infty$-functions, flat Whitney $\mathcal{C}^\infty$-functions and Whitney $\mathcal{C}^\infty$-functions on D, respectively (see [2] for a more detailed construction). There is an exact sequence

$$0 \to \rho_! \rho^{-1} \mathcal{C}^{\infty,w}_{X|D} \to \mathcal{C}^{\infty,w}_{X|D} \to \mathcal{C}^{\infty,w}_{X|D} / \rho_! \rho^{-1} \mathcal{C}^{\infty,w}_{X|D} \to 0.$$

Let $\mathcal{C}^{\infty,w}_{X|\cup D}$ be the kernel of the composition

$$\mathcal{C}^{\infty,w}_X \to \mathcal{C}^{\infty,w}_{X|D} \to \mathcal{C}^{\infty,w}_{X|D} / \rho_! \rho^{-1} \mathcal{C}^{\infty,w}_{X|D}.$$

A general result of homological algebra implies the exact sequence

$$0 \to \mathcal{C}^{\infty,w}_{X|X\setminus D} \to \mathcal{C}^{\infty,w}_{X|\cup D} \to \rho_! \rho^{-1} \mathcal{C}^{\infty,w}_{X|D} \to 0.$$

Define $\mathcal{O}^w_{X|\cup D}$ as the Dolbeaut complex with coefficients in $\mathcal{C}^{\infty,w}_{X|\cup D}$. We have a distinguished triangle

$$\mathcal{O}^w_{X|X\setminus D} \to \mathcal{O}^w_{X|\cup D} \to \rho_! \rho^{-1} \mathcal{O}^w_{X|D} \xrightarrow{+} .$$

Applying the multi-specialization functor, we get the triangle

$$\nu \mathcal{O}^w_{X|X\setminus D} \to \nu \mathcal{O}^w_{X|\cup D} \to \nu \rho_! \rho^{-1} \mathcal{O}^w_{X|D} \xrightarrow{+} .$$

The third term of the triangle is isomorphic to $(\mathcal{O}_{X\widehat{|}D})_0$. That is because, given a multi-direction ξ

$$\varinjlim_{\xi \in W} \mathcal{O}^w_{X|D}(\overline{W}) \simeq \varinjlim_{0 \in U} \mathcal{O}^w_{X|D}(U) \simeq (\mathcal{O}_{X\widehat{|}D})_0,$$

which is concentrated in degree zero. The last isomorphism follows from the definition of $\mathcal{O}^w_{X|D}$ and a result of [3]. Then $\nu \mathcal{O}^w_{X|\cup D}$ is concentrated in degree zero and the sequence

$$0 \to \nu \mathcal{O}^w_{X|X\setminus D} \to \nu \mathcal{O}^w_{X|\cup D} \to \nu \rho_! \rho^{-1} \mathcal{O}^w_{X|D} \to 0 \tag{2}$$

is exact. In the case of Majima's asymptotics, we have the isomorphisms (outside the zero section)

$$\mathcal{A}'_X \xrightarrow{\sim} \nu \mathcal{O}^w_{X|\cup D},$$

$$\mathcal{A}^{\leq 0}_X \xrightarrow{\sim} \nu \mathcal{O}^w_{X|X\setminus D},$$

$$(\mathcal{O}_{X\widehat{|}D})_0 \xrightarrow{\sim} \nu \rho_! \rho^{-1} \mathcal{O}^w_{X|D},$$

where as usual, we denote by $\mathcal{A}'_X$, $\mathcal{A}^{\leq 0}_X$ the sheaves of functions strongly asymptotically developable to $\mathcal{O}_{\widehat{X|D}}$ and flat asymptotics. So (2) is the Borel–Ritt exact sequence for minimal asymptotics.

References

1. F. Galindo and J. Sanz, On strongly asymptotically developable functions and the Borel-Ritt theorem. *Studia Math.*, **133**(1999), 231–248.
2. N. Honda and L. Prelli, Multi-specialization and multi-asymptotic expansions. *Adv. Math.*, **232**(2013), 432–498.
3. M. Kashiwara and P. Schapira, Moderate and formal cohomology associated with constructible sheaves. *Mémoires Soc. Math. France*, **64**(1996).
4. M. Kashiwara and P. Schapira, Ind-sheaves. *Astérisque*, **271**(2001).
5. H. Majima, *Asymptotic Analysis for Integrable Connections with Irregular Singular Points* (Lecture Notes in Math. 1075). Springer-Verlag, Berlin, 1984.
6. L. Prelli, Sheaves on subanalytic sites. *Rend. Sem. Mat. Univ. Padova*, **120**(2008), 167–216.

© 2022 World Scientific Publishing Europe Ltd.
https://doi.org/10.1142/9781800611368_0013

Chapter 13

On a Connection Problem for the Generalized Hypergeometric Equation

Shunya Adachi

Graduate School of Mathematical Science, Kumamoto University
2-39-1 Kurokami, Chuo-ku, Kumamoto, 860-8555, Japan
200d7101@st.kumamoto-u.ac.jp

We study a connection problem between the fundamental systems of solutions at singular points 0 and 1 for the generalized hypergeometric equation satisfied by the generalized hypergeometric series $_nF_{n-1}$. In general, the local solution space around $x = 1$ consists of one-dimensional singular solution space and $(n - 1)$-dimensional holomorphic solution space. Therefore, in the case of $n \geq 3$, the expression of connection matrix depends on the choice of the fundamental system of solutions at $x = 1$. On the study of connection problem for ordinary differential equations, Schäfke and Schmidt (LNM **810**, Springer, 1980) gave an impressive idea which focuses on the series expansion of fundamental system of solutions. We apply their idea to solve the connection problem for the generalized hypergeometric equation and derive the connection matrix.

1. Introduction

We consider the *generalized hypergeometric differential equation*:

$$\left[\delta \prod_{i=1}^{n-1} (\delta + \beta_i - 1) - x \prod_{i=1}^{n} (\delta + \alpha_i) \right] y = 0, \tag{E}$$

where $x, \alpha_i, \beta_i \in \mathbb{C}$ and $\delta = x\frac{d}{dx}$. This equation is an n-th order linear differential equation which has three regular singular points at $x = 0, 1, \infty$. The Riemann scheme of (E) is given by

$$\left\{ \begin{array}{ccc} x = 0 & x = 1 & x = \infty \\ 0 & 0 & \alpha_1 \\ 1 - \beta_1 & 1 & \alpha_2 \\ \vdots & \vdots & \vdots \\ 1 - \beta_{n-2} & n - 2 & \alpha_{n-1} \\ 1 - \beta_{n-1} & -\beta_n & \alpha_n \end{array} \right\}, \tag{1}$$

where β_n is defined by $\alpha_1 + \cdots + \alpha_n = \beta_1 + \cdots + \beta_n$. The generalized hypergeometric equation is rigid, i.e. the linear ordinary differential equation whose Riemann scheme is given by (1) is nothing but the generalized hypergeometric equation (E).

Throughout this chapter, we assume

$$\beta_i, \ \beta_i - \beta_j \notin \mathbb{Z} \tag{2}$$

for $1 \leq i, j \leq n$ with $i \neq j$. Under this assumption, a fundamental system of solutions (i.e. basis of local solution space) around the origin is given by

$$y_1^{[0]}(x) = {}_nF_{n-1}(\boldsymbol{\alpha}_0; \boldsymbol{\beta}_0; x),$$

$$y_{i+1}^{[0]}(x) = x^{1-\beta_i} {}_nF_{n-1}(\boldsymbol{\alpha}_i; \boldsymbol{\beta}_i; x),$$

where $\boldsymbol{\alpha}_0 = (\alpha_1, \alpha_2, \ldots, \alpha_n)$, $\boldsymbol{\beta}_0 = (\beta_1, \ldots, \beta_{n-1})$ and

$$\boldsymbol{\alpha}_i = (\alpha_1 + 1 - \beta_i, \alpha_2 + 1 - \beta_i, \ldots, \alpha_n + 1 - \beta_i),$$

$$\boldsymbol{\beta}_i = (\beta_1 + 1 - \beta_i, \ldots, 2 - \beta_i, \ldots, \beta_{n-1} + 1 - \beta_i)$$

for $1 \leq i \leq n - 1$. Here, the symbol ${}_nF_{n-1}$ denotes the *generalized hypergeometric series*:

$$_nF_{n-1}\left(\begin{array}{c} \alpha_1, \alpha_2, \ldots, \alpha_n \\ \beta_1, \ldots, \beta_{n-1} \end{array}; x\right) = \sum_{m \geq 0} \frac{(\alpha_1)_m (\alpha_2)_m \cdots (\alpha_n)_m}{(\beta_1)_m \cdots (\beta_{n-1})_m m!} x^m. \tag{3}$$

This series converges on $x \in D_0 := \{x \in \mathbb{C}; |x| < 1\}$ and the symbol $(a)_m$ is defined by

$$(a)_m = \frac{\Gamma(a + m)}{\Gamma(a)} = \begin{cases} 1 & m = 0, \\ a(a + 1) \cdots (a + m - 1) & m \geq 1. \end{cases}$$

On the other hand, a fundamental system of solutions around $x = 1$ is given by

$$y_i^{[1]}(x) = (1 - x)^{i-1} \sum_{m \geq 0} d_m^{(i)} (1 - x)^m, \quad d_0^{(i)} \neq 0,$$

$$y_n^{[1]}(x) = (1 - x)^{-\beta_n} \sum_{m \geq 0} d_m^{(n)} (1 - x)^m, \quad d_0^{(n)} \neq 0 \tag{4}$$

for $1 \leq i \leq n - 1$. Here, $y_1^{[1]}(x), y_2^{[1]}(x), \ldots, y_{n-1}^{[1]}(x)$ and the power series in $y_n^{[1]}$ are convergent on $D_1 = \{x \in \mathbb{C}\,;\, |1 - x| < 1\}$.

Remark 1.

(1) To determine the fundamental system of solutions $y_i^{[1]}(x)\,(1 \leq i \leq n - 1)$, we have to fix the coefficients $\{d_j^{(i)}\}_{j=0}^{n-1-i}$. Then the other coefficients $\{d_j^{(i)}\}_{j \geq n-i}$ are determined uniquely. On the other hand, the solution $y_n^{[1]}(x)$ is determined by only fixing $d_n^{(0)}$.

(2) In the case of $n = 2$, the fundamental system of solutions (4) is expressed by Gauss hypergeometric series ${}_2F_1$.

In this chapter, we consider the connection problem for the fundamental systems of solutions at singular points $x = 0$ and 1. That is, we consider the following problem.

Problem 1 (Connection problem). *We set n-row vectors as*

$$\mathcal{Y}^{[0]} = (y_1^{[0]}, y_2^{[0]}, \ldots, y_n^{[0]}), \quad \mathcal{Y}^{[1]} = (y_1^{[1]}, y_2^{[1]}, \ldots, y_n^{[1]}).$$

Then determine the connection matrix $C \in \mathrm{GL}(n, \mathbb{C})$ such that

$$\mathcal{Y}^{[0]} = \mathcal{Y}^{[1]} C, \quad x \in D_0 \cap D_1 \tag{5}$$

with $\arg x = \arg(1 - x) = 0$ on $0 < x < 1$.

Connection problem of the generalized hypergeometric equation between $x = 0$ and ∞ is studied well (cf. [1,3,5,6,9]). Recently, Oshima [7] studied the connection problem for more general Fuchsian equations from the viewpoint of the theory of middle convolution.

On the connection problem of the generalized hypergeometric equation between $x = 0$ and 1, there is a difficulty that did not appear in the case of between $x = 0$ and ∞: As is seen in (4), the solution space around $x = 1$ of (E) consists of one-dimensional singular solution space and $(n - 1)$-dimensional holomorphic solution space. Therefore, in the case of $n \geq 3$, the expression of the connection matrix depends on the choice of the fundamental system of solutions at $x = 1$.

Mimachi [5] considered the connection problem between singular solution at $x = 1$ and the fundamental system of solutions at $x = 0$ in the case of $n = 3$, and later he solved the same problem in more general case [6]. Matsuhira–Nagoya [4] considered the connection problem between the fundamental systems of solutions and derived the connection matrix.

In the above studies, the connection problem was solved by using the integral representation of solutions and the notion of twisted cycle. That is, Matsuhira–Nagoya's result gives an answer to Problem 1 in terms of the integral representation of solutions.

In this chapter, we solve Problem 1 from the viewpoints of series representation of local solutions. In doing this, Schäfke–Schmidt's work [8] plays an important role: They focused on the series representation of local solutions, and gave an impressive idea about connection problem for general linear differential equations.

Theorem 1 ([8]). *We consider the linear differential equation*

$$p_0(x)y^{(n)} + p_1(x)y^{(n-1)} + \cdots + p_n(x)y = 0,$$

which has regular singular points $x = 0, 1$ and doesn't have any other singular points on $|x| \leq 1$. If this equation has local solutions

$$y^{[0]}(x) = x^\alpha \sum_{m \geq 0} a_m x^m$$

on $D_0 \setminus \{0\}$ and

$$y_j^{[1]}(x) = (1 - x)^{\alpha_j} \sum_{m \geq 0} d_m^{(j)} (1 - x)^m, \quad 1 \leq j \leq n,$$

on $D_1 \setminus \{1\}$, then the following asymptotic formula for a_m holds:

$$a_m = \sum_{j=1}^{n} \frac{\Gamma(m + \alpha - \alpha_j)}{\Gamma(m + \alpha + 1)} \left\{ \sum_{\ell=0}^{k} \left(\prod_{s=1}^{\ell} \frac{-s - \alpha_j}{m + \alpha - s - \alpha_j} \right) d_\ell^{(j)} \right\} \frac{c_j}{\Gamma(-\alpha_j)} + O(m^{-\alpha - -k - 2})$$

as $m \to \infty$, where $c_1, c_2, \ldots, c_n \in \mathbb{C}$ are the connection coefficients such that

$$y^{[0]}(x) = \sum_{j=1}^{n} c_j y_j^{[1]}(x), \quad 0 < x < 1$$

with $\arg x = \arg(1 - x) = 0$, $k \in \mathbb{N}$ is an arbitrary number and

$$\alpha_- = \min \{ \operatorname{Re} \alpha_j \,;\, j \text{ s.t. } \alpha_j \notin \mathbb{N} \}.$$

Remark 2. Schäfke and Schmidt considered linear systems of first-order ordinary differential equations [8]. But the above statement can be shown by a similar way.

Theorem 1 says that the connection coefficients appear in the asymptotic behavior of the coefficients of the local solutions. We use this theorem to solve the connection problem for the generalized hypergeometric equation and derive the connection matrix explicitly. Our main result gives an example which shows Schäfke and Schmidt's result is useful for derivation of the connection matrix of the fundamental systems of solutions for Fuchsian differential equations.

This chapter is organized as follows. In Section 2, we give the main result Theorem 2 and the strategy of the proof. Since Theorem 2 is obtained from Proposition 1, we show Proposition 1 in Section 3. Section 4 is devoted to the proofs of some propositions which are necessary to show Proposition 1.

2. Main Result

In this section, we state our main theorem. Let us set

$$D = \begin{pmatrix}
d_0^{(1)} & & & & & \\
d_1^{(1)} & d_0^{(2)} & & & & \\
d_2^{(1)} & d_1^{(2)} & d_0^{(3)} & & & \\
\vdots & \vdots & \vdots & \ddots & & \\
d_{n-2}^{(1)} & d_{n-3}^{(2)} & d_{n-4}^{(3)} & \cdots & d_0^{(n-1)} & \\
0 & 0 & 0 & \cdots & 0 & d_0^{(n)}
\end{pmatrix}.$$

Since we assumed $d_0^{(1)} d_0^{(2)} \cdots d_0^{(n)} \neq 0$, the matrix D is invertible. Our main result is as follows.

Theorem 2. *Assume that the condition* (2) *and* $\operatorname{Re} \beta_n < -n + 2$ *hold. Then the connection matrix C in* (5) *is given by*

$$C = D^{-1} P.$$

Here, $P = (p_{ij})_{1 \leq i,j \leq n}$ *is the $n \times n$ matrix whose entries are*

$$p_{ij} = \begin{cases} \dfrac{(-1)^{i-1}}{(i-1)!} \dfrac{(\boldsymbol{\alpha}_{j-1})_{i-1}}{(\boldsymbol{\beta}_{j-1})_{i-1}} {}_n F_{n-1} \left(\boldsymbol{\alpha}_{j-1} + i - 1; \boldsymbol{\beta}_{j-1} + i - 1; 1\right) & 1 \leq i \leq n-1, \\[4mm] \dfrac{\Gamma(\boldsymbol{\beta}_{j-1}, \beta_n)}{\Gamma(\boldsymbol{\alpha}_{j-1})} & i = n. \end{cases}$$

The symbols $(\boldsymbol{\alpha})_m$ and $\Gamma(\boldsymbol{\alpha})$ are defined by

$$(\boldsymbol{\alpha})_m := \prod_{i=1}^{n} (\alpha_i)_m, \quad \Gamma(\boldsymbol{\alpha}) := \prod_{i=1}^{n} \Gamma(\alpha_i)$$

and

$$\boldsymbol{\alpha} + a := (\alpha_1 + a, \alpha_2 + a, \ldots, \alpha_n + a)$$

for $m \in \mathbb{N}$, $a \in \mathbb{C}$ *and* $\boldsymbol{\alpha} = (\alpha_1, \alpha_2, \ldots, \alpha_n) \in \mathbb{C}^n$.

Remark 3. We note that this theorem holds even in the cases

(1) $\alpha_i \in \mathbb{Z}$ for some $1 \leq i \leq n$,
(2) $\alpha_i - \beta_j \in \mathbb{Z}$ for some $1 \leq i \leq n$ and $1 \leq j \leq n-1$,
(3) $\alpha_i - \alpha_j \in \mathbb{Z}$ for some $1 \leq i, j \leq n$.

These cases were not treated in Matsuhira–Nagoya [4], and it is known that in the cases (1) and (2), our equation (E) is reducible (cf. [2]).

Example 1. In the case of $n = 3$,

$$D = \begin{pmatrix} d_0^{(1)} & & \\ d_1^{(1)} & d_0^{(2)} & \\ 0 & 0 & d_0^{(3)} \end{pmatrix}$$

and the connection matrix C is given by

$$
D^{-1}\begin{pmatrix}
{}_3F_2(\boldsymbol{\alpha}_0;\boldsymbol{\beta}_0;1) & {}_3F_2(\boldsymbol{\alpha}_1;\boldsymbol{\beta}_1;1) \\[2mm]
-\dfrac{(\boldsymbol{\alpha}_0)_1}{(\boldsymbol{\beta}_0)_1}{}_3F_2(\boldsymbol{\alpha}_0+1;\boldsymbol{\beta}_0+1;1) & -\dfrac{(\boldsymbol{\alpha}_1)_1}{(\boldsymbol{\beta}_1)_1}{}_3F_2(\boldsymbol{\alpha}_1+1;\boldsymbol{\beta}_1+1;1) \\[2mm]
\dfrac{\Gamma(\boldsymbol{\beta}_0,\beta_3)}{\Gamma(\boldsymbol{\alpha}_0)} & \dfrac{\Gamma(\boldsymbol{\beta}_1,\beta_3)}{\Gamma(\boldsymbol{\alpha}_1)}
\end{pmatrix}
$$

$$
\begin{pmatrix}
{}_3F_2(\boldsymbol{\alpha}_2;\boldsymbol{\beta}_2;1) \\[2mm]
-\dfrac{(\boldsymbol{\alpha}_2)_1}{(\boldsymbol{\beta}_2)_1}{}_3F_2(\boldsymbol{\alpha}_2+1;\boldsymbol{\beta}_2+1;1) \\[2mm]
\dfrac{\Gamma(\boldsymbol{\beta}_2,\beta_3)}{\Gamma(\boldsymbol{\alpha}_2)}
\end{pmatrix}.
$$

Remark 4. In this case, the connection matrix is complicated because D is not a diagonal matrix in general. But if we take fundamental system of solutions (4) satisfies $d_1^{(1)} = 0$, or retake $y_2(x)$ as $y_2(x) - d_1^1/d_0^1(1-x)y_1(x)$, we can make the matrix D into a diagonal matrix and the expression of connection matrix simple.

To give a proof of Theorem 2, it is sufficient to show the following proposition.

Proposition 1. *Assume that the condition* (2) *and* $\operatorname{Re}\beta_n < -n+2$ *hold. Then the connection coefficients* $c_1, c_2, \ldots, c_n \in \mathbb{C}$ *such that*

$$
y_1^{[0]}(x) = \sum_{j=1}^{n} c_j y_j^{[1]}(x), \quad x \in D_0 \cap D_1 \tag{6}
$$

with $\arg x = \arg(1-x) = 0$ *on* $0 < x < 1$ *are given by*

$$
\begin{pmatrix} c_1 \\ c_2 \\ \vdots \\ c_{n-1} \\ c_n \end{pmatrix} = D^{-1}
\begin{pmatrix}
{}_nF_{n-1}(\boldsymbol{\alpha}_0;\boldsymbol{\beta}_0;1) \\[2mm]
-\dfrac{(\boldsymbol{\alpha}_0)_1}{(\boldsymbol{\beta}_0)_1}{}_nF_{n-1}(\boldsymbol{\alpha}_0+1;\boldsymbol{\beta}_0+1;1) \\[2mm]
\vdots \\[2mm]
\dfrac{(-1)^{n-2}}{(n-2)!}\dfrac{(\boldsymbol{\alpha}_0)_{n-2}}{(\boldsymbol{\beta}_0)_{n-2}}{}_nF_{n-1}(\boldsymbol{\alpha}_0+n-2;\boldsymbol{\beta}_0+n-2;1) \\[2mm]
\dfrac{\Gamma(\boldsymbol{\beta}_0,\beta_n)}{\Gamma(\boldsymbol{\alpha}_0)}
\end{pmatrix}. \tag{7}
$$

The expression (7) gives the first column of the connection matrix C and other columns are obtained from the first column. This is seen from the fact that the Riemann scheme (1) determines the generalized hypergeometric equation (E). Let us explain this in detail. To clarify the dependency for the parameters, we rewrite (6) as

$$y_1^{[0]}(x) = y_1^{[0]}(\boldsymbol{\alpha}_0; \boldsymbol{\beta}_0; x), \quad y_j^{[1]}(x) = y_j^{[1]}(\boldsymbol{\alpha}_0; \boldsymbol{\beta}_0; x),$$

$$c_j = c_j(\boldsymbol{\alpha}_0; \boldsymbol{\beta}_0), \quad 1 \le j \le n$$

for a while. We note that the other solutions at $x = 0$ are written as

$$y_{i+1}^{[0]}(\boldsymbol{\alpha}_0; \boldsymbol{\beta}_0; x) = x^{1-\beta_i} y_1^{[0]}(\boldsymbol{\alpha}_i; \boldsymbol{\beta}_i; x)$$

for $1 \le i \le n - 1$. In the following, we focus on the fundamental system of solutions at $x = 1$ and show

$$x^{1-\beta_i} y_j^{[1]}(\boldsymbol{\alpha}_i; \boldsymbol{\beta}_i; x) = y_j^{[1]}(\boldsymbol{\alpha}_0; \boldsymbol{\beta}_0; x), \quad 1 \le j \le n. \tag{8}$$

Let us exchange the parameter

$$(\boldsymbol{\alpha}_0, \boldsymbol{\beta}_0) \longmapsto (\boldsymbol{\alpha}_i, \boldsymbol{\beta}_i) \tag{9}$$

in our equation (E). Then we have the equation which is satisfied by the functions $y_1^{[0]}(\boldsymbol{\alpha}_i; \boldsymbol{\beta}_i; x)$ and $y_j^{[1]}(\boldsymbol{\alpha}_i; \boldsymbol{\beta}_i; x)$. The Riemann scheme of that equation is given by

$$\left\{ \begin{array}{ccc} x = 0 & x = 1 & x = \infty \\ 0 & 0 & \alpha_1 + 1 - \beta_i \\ \beta_i - \beta_1 & 1 & \alpha_2 + 1 - \beta_i \\ \vdots & \vdots & \vdots \\ \beta_i - 1 & i & \alpha_{i+1} + 1 - \beta_i \\ \vdots & \vdots & \vdots \\ \beta_i - \beta_{n-2} & n - 2 & \alpha_{n-1} + 1 - \beta_i \\ \beta_i - \beta_{n-1} & -\beta_n & \alpha_n + 1 - \beta_i \end{array} \right\}. \tag{10}$$

Next we consider the transform of unknown function

$$y(x) \longmapsto x^{1-\beta_i} y(x). \tag{11}$$

Then the Riemann scheme (10) turns into

$$
\left\{
\begin{array}{ccc}
x = 0 & x = 1 & x = \infty \\
1 - \beta_i & 0 & \alpha_1 \\
1 - \beta_1 & 1 & \alpha_2 \\
\vdots & \vdots & \vdots \\
0 & i & \alpha_{i+1} \\
\vdots & \vdots & \vdots \\
1 - \beta_{n-2} & n - 2 & \alpha_{n-1} \\
1 - \beta_{n-1} & -\beta_n & \alpha_n
\end{array}
\right\}.
$$

The equation which corresponds to this scheme is nothing but (E). This means that the relation (8) holds. Therefore, we see that the connection formula (6) is transformed into

$$
y_{i+1}^{[0]}(x) = c_1(\boldsymbol{\alpha}_i; \boldsymbol{\beta}_i) y_1^{[1]}(\boldsymbol{\alpha}_0; \boldsymbol{\beta}_0; x) + c_2(\boldsymbol{\alpha}_i; \boldsymbol{\beta}_i) y_2^{[1]}(\boldsymbol{\alpha}_0; \boldsymbol{\beta}_0; x) + \cdots
$$
$$
+ c_n(\boldsymbol{\alpha}_i; \boldsymbol{\beta}_i) y_n^{[1]}(\boldsymbol{\alpha}_0; \boldsymbol{\beta}_0; x)
$$

by following the procedures (9) and (11). This gives the i-th column of the connection matrix C.

3. Proof of Proposition 1

We shall show Proposition 1 by dividing the proof into the following two steps.

Step 1. To derive

$$
d_0^{(n)} c_n = \frac{\Gamma(\boldsymbol{\beta}_0, \beta_n)}{\Gamma(\boldsymbol{\alpha}_0)}. \tag{12}
$$

Step 2. To derive

$$
\sum_{j=1}^{i} d_{i-j}^{(j)} c_j = \frac{(-1)^{i-1}}{(i-1)!} \frac{(\boldsymbol{\alpha}_0)_{i-1}}{(\boldsymbol{\beta}_0)_{i-1}} {}_n F_{n-1}(\boldsymbol{\alpha}_0 + i - 1, \boldsymbol{\beta}_0 + i - 1; 1)
$$

$$
\text{for } 1 \le i \le n - 1. \tag{13}
$$

The relation (13) gives the relation between $c_1, c_2, \ldots, c_{n-1}$ in (7): The expression (13) is rewritten as

$$d_0^{(1)} c_1 = {}_nF_{n-1}(\boldsymbol{\alpha}_0, \boldsymbol{\beta}_0; 1),$$

$$d_0^{(2)} c_2 + d_1^{(1)} c_1 = -\frac{(\boldsymbol{\alpha}_0)_1}{(\boldsymbol{\beta}_0)_1} {}_nF_{n-1}(\boldsymbol{\alpha}_0 + 1, \boldsymbol{\beta}_0 + 1; 1),$$

$$\vdots$$

$$d_0^{(n-1)} c_{n-1} + \cdots + d_{n-2}^{(1)} c_1 = \frac{(-1)^{n-2}}{(n-2)!} \frac{(\boldsymbol{\alpha}_0)_{n-2}}{(\boldsymbol{\beta}_0)_{n-2}} {}_nF_{n-1}$$
$$\times (\boldsymbol{\alpha}_0 + n - 2; \boldsymbol{\beta}_0 + n - 2; 1).$$

By expressing them as a matrix form, we have

$$\begin{pmatrix} d_0^{(1)} & & & \\ d_1^{(1)} & d_0^{(2)} & & \\ \vdots & \vdots & \ddots & \\ d_{n-2}^{(1)} & d_{n-3}^{(2)} & \cdots & d_0^{(n-1)} \end{pmatrix} \begin{pmatrix} c_1 \\ c_2 \\ \vdots \\ c_{n-1} \end{pmatrix}$$

$$= \begin{pmatrix} {}_nF_{n-1}(\boldsymbol{\alpha}_0; \boldsymbol{\beta}_0; 1) \\ -\frac{(\boldsymbol{\alpha}_0)_1}{(\boldsymbol{\beta}_0)_1} {}_nF_{n-1}(\boldsymbol{\alpha}_0 + 1; \boldsymbol{\beta}_0 + 1; 1) \\ \vdots \\ \frac{(-1)^{n-2}}{(n-2)!} \frac{(\boldsymbol{\alpha}_0)_{n-2}}{(\boldsymbol{\beta}_0)_{n-2}} {}_nF_{n-1}(\boldsymbol{\alpha}_0 + n - 2; \boldsymbol{\beta}_0 + n - 2; 1) \end{pmatrix}.$$

3.1. *Step 1: Derivation of* (12)

In this subsection, we derive (12). By setting

$$y_1^{[0]}(x) = {}_nF_{n-1}(\boldsymbol{\alpha}_0; \boldsymbol{\beta}_0; x) = \sum_{m \geq 0} a_m x^m,$$

$$a_m = \frac{(\alpha_1)_m (\alpha_2)_m \cdots (\alpha_n)_m}{(\beta_1)_m (\beta_2)_m \cdots (\beta_{n-1})_m m!}$$

and applying Theorem 1, we have

$$a_m = \frac{\Gamma(m+\beta_n)}{\Gamma(m+1)}\left\{\sum_{\ell=0}^{k}\left(\prod_{s=1}^{\ell}\frac{-s+\beta_n}{m-s+\beta_n}\right)d_\ell^{(n)}\right\}$$
$$\times\frac{c_n}{\Gamma(\beta_n)}+O(m^{\operatorname{Re}\beta_n-k-2}),\quad m\to\infty.$$

Taking $k=0$ in this expression, we get

$$a_m = \frac{\Gamma(m+\beta_n)}{\Gamma(m+1)}d_0^n\frac{c_n}{\Gamma(\beta_n)}+O(m^{\operatorname{Re}\beta_n-2}).$$

From this and the formula

$$\frac{\Gamma(m+1)}{\Gamma(m+\beta_n)} = m^{1-\beta_n}(1+O(m^{-1}))$$

which is obtained by Stirling's formula, we obtain

$$d_0^{(n)}c_n = \frac{\Gamma(\beta_n)\Gamma(m+1)}{\Gamma(m+\beta_n)}a_m + O(m^{-1}). \tag{14}$$

By substituting

$$a_m = \frac{(\alpha_1)_m(\alpha_2)_m\cdots(\alpha_n)_m}{(\beta_1)_m(\beta_2)_m\cdots(\beta_{n-1})_m m!}$$
$$= \frac{\Gamma(\alpha_1+m)\Gamma(\alpha_2+m)\cdots\Gamma(\alpha_n+m)}{\Gamma(\beta_1+m)\Gamma(\beta_2+m)\cdots\Gamma(\beta_{n-1}+m)\Gamma(1+m)}$$
$$\times\frac{\Gamma(\beta_1)\Gamma(\beta_2)\cdots\Gamma(\beta_{n-1})}{\Gamma(\alpha_1)\cdots\Gamma(\alpha_n)}$$

into (14) we have

$$d_0^{(n)}c_n = \frac{\Gamma(\boldsymbol{\beta}_0,\beta_n)}{\Gamma(\boldsymbol{\alpha}_0)}\frac{\Gamma(\alpha_1+m,\alpha_2+m,\ldots,\alpha_n+m)}{\Gamma(\beta_1+m,\ldots,\beta_{n-1}+m,\beta_n+m)}+O(m^{-1}). \tag{15}$$

From Stirling's formula, it holds that

$$\frac{\Gamma(\alpha_1+m,\alpha_2+m,\ldots,\alpha_n+m)}{\Gamma(\beta_1+m,\ldots,\beta_{n-1}+m,\beta_n+m)}$$
$$\to m^{\alpha_1+\alpha_2+\cdots+\alpha_n-(\beta_1+\cdots+\beta_{n-1}+\beta_n)} = 1$$

as $m\to\infty$. Therefore, by taking a limit $m\to\infty$ for (15), we get the conclusion (12).

3.2.   *Step 2: Derivation of* (13)

The relation (13) is obtained from the following two propositions.

Proposition 2. *It holds that*

$$\sum_{j=1}^{i} d_{i-j}^{(j)} c_j = \frac{(-1)^{i-1}}{(i-1)!} \sum_{h=0}^{m} [h]_{i-1} a_h + O(m^{\operatorname{Re}\beta_n + i - 1}), \quad m \to \infty$$

$$\tag{16}$$

for $1 \leq i \leq n-1$. Here, we set

$$[h]_j = \begin{cases} 1 & j = 0, \\ h(h-1)(h-2)\cdots(h-j+1) & j \geq 1. \end{cases}$$

Proposition 3. *Assume* $\operatorname{Re}\beta_n < -n+2$. *Then we have*

$$\lim_{m\to\infty} \sum_{h=0}^{m} [h]_{i-1} a_h = \frac{(\alpha_0)_{i-1}}{(\beta_0)_{i-1}} \, _n F_{n-1}(\alpha_0 + i - 1, \beta_0 + i - 1; 1)$$

$$for\ 1 \leq i \leq n-1. \tag{17}$$

Proofs of these propositions are given in the following section. We admit these propositions and finish the proof of Proposition 1. From the assumption, for any $1 \leq i \leq n-1$, we obtain

$$\operatorname{Re}\beta_n + i - 1 < -n+2+i-1 \leq -n+2+(n-1)-1 = 0.$$

Therefore, we see that $O(m^{\operatorname{Re}\beta_n + i - 1}) \to 0$ as $m \to \infty$ for each $1 \leq i \leq n-1$. Then by taking a limit $m \to \infty$ in (16) and using (17), we get the desired expression (13).

4.   Proofs of Propositions 2 and 3

4.1.   *Proof of Proposition* 2

We prove Proposition 2 by induction on i. At first, we show the case of $i = 1$, i.e.

$$d_0^{(1)} c_1 = \sum_{h=0}^{m} a_h + O(m^{\operatorname{Re}\beta_n}), \quad m \to \infty. \tag{18}$$

Let us consider the gauge transform

$$z(x) = (1 - x)^{-1} y(x)$$

in our equation (E). Then the equation satisfied by $z(x)$ is given by

$$\left\{ \begin{array}{ccc} x = 0 & x = 1 & x = \infty \\ 0 & -1 & \alpha_1 + 1 \\ 1 - \beta_1 & 0 & \alpha_2 + 1 \\ \vdots & \vdots & \vdots \\ 1 - \beta_{n-2} & n - 3 & \alpha_{n-1} + 1 \\ 1 - \beta_{n-1} & -\beta_n - 1 & \alpha_n + 1 \end{array} \right\}. \tag{19}$$

The fundamental system of solutions of this equation is obtained by considering the gauge transform for each solution of equation (E) (given in (4)). We write them as

$$z^{[0]}(x) = (1 - x)^{-1} y_1^{[0]}(x) = \sum_{m \geq 0} \left(\sum_{h=0}^{m} a_h \right) x^m,$$

$$z_j^{[1]}(x) = (1 - x)^{-1} y_j^{[1]}(x) = (1 - x)^{j-2} \sum_{m \geq 0} d_m^{(j)} (1 - x)^m,$$

$$1 \leq j \leq n - 1,$$

$$z_n^{[1]}(x) = (1 - x)^{-1} y_n^{[1]}(x) = (1 - x)^{-\beta_n - 1} \sum_{m \geq 0} d_m^{(n)} (1 - x)^m.$$

Then the connection formula (6) turns into

$$z^{[0]}(x) = c_1 z_1^{[1]}(x) + c_2 z_2^{[1]}(x) + \cdots + c_n z_n^{[1]}(x).$$

Here, we remark that $z^{[0]}(x)$ is a holomorphic solution and the function $z_1^{[1]}(x)$ is a singular solution whose characteristic exponent is -1.

Now we use Theorem 1 for equation (19). Then we have

$$\sum_{h=0}^{m} a_h = \left\{ \sum_{\ell=0}^{k} \left(\prod_{s=1}^{\ell} \frac{-s+1}{m-s+1} \right) d_\ell^{(1)} \right\} \frac{c_1}{\Gamma(1)}$$

$$+ \frac{\Gamma(m+\beta_n+1)}{\Gamma(m+1)} \left\{ \sum_{\ell=0}^{k} \left(\prod_{s=1}^{\ell} \frac{-s+\beta_n+1}{m-s+\beta_n+1} \right) d_\ell^{(n)} \right\}$$

$$\times \frac{c_n}{\Gamma(\beta_n+1)} + O(m^{-\alpha_- -k-2})$$

for any $k \in \mathbb{N}$. Since

$$\prod_{s=1}^{\ell} \frac{-s+1}{m-s+1} = 0$$

for all $\ell \geq 1$, it holds that

$$\sum_{\ell=0}^{k} \left(\prod_{s=1}^{\ell} \frac{-s+1}{m-s+1} \right) d_\ell^{(1)} = \sum_{\ell=0}^{0} \left(\prod_{s=1}^{\ell} \frac{-s+1}{m-s+1} \right) d_\ell^{(1)} = d_0^{(1)}.$$

Therefore, we have

$$\sum_{h=0}^{m} a_h = d_0^{(1)} c_1 + \frac{\Gamma(m+\beta_n+1)}{\Gamma(m+1)} \left\{ \sum_{\ell=0}^{k} \left(\prod_{s=1}^{\ell} \frac{-s+\beta_n+1}{m-s+\beta_n+1} \right) d_\ell^{(n)} \right\}$$

$$\times \frac{c_n}{\Gamma(\beta_n+1)} + O(m^{-\alpha_- -k-2}).$$

Here, $\alpha_- = \min\{-1, -\operatorname{Re}\beta_n - 1\}$. Since we assumed $\operatorname{Re}\beta_n < -n+2$ and $n \geq 2$, we have

$$-\operatorname{Re}\beta_n - 1 > n - 3 \geq -1.$$

Therefore, we see that $\alpha_- = -1$. Now we fix $k \in \mathbb{N}$, which satisfies $k \geq -\operatorname{Re}\beta_n - 1$. Then it holds

$$-\alpha_- - k - 2 = -k - 1 \leq \operatorname{Re}\beta_n.$$

From this we have

$$O(m^{-\alpha_- -k-2}) = O(m^{-k-1}) = O(m^{\operatorname{Re}\beta_n}).$$

Next, we focus on the second term of the right-hand side. From Stirling's formula, it holds that

$$\frac{\Gamma(m + \beta_n + 1)}{\Gamma(m + 1)} = O(m^{\operatorname{Re}\beta_n}).$$

Therefore, we have the desired relation (18).

Next, let $i \in \mathbb{N}$ with $2 \le i \le n - 1$, and suppose that (16) (with i replaced by p) is already proved for all $p \in \mathbb{N}$ with $1 \le p \le i - 1$. We consider the gauge transform

$$z(x) = (1 - x)^{-i} z(x)$$

in equation (E). Then the equation satisfied by $z(x)$ is given by

$$\left\{\begin{array}{ccc} x = 0 & x = 1 & x = \infty \\ 0 & -i & \alpha_1 + i \\ 1 - \beta_1 & -i + 1 & \alpha_2 + i \\ \vdots & \vdots & \vdots \\ 1 - \beta_{i-1} & -1 & \alpha_i + i \\ 1 - \beta_i & 0 & \alpha_{i+1} + i \\ \vdots & \vdots & \vdots \\ 1 - \beta_{n-1} & -\beta_n - i & \alpha_n + i \end{array}\right\}. \tag{20}$$

The fundamental system of solutions of this equation is obtained by considering the gauge transform for each solution of the equation (E) (given in (4)). We write them as

$$z^{[0]}(x) = (1 - x)^{-i} y_1^{[0]}(x) = \sum_{m \ge 0} \left(\frac{1}{(i - 1)!} \sum_{h=0}^{m} (m - h + 1)_{i-1} a_h\right) x^m,$$

$$z_j^{[1]}(x) = (1 - x)^{-i} y_j^{[1]}(x) = (1 - x)^{j-i-1} \sum_{m \ge 0} d_m^{(j)} (1 - x)^m,$$

$$1 \le j \le n - 1,$$

$$z_n^{[1]}(x) = (1 - x)^{-i} y_n^{[1]}(x) = (1 - x)^{-\beta_n - i} \sum_{m \ge 0} d_m^{(n)} (1 - x)^m.$$

Then the connection formula (6) turns into

$$z^{[0]}(x) = c_1 z_1^{[1]}(x) + c_2 z_2^{[1]}(x) + \cdots + c_n z_n^{[1]}(x).$$

Here, we remark that $z^{[0]}(x)$ is a holomorphic solution of (20) and the functions $z_j^{[1]}(x)\,(1 \le j \le i)$ are singular solutions of (20) whose characteristic exponent of each solution is $j - i - 1$.

Now we use Theorem 1 for the equation (20). Then we have

$$\frac{1}{(i-1)!} \sum_{h=0}^{m} (m - h + 1)_{i-1} a_h$$

$$= \sum_{j=1}^{i} \frac{\Gamma(m + i - j + 1)}{\Gamma(m + 1)} \left\{ \sum_{\ell=0}^{k} \left(\prod_{s=1}^{\ell} \frac{-s + i - j + 1}{m - s + i - j + 1} \right) d_\ell^{(j)} \right\}$$

$$\times \frac{c_j}{\Gamma(i - j + 1)} + \frac{\Gamma(\beta_n + i + m)}{\Gamma(m + 1)} \left\{ \sum_{\ell=0}^{k} \left(\prod_{s=1}^{\ell} \frac{-s + \beta_n + i}{m - s + \beta_n + i} \right) d_\ell^{(n)} \right\}$$

$$\times \frac{c_n}{\Gamma(\beta_n + i)} + O(m^{-\alpha_- - k - 2})$$

for any $k \in \mathbb{N}$. Next, we fix $k \in \mathbb{N}$, which satisfies $k \ge i-1\,(\ge i-j+1)$. Since

$$\prod_{s=1}^{\ell} \frac{-s + i - j + 1}{m - s + i - j + 1} = 0$$

for all $\ell \ge i - j + 1$, it holds that

$$\sum_{\ell=0}^{k} \left(\prod_{s=1}^{\ell} \frac{-s + i - j + 1}{m - s + i - j + 1} \right) d_\ell^{(j)} = \sum_{\ell=0}^{i-j} \left(\prod_{s=1}^{\ell} \frac{-s + i - j + 1}{m - s + i - j + 1} \right) d_\ell^{(j)}.$$

Therefore, we have

$$\frac{1}{(i-1)!} \sum_{h=0}^{m} (m - h + 1)_{i-1} a_h$$

$$= \sum_{j=1}^{i} \frac{\Gamma(m + i - j + 1)}{\Gamma(m + 1)} \left\{ \sum_{\ell=0}^{i-j} \left(\prod_{s=1}^{\ell} \frac{-s + i - j + 1}{m - s + i - j + 1} \right) d_\ell^{(j)} \right\}$$

$$\times \frac{c_j}{\Gamma(i - j + 1)} + \frac{\Gamma(\beta_n + i + m)}{\Gamma(m + 1)}$$

$$\times \left\{ \sum_{\ell=0}^{k} \left(\prod_{s=1}^{\ell} \frac{-s + \beta_n + i}{m - s + \beta_n + i} \right) d_\ell^{(n)} \right\} \frac{c_n}{\Gamma(\beta_n + i)} + O(m^{-\alpha_- - k - 2})$$

$$= \sum_{j=1}^{i} \frac{\Gamma(m + i - j + 1)}{\Gamma(m + 1)} \left(\prod_{s=1}^{i-j} \frac{-s + i - j + 1}{m - s + i - j + 1} \right) d_{i-j}^{(j)} \frac{c_j}{\Gamma(i - j + 1)}$$

$$+ \sum_{j=1}^{i} \frac{\Gamma(m + i - j + 1)}{\Gamma(m + 1)} \left\{ \sum_{\ell=0}^{i-j-1} \left(\prod_{s=1}^{\ell} \frac{-s + i - j + 1}{m - s + i - j + 1} \right) d_\ell^{(j)} \right\}$$

$$\times \frac{c_j}{\Gamma(i - j + 1)} + \frac{\Gamma(\beta_n + i + m)}{\Gamma(m + 1)}$$

$$\times \left\{ \sum_{\ell=0}^{k} \left(\prod_{s=1}^{\ell} \frac{-s + \beta_n + i}{m - s + \beta_n + i} \right) d_\ell^{(n)} \right\} \frac{c_n}{\Gamma(\beta_n + i)} + O(m^{-\alpha_- - k - 2}).$$

$$(21)$$

Here, it holds that

$$\alpha_- = \min\{-i, -i + 1, \ldots, -1, -\operatorname{Re}\beta_n - i\} = -i$$

from the assumptions $n \geq 2$ and $\operatorname{Re}\beta_n < -n + 2$. Now we retake $k \in \mathbb{N}$ sufficiently large such that it satisfies $k \geq -\operatorname{Re}\beta_n - 1$ if needed. Then it holds that

$$-\alpha_- - k - 2 = i - k - 2 \leq \operatorname{Re}\beta_n + i - 1.$$

From this we have

$$O(m^{-\alpha_- - k - 2}) = O(m^{i-k-2}) = O(m^{\operatorname{Re}\beta_n + i - 1}).$$

In addition, from Stirling's formula, we have

$$\frac{\Gamma(\beta_n + i + m)}{\Gamma(m + 1)} \left\{ \sum_{\ell=0}^{k} \left(\prod_{s=1}^{\ell} \frac{-s + \beta_n + i}{m - s + \beta_n + i} \right) d_\ell^{(n)} \right\} \frac{c_n}{\Gamma(\beta_n + i)}$$

$$= O(m^{\operatorname{Re}\beta_n + i - 2}).$$

Let us consider the other terms. On the first term of the right-hand side of (21),

$$\frac{\Gamma(m+i-j+1)}{\Gamma(m+1)}\left(\prod_{s=1}^{i-j}\frac{-s+i-j+1}{m-s+i-j+1}\right)\frac{1}{\Gamma(i-j+1)}$$

$$=\frac{\Gamma(m+i-j+1)}{\Gamma(m+1)}\left(\frac{(i-j)(i-j-1)\cdots\cdot 1}{(m+i-j)(m+i-j-1)\cdots\cdot(m+1)}\right)$$

$$\times\frac{1}{\Gamma(i-j+1)}=1$$

holds from the property of gamma function.

In the second term of the right-hand side of (21), by changing the order of summation, we have

$$\sum_{j=1}^{i}\frac{\Gamma(m+i-j+1)}{\Gamma(m+1)}\left\{\sum_{\ell=0}^{i-j-1}\left(\prod_{s=1}^{\ell}\frac{-s+i-j+1}{m-s+i-j+1}\right)d_\ell^{(j)}\right\}$$

$$\times\frac{c_j}{\Gamma(i-j+1)}=\frac{1}{\Gamma(m+1)}\sum_{p=1}^{i-1}\frac{\Gamma(m+p+1)}{\Gamma(p+1)}\sum_{j=1}^{i-p}d_{i-p-j}^{(j)}c_j.$$

Therefore, we have

$$\frac{1}{(i-1)!}\sum_{h=0}^{m}(m-h+1)_{i-1}a_h$$

$$=\sum_{j=1}^{i}d_{i-j}^{(j)}c_j+\frac{1}{\Gamma(m+1)}\sum_{p=1}^{i-1}\frac{\Gamma(m+p+1)}{\Gamma(p+1)}\sum_{j=1}^{i-p}d_{i-p-j}^{(j)}c_j$$

$$+O(m^{\operatorname{Re}\beta_n+i-1}). \tag{22}$$

Next, we consider the left-hand side of (22). We give a lemma.

Lemma 1. *The equality*

$$(m-h+1)_\ell=\sum_{p=0}^{\ell}(-1)^{\ell-p}\binom{\ell}{p}[h]_{\ell-p}(m+1)_p$$

holds for $\ell\in\mathbb{N}$.

This lemma can be proved by induction. From this lemma, we have

$$\frac{1}{(i-1)!}\sum_{h=0}^{m}(m-h+1)_{i-1}a_h$$

$$=\frac{1}{(i-1)!}\sum_{h=0}^{m}\left(\sum_{p=0}^{i-1}(-1)^{i-1-p}\binom{i-1}{p}[h]_{i-1-p}(m+1)_p\right)a_h$$

$$=\frac{1}{(i-1)!}\sum_{p=0}^{i-1}(-1)^{i-1-p}\binom{i-1}{p}(m+1)_p\sum_{h=0}^{m}[h]_{i-1-p}a_h$$

$$=\frac{(-1)^{i-1}}{(i-1)!}\sum_{h=0}^{m}[h]_{i-1}a_h+\frac{1}{(i-1)!}\sum_{p=1}^{i-1}(-1)^{i-1-p}\binom{i-1}{p}(m+1)_p$$

$$\times\sum_{h=0}^{m}[h]_{i-1-p}a_h$$

$$=\frac{(-1)^{i-1}}{(i-1)!}\sum_{h=0}^{m}[h]_{i-1}a_h+\frac{1}{\Gamma(m+1)}\sum_{p=1}^{i-1}(-1)^{i-1-p}\frac{\Gamma(m+p+1)}{\Gamma(p+1)\Gamma(i-p)}$$

$$\times\sum_{h=0}^{m}[h]_{i-1-p}a_h.$$

Here, we used the assumption of induction in the last term and Stirling's formula. Then we obtain

$$\frac{1}{(i-1)!}\sum_{h=0}^{m}(m-h+1)_{i-1}a_h$$

$$=\frac{(-1)^{i-1}}{(i-1)!}\sum_{h=0}^{m}[h]_{i-1}a_h+\frac{1}{\Gamma(m+1)}\sum_{p=1}^{i-1}\frac{\Gamma(m+p+1)}{\Gamma(p+1)}$$

$$\times\left(\sum_{j=1}^{i-p}d_{i-p-j}^{(j)}c_j+O(m^{\operatorname{Re}\beta_n+i-p-1})\right)$$

$$= \frac{(-1)^{i-1}}{(i-1)!} \sum_{h=0}^{m} [h]_{i-1} a_h + \frac{1}{\Gamma(m+1)} \sum_{p=1}^{i-1} \frac{\Gamma(m+p+1)}{\Gamma(p+1)}$$

$$\times \sum_{j=1}^{i-p} d_{i-p-j}^{(j)} c_j + O(m^{\mathrm{Re}\,\beta_n + i - 1}).$$

Substituting this into (22), we have

$$\frac{(-1)^{i-1}}{(i-1)!} \sum_{h=0}^{m} [h]_{i-1} a_h = \sum_{j=1}^{i} d_{i-j}^{(j)} c_j + O(m^{\mathrm{Re}\,\beta_n + i - 1}).$$

This completes the proof of Proposition 2.

4.2. *Proof of Proposition 3*

In the following, we give a proof of Proposition 3. From the definition of $[h]_{i-1}$, we obtain

$$\sum_{h=0}^{m} [h]_{i-1} a_h = \sum_{h=0}^{m} h(h-1)\cdots(h-i+2) \frac{(\boldsymbol{\alpha}_0)_h}{(\boldsymbol{\beta}_0)_h h!}$$

$$= \sum_{h=i-1}^{m} \frac{(\boldsymbol{\alpha}_0)_h}{(\boldsymbol{\beta}_0)_h (h-i+1)!}$$

for $1 \le i \le n-1$. We set $\ell = h - i + 1$. Then we have

$$\sum_{h=i-1}^{m} \frac{(\boldsymbol{\alpha}_0)_h}{(\boldsymbol{\beta}_0)_h (h-i+1)!} = \sum_{\ell=0}^{m-i+1} \frac{(\boldsymbol{\alpha}_0)_{\ell+i-1}}{(\boldsymbol{\beta}_0)_{\ell+i-1} \ell!}.$$

Since

$$(a)_{\ell+i-1} = a(a+1)\cdots(a+i-2)(a+i-1)\cdots(a+i+\ell-2)$$
$$= (a)_{i-1}(a+i-1)_\ell$$

holds for $a \in \mathbb{C}$, we obtain

$$\frac{(\boldsymbol{\alpha}_0)_{\ell+i-1}}{(\boldsymbol{\beta}_0)_{\ell+i-1}} = \frac{(\boldsymbol{\alpha}_0)_{i-1}}{(\boldsymbol{\beta}_0)_{i-1}} \frac{(\boldsymbol{\alpha}_0 + i - 1)_\ell}{(\boldsymbol{\beta}_0 + i - 1)_\ell}.$$

Then we have

$$\sum_{h=0}^{m}[h]_{i-1}a_h = \frac{(\boldsymbol{\alpha}_0)_{i-1}}{(\boldsymbol{\beta}_0)_{i-1}}\sum_{\ell=0}^{m-i+1}\frac{(\boldsymbol{\alpha}_0+i-1)_\ell}{(\boldsymbol{\beta}_0+i-1)_\ell\ell!}.$$

From Raabe's ratio test, we see that the power series

$$\sum_{\ell=0}^{\infty}\frac{(\boldsymbol{\alpha}_0+i-1)_\ell}{(\boldsymbol{\beta}_0+i-1)_\ell\ell!}$$

is convergent when $\operatorname{Re}\beta_n < -n+2$, and its value is nothing but $_nF_{n-1}(\boldsymbol{\alpha}_0+i-1,\boldsymbol{\beta}_0+i-1;1)$. In conclusion, we have

$$\lim_{m\to\infty}\sum_{h=0}^{m}[h]_{i-1}a_h = \frac{(\boldsymbol{\alpha}_0)_{i-1}}{(\boldsymbol{\beta}_0)_{i-1}}\,_nF_{n-1}(\boldsymbol{\alpha}_0+i-1,\boldsymbol{\beta}_0+i-1;1).$$

Acknowledgment

The author would like to thank Professor Yoshishige Haraoka for his valuable comments and suggestions.

References

1. E. W. Barnes, A new development of the theory of the hypergeometric functions. *Proc. London Math. Soc.*, **6**(2) (1908), 141–177.
2. F. Beukers and G. Heckman, Monodromy for the hypergeometric function $_nF_{n-1}$. *Invent. Math.*, **95**(1989), 325–354.
3. Y. Kawahata, Connection problems associated with the Fuchsian differential equation of rank n with three regular singularities, II (in Japanese). *Proc. Tsuda College*, **10**(1978), 45–55.
4. Y. Matsuhira and H. Nagoya, Connection problem for the generalized hypergeometric function, arXiv:1904.02935.
5. K. Mimachi, Connection matrices associated with the generalized hypergeometric function $_3F_2$. *Funkcialaj Ekvacioj*, **51**(2008), 107–133.
6. K. Mimachi, Intersection numbers for twisted cycles and the connection problem associated with the generalized hypergeometric function $_{n+1}F_n$. *Int. Math. Res. Notices*, **2011**(8) (2011), 1757–1781.
7. T. Oshima, *Fractional Calculus of Weyl Algebra and Fuchsian Differential Equations*, MSJ Memoirs **28**. *Math. Society of Japan*, 2012.

8. R. Schäfke and D. Schmidt, Connection problems for linear ordinary differential equations in the complex domain. In: *Geometrical Approaches to Differential Equations*, Martini R. (ed.), Lecture Notes in Mathematics, Vol. **810**. Springer, Berlin, Heidelberg, 1980.

9. F. C. Smith, Relations among the fundamental solutions of the generalized hypergeometric equation when $p = q + 1$. Non-logarithmic cases. *Bull. Amer. Math. Soc.*, **44**(1938), 429–433.

© 2022 World Scientific Publishing Europe Ltd.
https://doi.org/10.1142/9781800611368_0014

Chapter 14

The Annihilation Operator for Certain Family of q-Hermite Sobolev-Type Orthogonal Polynomials

Carlos Hermoso[*,‡] and Anier Soria-Lorente[†,§]

*Department of Physics and Mathematics, Universidad de Alcalá
Ctra. Madrid-Barcelona
Km. 33,600. C.P.: 28805 Alcalá de Henares
Madrid, Spain
†Department of Basic Sciences, Universidad de Granma
Carretera de Bayamo a Manzanillo
Km. 17,500. Bayamo 85100, Cuba
‡ carlos.hermoso@uah.es
§ asorial@udg.co.cu

We present a new family $\{S_n(x;q)\}_{n\geq 0}$ of monic polynomials in x, orthogonal with respect to a Sobolev-type inner product related to the q-Hermite I orthogonal polynomials, involving a first-order q-derivative on a mass-point $\alpha \in \mathbb{R}$ located out of the corresponding orthogonality interval $[-1, 1]$, for some fixed real number $q \in (0, 1)$. We present connection formulas, and the annihilation operator for this non-standard orthogonal polynomial family.

1. Introduction

Let $\{H_n(x;q)\}_{n\geq 0}$ be the sequence of *monic q-Hermite I polynomials* of degree n, orthogonal with respect to the inner product

$$\langle f,g\rangle = \int_{-1}^{1} f(x;q)g(x;q)(qx,-qx;q)_\infty d_q x,\ 0<q<1,\quad f,g\in\mathbb{P}$$

(1)

in the linear space $\mathbb{P}$ of polynomials of real coefficients. This family of q-hypergeometric polynomials was introduced at the end of the 19th century by Rogers (see [14–16]) and studied throughout the 20th century by Szegő (see [18]) and Carlitz (see [4–6]), currently playing an important role in different fields such as, for instance, classical and non-commutative probability theory, combinatory or quantum-physics. The monic q-Hermite I polynomials can also be described as the family of polynomials satisfying the orthogonality relation

$$\int_{-1}^{1} H_m(x;q)H_n(x;q)(qx,-qx;q)_\infty d_q x = (1-q)(q;q)_n$$

$$\times\, (q,-1,-q;q)_\infty q^{\binom{n}{2}}\delta_{m,n},$$

(2)

where $\delta_{m,n}$ is the well-known Kronecker-delta symbol (see [3]). $\{H_n(x;q)\}_{n\geq 0}$ form a system of polynomials, orthogonal with respect to the measure $(qx,-qx;q)_\infty d_q x$, where $q\in\mathbb{R}$ stands for its unique fixed parameter, for which we assume that $0<q<1$. This last condition means that they belong to the class of orthogonal polynomial solutions of certain second-order q-difference equations, known in the literature as the Hahn class (see [12,13]). It is also important to highlight that, for $q=1$ one recovers the classical Hermite polynomials and, for $q=0$ one recovers a re-scaled version of the Chebyshev polynomials of the second kind.

On the other hand, the introduction of derivatives or q-derivatives acting on orthogonal polynomials constitutes a powerful tool to obtain new families of orthogonal polynomials. The polynomials resulting under the action of the q-derivatives on the q-Hermite family, as well as their orthogonality properties, have recently been studied (see [1]). Furthermore, q-derivatives can also appear involved in the discrete part of the measure, thus modifying the continuous part and giving rise to Sobolev-type perturbations of the q-Hermite I polynomials. For a recent and comprehensive study on general discrete Sobolev orthogonal polynomials including difference operators, see [11].

In this contribution, we inquired on the following Sobolev-type modification:

$$\langle f, g \rangle_\lambda = \int_{-1}^{1} f(x; q) g(x; q)(qx, -qx; q)_\infty d_q x$$
$$+ \lambda (\mathscr{D}_q f)(\alpha; q)(\mathscr{D}_q g)(\alpha; q) \tag{3}$$

of (1), where $\alpha \in \mathbb{R} \setminus [-1, 1]$, $\lambda \in \mathbb{R}^+$ and $\mathscr{D}_q$ denotes the q-derivative operator, as will be defined in what follows. In the sequel, we will denote by $\{S_n(x; q)\}_{n \geq 0}$ the sequence of monic polynomials orthogonal with respect to (3).

The perturbation above, induced by discrete measures with a single mass-point and first-order q-derivatives, addresses a first step in the analysis of the Sobolev-type modifications involving q-derivatives. Thus, the main goal of the chapter will be to study this type of perturbations by providing the q-Hermite I-Sobolev-type polynomials $S_n(x; q)$, and to establish some interesting results concerning this family. We will assume here that the discrete mass λ is located at a point α outside the support $[-1, 1]$ of the q-discrete measure $(qx, -qx; q)_\infty d_q x$.

The structure of the manuscript is as follows. In Section 2, we summarize few preliminaries of the q-calculus which will be useful throughout the manuscript. In Section 3, we obtain a structure relation for the q-derivatives of $S_n(x; q)$. In Section 4, we provide connection formulas between orthogonal polynomials of q-Hermite I and q-Hermite Sobolev-type families. Finally, in Section 5, we obtain an explicit expression for the annihilation operator for the family $S_n(x; q)_n \geq 0$, in the same fashion as in [7,8].

2. Preliminary Results from q-Calculus

In this section, we summarize few concepts and definitions from q-calculus, which will be needed in the sequel.

The q-number $[n]_q$ is defined by [13]

$$[n]_q = \begin{cases} 0, & \text{if } n = 0, \\ \dfrac{1-q^n}{1-q} = \displaystyle\sum_{0 \le k \le n-1} q^k, & \text{if } n \ge 1. \end{cases}$$

A q-analogue of the factorial of n is given by

$$[n]_q! = \begin{cases} 1, & \text{if } n = 0, \\ [n]_q[n-1]_q \cdots [2]_q[1]_q, & \text{if } n \ge 1. \end{cases}$$

Similarly, a q-analogue of the Pochhammer symbol, or shifted factorial, [13], is defined by

$$(a;q)_n = \begin{cases} 1, & \text{if } n = 0, \\ \displaystyle\prod_{0 \le j \le n-1} \left(1 - aq^j\right), & \text{if } n \ge 1, \\ (a;q)_\infty = \displaystyle\prod_{j \ge 0}(1 - aq^j), & \text{if } n = \infty \text{ and } |a| < 1. \end{cases}$$

Moreover, we will use the following notation

$$(a_1, \ldots, a_r; q)_k = \prod_{1 \le j \le r} (a_j; q)_k.$$

The q-falling factorial (see [2]) is defined by

$$[s]_q^{(n)} = \frac{(q^{-s};q)_n}{(q-1)^n} q^{ns - \binom{n}{2}}, \quad n \ge 1.$$

The q-binomial coefficient is given by (see [13])

$$\begin{bmatrix} k \\ n \end{bmatrix}_q = \frac{(q;q)_n}{(q;q)_k(q;q)_{n-k}} = \frac{[n]_q!}{[k]_q![n-k]_q!} = \begin{bmatrix} k \\ n-k \end{bmatrix}_q, \quad k = 0, 1, \ldots, n,$$

where n denotes a non-negative integer.

The Jackson–Hahn–Cigler q-subtraction (see for example [9] or [10, Def. 6]) and the references given there)

$$(x \boxminus_q y)^n = \prod_{0 \le j \le n-1} \left(x - y q^j\right) = x^n (y/x; q)_n$$

$$= \sum_{0 \le k \le n} \begin{bmatrix} k \\ n \end{bmatrix}_q q^{\binom{k}{2}} (-y)^k x^{n-k}.$$

The q-derivative [13] or the Euler–Jackson q-difference operator

$$(\mathscr{D}_q f)(z) = \begin{cases} \dfrac{f(qz) - f(z)}{(q-1)z}, & \text{if } z \ne 0, \ q \ne 1, \\[2ex] f'(z), & \text{if } z = 0, \ q = 1, \end{cases}$$

where $\mathscr{D}_q^0 f = f$, $\mathscr{D}_q^n f = \mathscr{D}_q(\mathscr{D}_q^{n-1} f)$, with $n \ge 1$, and

$$\lim_{q \to 1} \mathscr{D}_q f(z) = f'(z).$$

Furthermore, one has the following properties:

$$\mathscr{D}_q[f(\gamma z)] = \gamma (\mathscr{D}_q f)(\gamma z), \quad \forall \, \gamma \in \mathbb{C}, \tag{4}$$

$$\mathscr{D}_q f(z) = \mathscr{D}_{q^{-1}} f(qz) \Leftrightarrow \mathscr{D}_{q^{-1}} f(z) = \mathscr{D}_q f(q^{-1} z), \tag{5}$$

$$\mathscr{D}_q[f(z)g(z)] = f(qz)\mathscr{D}_q g(z) + g(z)\mathscr{D}_q f(z) \tag{6}$$

$$= f(z)\mathscr{D}_q g(z) + g(qz)\mathscr{D}_q f(z),$$

and the following interesting property:

$$\mathscr{D}_{q^{-1}}(\mathscr{D}_q f)(z) = q \mathscr{D}_q(\mathscr{D}_{q^{-1}} f)(z). \tag{7}$$

The q-Taylor formula (see [17, Th. 6.3]), with the Cauchy remainder term, is defined by

$$f(x) = \sum_{k=0}^{n} \frac{(\mathscr{D}_q^k f)(a)}{[k]_q!} (x \boxminus_q a)^k + \frac{1}{[n]_q!} \int_a^x (\mathscr{D}_q^{n+1} f)(t) \cdot (x \boxminus_q qt)^n d_q t.$$

                    C. Hermoso & A. Soria-Lorente

The Jackson q-integral is given by

$$\int_0^z f(x)d_qx = (1-q)z\sum_{k=0}^{\infty} q^k f(q^k z),$$

which in a generic interval $[a,b]$ is given by

$$\int_a^b f(x)d_qx = \int_0^b f(x)d_qx - \int_0^a f(x)d_qx.$$

3. q-Hermite I Orthogonal Polynomials

After the above q-calculus introduction, we continue by giving several aspects and properties of the q-Hermite I polynomials $\{H_n(x;q)\}_{n\geq 0}$.

The monic q-Hermite I polynomials can be given by means of their generating function:

$$\sum_{n=0}^{\infty} H_n(x;q)\frac{t^n}{(q;q)_n} = \prod_{n=0}^{\infty} \frac{1}{1-2xtq^n+t^2q^{2n}}.$$

We also have the following classic and well-known results in the literature for this family of polynomials:

Proposition 1. *Let* $\{H_n(x;q)\}_{n\geq 0}$ *be the sequence of q-Hermite I polynomials of degree n. Then the following statements hold.*

(1) *The recurrence relation* [13]

$$xH_n(x;q) = H_{n+1}(x;q) + \gamma_n H_{n-1}(x;q), \tag{8}$$

with initial conditions $H_{-1}(x;q) = 0$ and $H_0(x;q) = 1$. Here, $\gamma_n = q^{n-1}(1-q^n)$.

(2) *Squared norm* [13]. *For every $n \in \mathbb{N}$,*

$$\|H_n\|^2 = (1-q)(q;q)_n(q,-1,-q;q)_\infty q^{\binom{n}{2}}.$$

(3) *Forward shift operator* [13]

$$\mathscr{D}_q^k H_n(x;q) = [n]_q^{(k)} H_{n-k}(x;q),\qquad(9)$$

where

$$[n]_q^{(k)} = \frac{(q^{-n};q)_k}{(q-1)^k} q^{kn-\binom{k}{2}},$$

denote the q-falling factorial [2]. *Observe that* $[n]_q^{(1)} = [n]_q$.

(4) *Second-order q-difference equation* [13]

$$\sigma(x)\mathscr{D}_q\mathscr{D}_{q^{-1}} H_n(x;q) + \tau(x)\mathscr{D}_q H_n(x;q) + \lambda_{n,q} H_n(x;q) = 0,$$

where $\sigma(x) = x^2 - 1$, $\tau(x) = (1-q)^{-1}x$ *and* $\lambda_{n,q} = [n]_q([1 - n]_q\sigma''/2 - \tau')$.

A Christoffel–Darboux formula for this family can also be established:

Proposition 2 (Christoffel–Darboux formula). *Let* $\{H_n(x;q)\}_{n\geq 0}$ *be the sequence of q-Hermite I polynomials. If we denote the nth reproducing kernel by*

$$K_{n,q}(x,y) = \sum_{k=0}^{n} \frac{H_k(x;q)H_k(y;q)}{||H_k||^2}.$$

Then, for all $n \in \mathbb{N}$, *it holds that*

$$K_{n,q}(x,y) = \frac{H_{n+1}(x;q)H_n(y;q) - H_{n+1}(y;q)H_n(x;q)}{(x-y)\,||H_n||^2}.\qquad(10)$$

Concerning the partial q-derivatives of $K_{n,q}(x,y)$, we use the following notation:

$$K_{n,q}^{(i,j)}(x,y) = \mathscr{D}_{q,y}^j(\mathscr{D}_{q,x}^i K_{n,q}(x,y))$$

$$= \sum_{k=0}^{n} \frac{\mathscr{D}_q^i H_k(x;q)\mathscr{D}_q^j H_k(y;q)}{||H_k||^2}.$$

In order to establish connection formulas between the polynomials of $\{S_n(x;q)\}_{n\geq 0}$ and the polynomials of $\{H_n(x;q)\}_{n\geq 0}$, we first need to relate the first-order partial derivatives of the kernel polynomials of $\{H_n(x;q)\}_{n\geq 0}$ to the polynomials of this family. The next result is, in fact, a connection formula between the partial q-derivative with respect to one variable of the kernel polynomials and the polynomials of $\{H_n(x;q)\}_{n\geq 0}$. The proof is quite involved, and follows from the techniques recently presented in [8].

Proposition 3. *Let $\{H_n(x;q)\}_{n\geq 0}$ be the sequence of q-Hermite I polynomials of degree n. Then following statements hold, for all $n \in \mathbb{N}$,*

$$K_{n-1,q}^{(0,1)}(x,y) = \mathcal{A}_n^{(1)}(x,y)H_n(x;q) + \mathcal{B}_n^{(1)}(x,y)H_{n-1}(x;q), \qquad (11)$$

where

$$\mathcal{A}_n^{(1)}(x,y) = \frac{[1]_q!}{\|H_{n-1}\|^2\,(x \boxminus_q y)^2} \sum_{k=0}^{1} \frac{\mathscr{D}_q^k H_{n-1}(y;q)}{[k]_q!}(x \boxminus_q y)^k$$

$$= \frac{1}{\|H_{n-1}\|^2\,(x \boxminus_q y)^2}\,(H_{n-1}(y;q) + \mathscr{D}_q H_{n-1}(y;q)(x \boxminus_q y)),$$

and

$$\mathcal{B}_n^{(1)}(x,y) = -\frac{[j]_1!}{\|H_{n-1}\|^2\,(x \boxminus_q y)^2} \sum_{k=0}^{1} \frac{\mathscr{D}_q^k H_n(y;q)}{[k]_q!}(x \boxminus_q y)^k$$

$$= -\frac{1}{\|H_{n-1}\|^2\,(x \boxminus_q y)^2}\,(x \boxminus_q y)^2\,(H_n(y;q)$$

$$+ \mathscr{D}_q H_n(y;q)(x \boxminus_q y)).$$

Furthermore, we can state a connection formula between the partial q-derivatives with respect to both variables x,y of the kernel polynomials, and the polynomials of $\{H_n(x;q)\}_{n\geq 0}$:

Proposition 4. *Let $\{H_n(x;q)\}_{n\geq 0}$ be the sequence of q-Hermite I polynomials of degree n. Then following statement holds, for all $n \in \mathbb{N}$,*

$$K_{n-1,q}^{(1,1)}(x,y) = \mathcal{C}_{1,n}(x,y)H_n(x;q) + \mathcal{D}_{1,n}(x,y)H_{n-1}(x;q), \qquad (12)$$

where

$$\mathcal{C}_{1,n}(x,y) = \mathscr{D}_q \mathcal{A}_n^{(1)}(x,y) - [n-1]_q \gamma_{n-1}^{-1} \mathcal{B}_n^{(1)}(qx,y),$$

and

$$\mathcal{D}_{1,n}(x,y) = [n]_q \mathcal{A}_n^{(1)}(qx,y) + [n-1]_q \gamma_{n-1}^{-1} x \mathcal{B}_n^{(1)}(qx,y) + \mathscr{D}_q \mathcal{B}_n^{(1)}(x,y).$$

Proof. Applying the q-derivative operator $\mathscr{D}_q$ to (11), together with the property (6), we have

$$K_{n-1,q}^{(1,1)}(x,y) = \mathcal{A}_n^{(1)}(qx)\mathscr{D}_q H_n(x;q) + H_n(x;q)\mathscr{D}_q \mathcal{A}_n^{(1)}(x)$$

$$+ \mathcal{B}_n^{(1)}(qx)\mathscr{D}_q H_{n-1}(x;q) + H_{n-1}(x;q)\mathscr{D}_q \mathcal{B}_n^{(1)}(x).$$

Using (9) and (8) we easily deduce (12), and the proof is complete.

$$\square$$

4. Connection Formulas

In Section 1, we have presented the q-Hermite I-Sobolev-type orthogonal polynomials $\{S_n(x;q)\}_{n\geq 0}$, which are orthogonal with respect to Sobolev-type inner product

$$\langle f,g \rangle_\lambda = \int_{-1}^{1} f(x;q)g(x;q)(qx,-qx;q)_\infty d_q x$$

$$+ \lambda(\mathscr{D}_q f)(\alpha;q)(\mathscr{D}_q g)(\alpha;q),$$

where $\alpha \in \mathbb{R} \setminus [-1,1]$, $\lambda \in \mathbb{R}^+$ and $j \in \mathbb{N}$. Now, in this section, we will express these q-Hermite I-Sobolev-type orthogonal polynomials $\{S_n(x;q)\}_{n\geq 0}$ in terms of the q-Hermite polynomials $\{H_n(x;q)\}_{n\geq 0}$, the kernel polynomials and their corresponding derivatives. The main idea behind the identities presented here is that of expressing the new polynomials in terms of another with well-established properties, to infer, from these, those of the first one. As a consequence, we will obtain connection formulas between the polynomials of $\{S_n(x;q)\}_{n\geq 0}$ and $\{H_n(x;q)\}_{n\geq 0}$, as well as between the first-order q-derivatives of the polynomials of this new family and the polynomials of the old one.

We begin by relating the q-Hermite I-Sobolev-type polynomials and their q-derivatives to the q-Hermite polynomials and the kernel polynomials:

Proposition 5. *Let $\{S_n(x;q)\}_{n\geq 0}$ be the sequence of q-Hermite I-Sobolev-type orthogonal polynomials of degree n. Then, following statements hold,*

$$S_n(x;q) = H_n(x;q) - \lambda \frac{[n]_q^{(1)} H_{n-1}(\alpha;q)}{1 + \lambda K_{n-1,q}^{(1,1)}(\alpha,\alpha)} K_{n-1,q}^{(0,1)}(x,\alpha). \tag{13}$$

Proof. Taking into account the Fourier expansion

$$S_n(x;q) = H_n(x;q) + \sum_{0\leq k\leq n-1} a_{n,k} H_k(x;q).$$

Next, from (5) and considering the orthogonality properties of $H_n(x;q)$, the coefficients in the previous expansion are given by

$$a_{n,k} = -\frac{\lambda \mathscr{D}_q S_n(\alpha;q) \mathscr{D}_q H_k(\alpha;q)}{\|H_k\|^2}, \quad 0 \leq k \leq n-1.$$

Thus,

$$S_n(x;q) = H_n(x;q) - \lambda \mathscr{D}_q S_n(\alpha;q) K_{n-1,q}^{(0,1)}(x,\alpha).$$

Applying the operator $\mathscr{D}_q$ to the previous equation, we get

$$\mathscr{D}_q S_n(x;q) = [n]_q^{(1)} H_{n-1}(x;q) - \lambda \mathscr{D}_q S_n(\alpha;q) K_{n-1,q}^{(1,1)}(x,\alpha).$$

After some manipulations, we deduce

$$\mathscr{D}_q S_n(\alpha;q) = \frac{[n]_q^{(1)} H_{n-1}(\alpha;q)}{1 + \lambda K_{n-1,q}^{(1,1)}(\alpha,\alpha)}.$$

Therefore, we obtain (13). $\qquad\qquad\qquad\qquad\qquad\qquad\qquad\square$

As a consequence, we have the following result:

Corollary 1. *Let $\{S_n(x;q)\}_{n\geq 0}$ be the sequence of q-Hermite I-Sobolev–type orthogonal polynomials of degree n. Then, following statement holds,*

$$\mathscr{D}_q S_n(x;q) = [n]_q^{(1)} H_{n-1}(x;q) - \lambda \frac{[n]_q^{(1)} H_{n-1}(\alpha;q)}{1 + \lambda K_{n-1,q}^{(1,1)}(\alpha,\alpha)} K_{n-1,q}^{(1,1)}(x,\alpha).$$

Now, we are in a position to state connection formulas between the orthogonal polynomials of the q-Hermite I and q-Hermite I-Sobolev-type families. These lemmas provide the required results:

Lemma 1. *Let $\{S_n(x;q)\}_{n\geq 0}$ be the sequence of q-Hermite I-Sobolev-type orthogonal polynomials of degree n. Then, we have*

$$S_n(x;q) = \mathcal{E}_{1,n}(x)H_n(x;q) + F_{1,n}(x)H_{n-1}(x;q), \qquad (14)$$

where

$$\mathcal{E}_{1,n}(x) = 1 - \lambda \frac{[n]_q^{(1)} H_{n-1}(\alpha;q)}{1 + \lambda K_{n-1,q}^{(1,1)}(\alpha,\alpha)} \mathcal{A}_n^{(1)}(x,\alpha),$$

and

$$F_{1,n}(x) = -\lambda \frac{[n]_q^{(1)} H_{n-1}(\alpha;q)}{1 + \lambda K_{n-1,q}^{(1,1)}(\alpha,\alpha)} \mathcal{B}_n^{(1)}(x,\alpha).$$

Proof. From (13) and Proposition 3, the lemma holds. $\qquad\square$

On the other hand, from previous Lemma and recurrence relation (8), we obtain the following result:

$$S_{n-1}(x;q) = \mathcal{E}_{2,n}(x)H_n(x;q) + \mathcal{F}_{2,n}(x)H_{n-1}(x;q), \qquad (15)$$

where

$$\mathcal{E}_{2,n}(x) = -\frac{\mathcal{F}_{1,n-1}(x)}{\gamma_{n-1}},$$

and

$$\mathcal{F}_{2,n}(x) = \mathcal{E}_{1,n-1}(x) - x\mathcal{E}_{2,n}(x).$$

Lemma 2. *Let $\{S_n(x;q)\}_{n\geq 0}$ be the sequence of q-Hermite I-Sobolev-type orthogonal polynomials of degree n. Then, next statements hold,*

$$\Xi_{1,n}(x)H_n(x;q) = \begin{vmatrix} S_n(x;q) & S_{n-1}(x;q) \\ \mathcal{F}_{1,n}(x) & \mathcal{F}_{2,n}(x) \end{vmatrix}, \qquad (16)$$

and

$$\Xi_{1,n}(x)H_{n-1}(x;q) = -\begin{vmatrix} S_n(x;q) & S_{n-1}(x;q) \\ \mathcal{E}_{1,n}(x) & \mathcal{E}_{2,n}(x) \end{vmatrix}, \qquad (17)$$

where

$$\Xi_{1,n}(x) = \begin{vmatrix} \mathcal{E}_{1,n}(x) & \mathcal{E}_{2,n}(x) \\ \mathcal{F}_{1,n}(x) & \mathcal{F}_{2,n}(x) \end{vmatrix}.$$

Proof. Multiplying (14) by $\mathcal{F}_{2,n}(x)$ and (15) by $-\mathcal{F}_{1,n}(x)$, adding and simplifying the resulting equations, we deduce (16). In addition, we can now proceed analogously to get (17). $\qquad\square$

We end this section stating another connection formula between the first-order q-derivatives of the q-Hermite I-Sobolev-type orthogonal polynomials and the q-Hermite I polynomials:

Lemma 3. *Let $\{S_n(x;q)\}_{n\geq 0}$ be the sequence of q-Hermite I-Sobolev-type orthogonal polynomials of degree n. Then, following statements hold:*

$$\mathscr{D}_q S_n(x;q) = \mathcal{E}_{3,n}(x) H_n(x;q) + \mathcal{F}_{3,n}(x) H_{n-1}(x;q), \qquad (18)$$

where

$$\mathcal{E}_{3,n}(x) = -\lambda \frac{[n]_q^{(1)} H_{n-1}(\alpha;q)}{1 + \lambda K_{n-1,q}^{(1,1)}(\alpha,\alpha)} \mathcal{C}_{1,n}(x,\alpha),$$

and

$$\mathcal{F}_{3,n}(x) = [n]_q^{(1)} - \lambda \frac{[n]_q^{(1)} H_{n-1}(\alpha;q)}{1 + \lambda K_{n-1,q}^{(1,1)}(\alpha,\alpha)} \mathcal{D}_{1,n}(x,\alpha).$$

Proof. The statements easily follow by replacing the expansion of $K_{n-1,q}^{(1,1)}(x,y)$ as a linear combination of $H_n(x;q)$ and $H_{n-1}(x;q)$ (formula (12), Proposition 4) in the expression of $\mathscr{D}_q S_n(x;q)$ (Corollary 1):

$$\mathscr{D}_q S_n(x;q)$$

$$= [n]_q^{(1)} H_{n-1}(x;q) - \lambda \frac{[n]_q^{(1)} H_{n-1}(\alpha;q)}{1 + \lambda K_{n-1,q}^{(1,1)}(\alpha,\alpha)} (\mathcal{C}_{1,n}(x,y) H_n(x;q)$$

$$+ \mathcal{D}_{1,n}(x,y) H_{n-1}(x;q)),$$

and rearranging the terms keeping the linear combination of the two consecutive q-Hermite I polynomials of degrees n and $n-1$:

$$\mathscr{D}_q S_n(x;q)$$

$$= [n]_q^{(1)} H_{n-1}(x;q) - \lambda \frac{[n]_q^{(1)} H_{n-1}(\alpha;q)}{1 + \lambda K_{n-1,q}^{(1,1)}(\alpha,\alpha)} (\mathcal{C}_{1,n}(x,y) H_n(x;q)$$

$$+ \mathcal{D}_{1,n}(x,y) H_{n-1}(x;q))$$

$$= [n]_q^{(1)} H_{n-1}(x;q) - \lambda \frac{[n]_q^{(1)} H_{n-1}(\alpha;q)}{1 + \lambda K_{n-1,q}^{(1,1)}(\alpha,\alpha)} \mathcal{C}_{1,n}(x,y) H_n(x;q)$$

$$- \lambda \frac{[n]_q^{(1)} H_{n-1}(\alpha;q)}{1 + \lambda K_{n-1,q}^{(1,1)}(\alpha,\alpha)} \mathcal{D}_{1,n}(x,y) H_{n-1}(x;q)$$

$$= \left(-\lambda \frac{[n]_q^{(1)} H_{n-1}(\alpha;q)}{1 + \lambda K_{n-1,q}^{(1,1)}(\alpha,\alpha)} \mathcal{C}_{1,n}(x,y) \right) H_n(x;q)$$

$$+ \left([n]_q^{(1)} - \lambda \frac{[n]_q^{(1)} H_{n-1}(\alpha;q)}{1 + \lambda K_{n-1,q}^{(1,1)}(\alpha,\alpha)} \mathcal{D}_{1,n}(x,y) \right) H_{n-1}(x;q). \qquad \square$$

5. Annihilation Operator

In this section, we finally provide an expression of the annihilation operator for this new family of q-Hermite I-Sobolev-type polynomials:

Proposition 6 (Annihilation operator). *The annihilation operator of the q-Hermite I-Sobolev-type orthogonal polynomials $S_n(x;q)$, has the following explicit expression*

$$\mathfrak{a}_n S_n(x;q) = S_{n-1}(x;q),$$

where

$$\mathfrak{a}_n := \frac{1}{\mathcal{F}_{4,n}(x)} \left(\Xi_{1,n}(x) \mathscr{D}_q - \mathcal{E}_{4,n}(x) I \right),$$

being I the identity operator. One also has

$$\mathcal{E}_{4,n}(x) = - \begin{vmatrix} \mathcal{E}_{2,n}(x) & \mathcal{E}_{3,n}(x) \\ \mathcal{F}_{2,n}(x) & \mathcal{F}_{3,n}(x) \end{vmatrix},$$

and

$$\mathcal{F}_{4,n}(x) = \begin{vmatrix} \mathcal{E}_{1,n}(x) & \mathcal{E}_{3,n}(x) \\ \mathcal{F}_{1,n}(x) & \mathcal{F}_{3,n}(x) \end{vmatrix}.$$

Proof. Using the Lemmas 2 and 3, respectively, we get

$$\begin{vmatrix} S_n(x;q) & S_{n-1}(x;q) \\ \mathcal{F}_{1,n}(x) & \mathcal{F}_{2,n}(x) \end{vmatrix} \mathcal{E}_{3,n}(x) - \begin{vmatrix} S_n(x;q) & S_{n-1}(x;q) \\ \mathcal{E}_{1,n}(x) & \mathcal{E}_{2,n}(x) \end{vmatrix} \mathcal{F}_{3,n}(x)$$

$$= \mathcal{E}_{3,n}(x)\mathcal{F}_{2,n}(x)S_n(x;q) - \mathcal{E}_{3,n}(x)\mathcal{F}_{1,n}(x)S_{n-1}(x;q)$$

$$- \mathcal{E}_{2,n}(x)\mathcal{F}_{3,n}(x)S_n(x;q) + \mathcal{E}_{1,n}(x)\mathcal{F}_{3,n}(x)S_{n-1}(x;q)$$

$$= - \begin{vmatrix} \mathcal{E}_{2,n}(x) & \mathcal{E}_{3,n}(x) \\ \mathcal{F}_{2,n}(x) & \mathcal{F}_{3,n}(x) \end{vmatrix} S_n(x;q) + \begin{vmatrix} \mathcal{E}_{1,n}(x) & \mathcal{E}_{3,n}(x) \\ \mathcal{F}_{1,n}(x) & \mathcal{F}_{3,n}(x) \end{vmatrix} S_{n-1}(x;q).$$

The proof is complete after reordering terms in the above expression. $\square$

6.   Conclusions

In this chapter, a new family of q-Hermite I-Sobolev-type polynomials, orthogonal with respect to the q-discrete measure of the q-Hermite I polynomials modified by means of discrete measures with a single mass-point and first-order q-derivatives, has been presented. Interesting and intrincate results for this sequence of polynomials, such as several connection formulas and an explicit expression for its annihilation operator, have also been obtained.

A pending question is the expression for the corresponding creation operator, as well as a second-order q-difference equation satisfied by this q-Hermite I Sobolev-type polynomial. Thus, the results presented in this chapter can be the starting point to study the modified q-Hermite polynomials by adding more mass-points at the discrete part of the measure to later include higher-order q-derivatives. This study is not easy, since it involves a greater number of calculations as well as a greater complexity in the expressions, so new strategies are required.

Another important line of future research would deal with the analysis of the behavior of the zeros of these families and their possible physical interpretations.

Acknowledgments

The work of the first author (C.H.) was funded by Dirección General de Investigación e Innovación, Consejería de Educación e Investigación of the Comunidad de Madrid (Spain), and Universidad de Alcalá, under grant CM/JIN/2019-010, Proyectos de I+D para Jóvenes Investigadores de la Universidad de Alcalá 2019.

References

1. S. Alwhishi, R. S. Adigüzel and M. Turan, On the orthogonality of the q-derivatives of the discrete q-Hermite I polynomials. *Emerging Applications of Differential Equations and Game Theory*, IGI Global, Pennsilvania, USA (2020), 135–162.
2. J. Arvesú and A. Soria–Lorente, First-order non-homogeneous q-difference equation for Stieltjes function characterizing q-orthogonal polynomials. *J. Diff. Eqn. Appl.*, **19**(5), (2013), 814–838.
3. C. Berg, and M. E. H. Ismail, Q-Hermite polynomials and classical orthogonal polynomials. *Can. J. Math.*, **48**(1), (1996), 43–63.
4. L. Carlitz, Some polynomials related to theta functions. *Ann. Mat. Pura Appl.*, **41**(4), (1956), 359–373.
5. L. Carlitz, Some polynomials related to theta functions. *Duke Math. J.*, **24**(1957), 521–527.
6. L. Carlitz, Generating functions for certain q-orthogonal polynomials. *Collect. Math.*, **23**(1972), 91–104.
7. C. Hermoso, E. J. Huertas, and A. Lastra, Determinantal form for ladder operators in a problem concerning a convex linear combination of discrete and continuous measures. In: *FASdiff 2017 — Springer Proceedings in Mathematics and Statistics (PROMS) G. Filipuk et al.* (eds.), **256**, pp. 263–274 (2018).
8. C. Hermoso, E. J. Huertas, A. Lastra, and Anier Soria-Lorente, On second order q-difference equations satisfied by Al-Salam–Carlitz I-Sobolev type polynomials of higher order. *Mathematics*, **8**(8) (2020), 1300.
9. T. Ernst, q-Complex Numbers, A natural consequence of umbral calculus. *Uppsala University Department of Mathematics*, Report **44**(2007).

10. T. Ernst, q-Calculus as operational algebra. *Proc. Estonian Acad. Sci.*, **58**(2) (2009), 73–97.

11. G. Filipuk, J. F. Mañas-Mañas and J. J. Moreno-Balcázar, Ladders operators for general discrete Sobolev orthogonal polynomials, (2020), arXiv:2006.14391v1 [math.CA].

12. W. Hahn, Über orthogonalpolynome, die q-Differenzengleichungen genügen. *Math. Nachr.*, **2**(1949), 4–34.

13. R. Koekoek, P. A. Lesky, and R. F. Swarttouw, *Hypergeometric Orthogonal Polynomials and their q-Analogues.* Springer Science & Business Media, Berlín/Heidelberg, Germany, 2010.

14. L. J. Rogers, On the expansion of certain infinite products. *Proc. London Math. Soc.*, **24**(1893), 337–352.

15. L. J. Rogers, Second memoir on the expansion of certain infinite products. *Proc. London Math. Soc.*, **25**(1894), 318–343.

16. L. J. Rogers, Third memoir on the expansion of certain infinite products. *Proc. London Math. Soc.*, **26**(1895), 15–32.

17. H. M. Srivastava and J. Choi, *Zeta and q-Zeta Functions and Associated Series and Integrals*, Elsevier, 2001.

18. G. Szegő, Ein Beitrag zur Theorie der Thetafunktionen, *Sitz. Preuss. Akad. Wiss. Phys. Math. KI.*, **XIX**: (1926), 242–252.

© 2022 World Scientific Publishing Europe Ltd.
https://doi.org/10.1142/9781800611368_0015

Chapter 15

Divided-Difference Operators from the Geometric Point of View

Maria das Neves Rebocho

Departamento de Matemática, Universidade da Beira Interior
Rua Marquês D'Ávila e Bolama, 6201-001 Covilhã, Portugal
CMUC, Department of Mathematics, University of Coimbra
3001-501 Coimbra, Portugal
mneves@ubi.pt

It is presented a study of general divided-difference operators having the fundamental property of leaving a polynomial of degree $n - 1$ when applied to a polynomial of degree n.

1. Introduction

In the present chapter, it is presented a study on divided-difference operators having the fundamental property of leaving a polynomial of degree $n - 1$ when applied to a polynomial of degree n. Primarily, the focus is on the geometric interpretation, by analyzing the connection between the divided-difference operators and their relation with a corresponding conic, which, in turn, gives rise to a corresponding lattice of points that well-defines the operator (see [11]). Essentially, there are four primary classes of lattices and related divided-difference operators having the above-mentioned property: (i) the linear lattice, related to the forward difference operator [15, Chapter 2, Section 12]; (ii) the q-linear lattice, related to the q-difference operator [6]; (iii) the quadratic lattice, related to the Wilson operator [2];

301

(iv) the q-quadratic lattice, related to the Askey–Wilson operator [2]. This list gives a hierarchy of operators, as each of the operators in (i)–(iv) is an extension of the preceding one, which can be recovered as a special case and/or a limited case, up to a linear transformation of the variable.

The analysis of divided-difference operators (i)–(iv) is rather sparse in the literature. For instance, they are a fundamental machinery for the study of certain special functions appearing in problems from Mathematical-Physics, e.g. within the general theory of orthogonal polynomials (see [2,7,9,15]). Very often, when dealing with applications, final and combined formulae are given, together with a notation that may lead to a heavy reading for readers unaware of basic relations in the theory of divided-difference operators. With this idea in mind, the main goal of the present chapter is to give a concise but detailed study of some basic aspects of the divided-difference operators referred above, showing details of fundamental formulae that emerge from the geometric interpretation (given in the seminal paper [10]) and its connection with algebraic aspects of operator calculus. Here, the following topics are covered: the geometric interpretation — namely, the connection between the operator and a conic/lattice (cf. Section 2); the classification of operators in terms of a set of parameters in the given conic (cf. Section 3); the analysis of coalescences between the operators (cf. Section 4); basic and fundamental formulae in the divided-difference calculus (cf. Section 5).

2. The Conic and the Related Lattice

We start by following the approach from [10], where it is considered a divided-difference operator involving the values of a function at two points, with the property that it leaves a polynomial of degree $n - 1$ when applied to a polynomial of degree n. Let us take the divided-difference operator $\mathbb{D}_x$ as given in [10, Eq. (1.1)], defined on the space of arbitrary functions, by

$$\mathbb{D}_x f(x) = \frac{f(y_+(x)) - f(y_-(x))}{y_+(x) - y_-(x)}, \tag{1}$$

where, at this stage, y_+ and y_- are unknown functions. To define them, one starts by using the property that $\mathbb{D}_x f$ is a polynomial of degree $n-1$ whenever f is a polynomial of degree n. Then, applying $\mathbb{D}_x$ to $f(x) = x^2$ and $f(x) = x^3$, we obtain, respectively,

$$y_-(x) + y_+(x) = \text{polynomial of degree 1}, \qquad (2)$$

$$(y_-(x))^2 + y_-(x)y_+(x) + (y_+(x))^2 = \text{polynomial of degree 2}, \qquad (3)$$

the later condition being equivalent to $y_-(x)y_+(x) = $ polynomial of degree less or equal than two. From standard polynomial properties, the conditions (2)–(3) define y_- and y_+ as the two y-roots of a quadratic equation, say,

$$ay^2 + 2bxy + cx^2 + 2dy + 2ex + f = 0, \quad a \neq 0. \qquad (4)$$

The conic defined by the equation above plays an essential role in the sequel. The following identities, to be used later on, follow from the fact that y_-, y_+ are the y-roots of (4):

$$y_-(x) + y_+(x) = -2(bx + d)/a, \qquad (5)$$

$$y_-(x)y_+(x) = (cx^2 + 2ex + f)/a, \qquad (6)$$

$$y_-(x) = p(x) - \sqrt{r(x)}, \quad y_+(x) = p(x) + \sqrt{r(x)}, \qquad (7)$$

with p, r polynomials given by

$$p(x) = -\frac{b}{a}x - \frac{d}{a}, \quad r(x) = \frac{(b^2 - ac)}{a^2}x^2 + 2\frac{(bd - ae)}{a^2}x + \frac{(d^2 - af)}{a^2}. \qquad (8)$$

By virtue of (7), the operator $\mathbb{D}_x$ defined in (1) is given as

$$\mathbb{D}_x f(x) = \frac{f(p(x) + \sqrt{r(x)}) - f(p(x) - \sqrt{r(x)})}{2\sqrt{r(x)}}. \qquad (9)$$

Remark 1. The polynomials p, r will play a fundamental role in the sequel. Note that, from (7), it follows that

$$y_-(x) + y_+(x) = 2p(x), \quad (y_-(x) - y_+(x))^2 = 4r(x). \qquad (10)$$

Let us now look at the lattices.

Associated to each conic (4) two lattices are determined: the x-lattice and the y-lattice. The construction is based on the parametric representations of the conic, as follows (see [11]):

Let $\{x(s),\, y(s)\}$ be a parametric representation of the conic (4). For a given $x = x(s)$ value, the quadratic (4) defines two y-roots, say $y_s := y(s)$ and $y_{s+1} := y(s+1)$, which are the two ordinates associated to the abcissa $x(s)$. Then one starts from some point $\{x_1 = x(s_1),\, y_1 = y(s_1)\}$ on the conic, and one looks for the points $\{x_k = x(s_1+k),\, y_k = y(s_1+k)\}$, $k = 1, 2, \ldots$. This determines the so-called y-lattice, also known as the dual lattice. Conversely, if $c \neq 0$ in (4), then, for a given y-value, the quadratic (4) defines two x-roots, say $x_s := x(s)$, $x_{s+1} := x(s+1)$, which are consecutive points on the so-called x-lattice, also known as the direct lattice.

Remark 2. With the above notation, in terms of the operator $\mathbb{D}_x$ defined in (1), we have

$$y_s = y_-(x(s)), \quad y_{s+1} = y_+(x(s)).$$

2.1.   *The quadratic class of lattices — Explicit parameterizations*

The quadratic class of lattices appears when the conic (4) is such that $(b^2 - ac)(d^2 - af) - (bd - ae)^2 \neq 0$. Two subcases hold: the conic is a parabola — when $b^2 - ac = 0$ — this corresponds to the quadratic case; the conic is a hyperbola or an ellipse — when $b^2 - ac > 0$ or $b^2 - ac < 0$, respectively — this corresponds to the q-quadratic case.

For the quadratic class of lattices there is a parametric representation of the conic, say $\{x(s),\, y(s)\}$, such that the functions y_- and y_+ in (1) satisfy [11,14,16]

$$y_-(x(s)) = y(s) = x(s - 1/2), \quad y_+(x(s)) = y(s + 1) = x(s + 1/2). \tag{11}$$

Hence, the divided-difference operator (1) is given as

$$\mathbb{D}_x f(x(s)) = \frac{f(x(s + 1/2)) - f(x(s - 1/2))}{x(s + 1/2) - x(s - 1/2)}. \tag{12}$$

The parametrization on s is explicit [13], given by

$$x(s) = \tilde{\kappa}_2 s^2 + \tilde{\kappa}_1 s + \tilde{\kappa}_0, \tag{13}$$

where $\tilde{\kappa}_2 \neq 0$ in the quadratic case, and

$$x(s) = \kappa_1 q^s + \kappa_2 q^{-s} + \kappa_3, \tag{14}$$

where $\kappa_1\kappa_2 \neq 0$ in the q-quadratic case. Here, the κs and $\tilde{\kappa}$s are appropriate constants.

The parameterizations of the form (13) and (14) cover the whole set of canonical forms for the lattices. A formal deduction of formulae (13) and (14), based on properties of adjoint operators, will be given in Section 5.1.

Remark 3. Note that, in the account of (10) and (11), the polynomials p, r in (9) are then recovered under

$$x(s+1/2)+x(s-1/2) = 2p(x(s)), \quad (x(s+1/2) - x(s-1/2))^2 = 4r(x(s)).$$

Indeed, by writing $p(x) = p_1 x + p_0$, $r(x) = r_2 x^2 + r_1 x + r_0$, we get

$$p_1 = 1, \ p_0 = \tilde{\kappa}_2/4, \ r_2 = 0, \ r_1 = \tilde{\kappa}_2, \ r_0 = \tilde{\kappa}_1^2/4 - \tilde{\kappa}_2\tilde{\kappa}_0 \qquad (15)$$

in the case (13), and

$$p_1 = \frac{q^{1/2} + q^{-1/2}}{2}, \quad p_0 = \kappa_3 \left(1 - \frac{(q^{1/2} + q^{-1/2})}{2} \right), \qquad (16)$$

$$r_2 = \frac{(q^{1/2} - q^{-1/2})^2}{4}, \quad r_1 = -\kappa_3 \frac{(q^{1/2} - q^{-1/2})^2}{2}, \qquad (17)$$

$$r_0 = \left(-\kappa_1\kappa_2 + \frac{\kappa_3^2}{4}\right)(q^{1/2} - q^{-1/2})^2 \qquad (18)$$

in the case (14).

Magnus, in [11, p. 255], gives the following precise parameterizations.

Proposition 1. *Consider the conic* (4), $ay^2 + 2bxy + cx^2 + 2dy + 2ex + f = 0$, *with* $ac \neq 0$. *The following assertions hold.*

(a) *If the conic has a center* $\lambda := b^2 - ac \neq 0$, *then, with the center coordinates*

$$x_c = \frac{ae - bd}{\lambda}, \quad x_c = \frac{cd - be}{\lambda},$$

one has (4) *written in the form*

$$a(y - y_c)^2 + 2b(x - x_c)(y - y_c) + c(x - x_c)^2 + \tilde{f} = 0,$$

with

$$\tilde{f} = f - ay_c^2 - 2bx_cy_c - cx_c^2 = f + dy_c + ex_c = f + \frac{cd^2 - 2bde + ae^2}{\lambda}.$$

(a.1) *If $\tilde{f} \neq 0$, then*

$$x(s) = x_c + \xi\sqrt{a}(q^s + q^{-s}), \quad y(s) = y_c + \xi\sqrt{c}(q^{s-1/2} + q^{-s+1/2})$$

is a parametric representation of (4)*, where $\xi^2 = \tilde{f}/(4\lambda)$, and*

$$q^{1/2} + q^{-1/2} = -\frac{2b}{\sqrt{ac}}, \quad i.e. \quad q + q^{-1} = \frac{4b^2}{ac} - 2.$$

(a.2) *If $\tilde{f} = 0$, then one finds the parametric representation*

$$x(s) = x_c + X\sqrt{a}q^s, \quad y(s) = y_c + X\sqrt{c}q^{s\pm1/2},$$

for arbitrary parameters X.
(b) *If the conic has a center $\lambda := b^2 - ac = 0$, then*

$$\begin{cases} x(s) = \sqrt{a}\left(\dfrac{d^2 - af}{2a(d\sqrt{c} + e\sqrt{a})} - 2\dfrac{(d\sqrt{c} + e\sqrt{a})}{ac}s^2\right) \\[2em] y(s) = \sqrt{c}\left(\dfrac{e^2 - cf}{2c(d\sqrt{c} + e\sqrt{a})} - 2\dfrac{(d\sqrt{c} + e\sqrt{a})}{ac}(s - 1/2)^2\right) \end{cases}$$

is a parametric representation of (4)*.*

Remark 4. In the generic case q-quadratic case $|q| \neq 1$, the conic gives a hyperbola. In such a case, the asymptotes are given by $y = (c/a)^{1/2}q^{\pm1/2}x$, thus, q is precisely the ratio of the slopes of the asymptotes of the conic.

3.   Classification

There are four primary classes of lattices and related divided-difference operators:

(i)   the linear lattice, related to the forward difference operator [15, Chapter 2, Section 12];
(ii) the q-linear lattice, related to the q-difference operator [6];

(iii) the quadratic lattice, related to the Wilson operator [2];

(iv) the q-quadratic lattice, related to the Askey–Wilson operator [2].

Such a classification can be done according to the two parameters λ, τ defined in terms of the conic (4), $ay^2 + 2bxy + cx^2 + 2dy + 2ex + f = 0$, as follows:

$$\lambda = b^2 - ac, \quad \tau = \left((b^2 - ac)(d^2 - af) - (bd - ae)^2\right)/a, \qquad (19)$$

or, using the determinant notation,

$$\tau = \det \begin{bmatrix} a & b & d \\ b & c & e \\ d & e & f \end{bmatrix}.$$

Note that $\lambda \neq 0$ allows us to write the polynomial r in (8) as

$$r(x) = \frac{\lambda}{a^2}\left(x + \frac{bd - ae}{\lambda}\right)^2 + \frac{\tau}{a\lambda}. \qquad (20)$$

A detailed analysis of each case (i)–(iv), showing each of the operators in the form (9) with the corresponding polynomials p, r, is given in the following subsections.

3.1. *The linear lattice:* $\lambda = \tau = 0$ *in* (19)

If $\lambda = 0$ and $\tau = 0$, then, from (19), $bd - ae = 0$, thus, the polynomial r defined in (8) is constant, $r(x) = \frac{d^2 - af}{a^2}$. Hence, we have the polynomials p, r defined in (8) given by

$$p(x) = -\frac{b}{a}x - \frac{d}{a}, \quad r(x) = \frac{d^2 - af}{a^2}.$$

Recalling (7), it follows that

$$y_\pm(x) = -\frac{b}{a}x - \frac{d}{a} \pm \frac{\sqrt{d^2 - af}}{a^2}, \qquad (21)$$

that is, we have two parallel lines,

$$y_\pm(x) = mx \pm b_\pm,$$

with

$$m = -\frac{b}{a}, \quad b_{\pm} = -\frac{d}{a} \pm \frac{\sqrt{d^2 - af}}{a^2}.$$

Proposition 2. *The canonical divided-difference operator related to the linear lattices is the forward difference operator* $\mathbb{D}_x = \Delta_w$ *— the so-called Hahn operator* [6], *where*

$$\Delta_w f(x) = \frac{f(x + w) - f(x)}{w}, \quad w \neq 0, \tag{22}$$

for arbitrary functions f. *Hence, the operator* Δ_w *can be written in the form* (9), *with the polynomials* p, r *given by*

$$p(x) = x + \frac{w}{2}, \quad r(x) = \frac{w^2}{4}.$$

Proof. Combining (1) with (21), the operator (22) is recovered through the specialization

$$b = -a \ c = a, \quad d = -aw/2, \quad e = aw/2, \quad f = 0, \tag{23}$$

and it follows the assertion on the polynomials p, r. $\qquad\qquad\square$

Also, by using the values of (23) into (4), we get the conic with equation

$$y^2 - 2xy + x^2 - wy + wx = 0,$$

which can be factorized as

$$(y - x)(y - x - w) = 0.$$

The linear lattice, obtained via two parallel lines, is illustrated in ([11, p. 256, Fig. 2.d]).

3.2.   *The q-linear lattice:* $\lambda \neq 0, \ \tau = 0$ *in* (19)

If $\lambda \neq 0$ and $\tau = 0$, the polynomials p, r defined in (8) are given by

$$p(x) = -\frac{b}{a}x - \frac{d}{a}, \quad r(x) = \frac{\lambda}{a^2}\left(x + \frac{bd - ae}{\lambda}\right)^2.$$

Recalling (7), it follows that

$$y_\pm(x) = -\frac{b}{a}x - \frac{d}{a} \pm \frac{\sqrt{\lambda}}{a}\left(x + \frac{bd - ae}{\lambda}\right), \qquad (24)$$

that is, we have two intersecting lines,

$$y_+(x) = m_+ x + b_+, \quad y_-(x) = m_- x + b_-,$$

with

$$m_+ = -\frac{b}{a} + \frac{\sqrt{\lambda}}{a}, \quad m_- = -\frac{b}{a} - \frac{\sqrt{\lambda}}{a},$$

$$b_+ = \frac{\sqrt{\lambda}}{a}\left(\frac{bd - ae}{\lambda}\right) - \frac{d}{a}, \quad b_- = -\frac{\sqrt{\lambda}}{a}\left(\frac{bd - ae}{\lambda}\right) - \frac{d}{a}.$$

Proposition 3. *The canonical divided-difference operator related to the q-linear lattices is the q-linear difference operator, $\mathbb{D}_x = \Delta_{q,w}$ [6], where*

$$\Delta_{q,w}f(x) = \frac{f(qx + w) - f(x)}{(q - 1)x + w}, \quad q \neq 1, \qquad (25)$$

for arbitrary functions f. Hence, the operator $\Delta_{q,w}$ can be written in the form (9), with the polynomials p, r given by

$$p(x) = \frac{(q + 1)}{2}x + \frac{w}{2}, \quad r(x) = \frac{(q - 1)^2}{4}\left(x + \frac{w}{q - 1}\right)^2.$$

Proof. Combining (1) with (24), the operator (25) is recovered through the specialization

$$b = -\frac{(q + 1)}{2}a, \quad c = qa, \quad d = -\frac{w}{2}a, \quad e = \frac{w}{2}a, \quad f = 0, \qquad (26)$$

and it follows the assertion on the polynomials p, r. $\qquad\square$

Also, by using the values of (26) into (4), we get the conic with equation

$$y^2 - (q + 1)xy + qx^2 - wy + wx = 0,$$

which can be factorized as

$$(y - x)(y - qx - w) = 0.$$

The q-linear lattice, obtained via two intersecting lines, is illustrated in ([11, p. 256, Fig. 2.b]).

Remark 5. In [6, p. 6], it is shown that, whenever $q \neq 1$, the constant w in (25) can be eliminated through a linear transformation: by setting $x = \hat{a}z + \hat{b}$ and $f(x) = h(z)$, the operator $\Delta_{q,w}$ can be written as

$$\Delta_{q,w} f(x) = \frac{h\left(qz + \frac{(q-1)\hat{b}+w}{\hat{a}}\right) - h(z)}{(q-1)z + \frac{(q-1)\hat{b}+w}{\hat{a}}}.$$

Now, choosing $\hat{a} = 1$, $\hat{b} = \dfrac{w}{1-q}$, we get the operator

$$\mathcal{D}_q f(x) = \frac{f(qx) - f(x)}{(q-1)x}. \tag{27}$$

3.3.   *The quadratic lattice: $\lambda = 0$, $\tau \neq 0$ in (19)*

If $\lambda = 0$ and $\tau \neq 0$, the polynomials p, r defined in (8) are both of degree one, given by

$$p(x) = -\frac{b}{a}x - \frac{d}{a}, \quad r(x) = 2\frac{(bd - ae)}{a^2}x + \frac{(d^2 - af)}{a^2}.$$

Recalling (7), it follows that

$$y_{\pm}(x) = -\frac{b}{a}x - \frac{d}{a} \pm \frac{\sqrt{2(bd - ae)x + (d^2 - af)}}{a}. \tag{28}$$

Proposition 4. *The canonical divided-difference operator related to the quadratic lattices is the Wilson operator* [1,2], *$\mathbb{D}_x = \mathcal{W}$ where*

$$\mathcal{W}f(x) = \frac{f\left((\sqrt{x} + \frac{i}{2})^2\right) - f\left((\sqrt{x} - \frac{i}{2})^2\right)}{2i\sqrt{x}}, \tag{29}$$

for arbitrary functions f. Hence, the operator $\mathcal{W}$ can be written in the form (9), with the polynomials p, r given by

$$p(x) = x - \frac{1}{4}, \quad r(x) = -x.$$

Proof. Combining (1) with (28), the operator (29) is recovered through the specialization

$$b = -a, \quad c = a, \quad d = e = \frac{a}{4}, \quad f = \frac{a}{16}, \tag{30}$$

and it follows the assertion on the polynomials p, r. $\qquad\square$

Also, by using the values of (30) into (4), we get the conic with equation

$$y^2 - 2xy + x^2 + \frac{y}{2} + \frac{x}{2} + \frac{1}{16} = 0,$$

which is a parabola (we have $\lambda = 0$ and $\tau < 0$). The corresponding lattice, obtained via a parabola, is illustrated ([11, p. 256, Fig. 2.c]).

3.4. *The q-quadratic lattice:* $\lambda \neq 0, \quad \tau \neq 0$ *in* (19)

If $\lambda \neq 0$ and $\tau \neq 0$, the polynomials p, r defined in (8) are of degree one and two, respectively, given as

$$p(x) = -\frac{b}{a}x - \frac{d}{a}, \quad r(x) = r(x) = \frac{\lambda}{a^2}\left(x + \frac{bd - ae}{\lambda}\right)^2 + \frac{\tau}{a\lambda}.$$

Recalling (7), it follows that

$$y_\pm(x) = -\frac{b}{a}x - \frac{d}{a} \pm \sqrt{\frac{\lambda}{a^2}\left(x + \frac{bd - ae}{\lambda}\right)^2 + \frac{\tau}{a\lambda}}. \tag{31}$$

Under some specializations, by considering the centred and symmetrized forms of the lattice, one can recover the Askey–Wilson operator [1,2] (see also [7, Eq. (12.1.12)]), given by

$$\mathbb{D}_x f(x) = \frac{f(\frac{1}{2}(q^{1/2}z + q^{-1/2}z^{-1})) - f(\frac{1}{2}(q^{-1/2}z + q^{1/2}z^{-1}))}{\frac{1}{2}(q^{1/2} - q^{-1/2})(z - z^{-1})}. \tag{32}$$

Indeed, let us begin by defining the base $q = e^{2i\eta}$ and consider the projection map from the unit circle $\{z = e^{i\theta}, \ \theta \in [-\pi, \pi[\}$ onto

$[-1, 1]$ by

$$x = \frac{1}{2}(z + z^{-1}).$$

Note that we have

$$y_-(x) = \frac{1}{2}(q^{-1/2}z + q^{1/2}z^{-1}), \quad y_+(x) = \frac{1}{2}(q^{1/2}z + q^{-1/2}z^{-1}). \quad (33)$$

Proposition 5. *The canonical divided-difference operator related to the q-quadratic lattices, in the symmetrical form, is the Askey–Wilson operator* (32) [1,2]. *The operator* (32) *can be written in the form* (9), *with the polynomials* p, r *given by*

$$p(x) = \frac{(q^{1/2} + q^{-1/2})}{2}x, \quad r(x) = \frac{(q^{1/2} - q^{-1/2})}{4}(x^2 - 1).$$

Proof. Combining (1) with (33), we have, after basic computations,

$$y_-(x) + y_+(x) = 2\cos(\eta)x = (q^{1/2} + q^{-1/2})x, \quad (34)$$

$$(y_-(x) - y_+(x))^2 = (q^{1/2} - q^{-1/2})(x^2 - 1). \quad (35)$$

In the account of (10), that is, $y_-(x) + y_+(x) = 2p(x)$ and $(y_-(x) - y_+(x))^2 = 4r(x)$, there follow the polynomials p, r as stated.

The operator (32) is recovered through the specialization

$$a = c, \text{ arbitrary and non-zero}, \quad b = -a\cos(\eta), \quad d = e = 0,$$

$$f = -a\sin^2(\eta). \qquad\qquad \square$$

In the q-quadratic case, the conic is an hyperbola (when $\lambda > 0$ and $\tau < 0$), or an ellipse (when $\lambda < 0$ and $\tau < 0$, respectively). The corresponding lattice, obtained via an hyperbola or an ellipse, is illustrated) in [11, p. 256], Figs. 1 and 2.a).

4.   Coalescence

The set of lattices previously defined can be classified through specifications on the constants in the parametrization formulae (13) and (14), that is, in

$$x(s) = \tilde{\kappa}_2 s^2 + \tilde{\kappa}_1 s + \tilde{\kappa}_0$$

and

$$x(s) = \kappa_1 q^s + \kappa_2 q^{-s} + \kappa_3,$$

respectively. Indeed, depending on the constants κs and $\tilde{\kappa}$s, we recover the four primary classes for the lattices $x(s)$:

(i) Linear lattices: $\tilde{\kappa}_2 = 0$ and $\tilde{\kappa}_1 \neq 0$ in (13);
(ii) q-linear lattices: $\kappa_2 = 0$ and $\kappa_1 \neq 0$ in (14);
(iii) Quadratic lattices: $\tilde{\kappa}_2 \neq 0$ in (13);
(iv) q-Quadratic lattices: $\kappa_1 \kappa_2 \neq 0$ in (14).

The q-quadratic lattice, in its general non-symmetrical form, is the most general case and the other lattices can be found from this by limiting processes.

It turns out that each of the operators listed in (i)–(iii) of the previous section, specified in Sections 3.1–3.3, can be recovered as a particular case or as a limit case, up to a linear transformation of the variable, from one of the operators in the list. Details are given as follows.

Recall the polynomials p, r in (8): by writing $p(x) = p_1 x + p_0$, $r(x) = r_2 x^2 + r_1 x + r_0$, we have

$$p_1 = -\frac{b}{a}, \quad p_0 = -\frac{d}{a}, \tag{36}$$

$$r_2 = \frac{b^2 - ac}{a^2}, \quad r_1 = 2\frac{(bd - ae)}{a^2}, \quad r_0 = \frac{d^2 - af}{a^2}. \tag{37}$$

4.1. *From q-quadratic to quadratic*

Taking limits $q \to 1$ in (16), as well as in (17), we get $p_1 = 1$ and $r_2 = 0$. In the account of (37), $r_2 = 0$ yields $b^2 - ac = 0$. Furthermore, in the account of (37), note that $\tau \neq 0$ in (19) if, and only if, $r_0 r_2 - (r_1/2)^2 \neq 0$. As we have $r_2 = 0$, then $\tau \neq 0$ if, and only if, $r_1 \neq 0$, which must hold upon a suitable choice of κ_3. Thus, we get the quadratic case: $\lambda = 0$ and $\tau \neq 0$ (cf. Section 3.3).

4.2. *From q-quadratic to q-linear*

Recalling Remark 5, let us take the operator $\mathcal{D}_q$ defined by (27),

$$\mathcal{D}_q f(x) = \frac{f(qx) - f(x)}{(q-1)x}.$$

We begin by fixing the parameter $q \neq 1$. Taking limits $\kappa_2 \to 0$, $\kappa_3 \to 0$, and fixing $q \neq 1$ in (14) we get $r_2 \neq 0$, $r_1 = 0$, $r_0 = 0$ in (17)–(18), that, in the account of (37), yields $b^2 - ac \neq 0$, $bd - ae = 0$, $d^2 - af = 0$. Thus, we get the q-linear case: $\lambda \neq 0$ and $\tau = 0$ (cf. Section 3.2).

Note that, in such a situation, the operator $\mathcal{D}_q$ obtained via the above limiting process is given by

$$\mathcal{D}_q f(x(s)) = \frac{f(\kappa_1 q^{s+1/2}) - f(\kappa_1 q^{s-1/2})}{\kappa_1 (q^{s+1/2} - q^{s-1/2})},$$

which can be easily written as (27) through the change of variable $x(s) = \kappa_1 q^{s-1/2}$.

4.3. *From q-linear to linear*

The linear case follows easily by taking limits $q \to 1$ in (25). Indeed, we get the coefficients of the polynomials p, r as given in Proposition 2, thus, in the account of (37), we have $\lambda = 0$ and $\tau = 0$ (cf. Section 3.1).

5. Divided-Difference Operator Calculus

Recall the operator $\mathbb{D}_x$ in its general form given by (1), together with the corresponding conic (4) and the polynomials p, r defined in (8). In the sequel, we shall take $\Delta_y = y_+ - y_-$. From (7), there follows:

$$\Delta_y = 2\sqrt{r}. \tag{38}$$

In order to deduce further properties, let us now introduce the operators $\mathbb{E}_x^+$ and $\mathbb{E}_x^-$ (see [10]), acting on arbitrary functions f, as

$$\mathbb{E}^\pm f(x) = f(y_\pm(x)).$$

With this notation, (1) is also given by

$$\mathbb{D}_x f(x) = \frac{\mathbb{E}_x^+ f - \mathbb{E}_x^- f}{\mathbb{E}_x^+ x - \mathbb{E}_x^- x} .$$

The companion operator of $\mathbb{D}$ is then defined as (see [10])

$$\mathbb{M}_x f(x) = \frac{\mathbb{E}_x^+ f(x) + \mathbb{E}_x^- f(x)}{2} . \tag{39}$$

Note that $\mathbb{M}_x f$ is a polynomial whenever f is a polynomial. Furthermore, if $\deg(f) = n$, then $\deg(\mathbb{M}_x f) = n$.

The operators $\mathbb{D}_x$ and $\mathbb{M}_x$ satisfy the product and quotient rules listed as follows (see [10]):

$$\mathbb{D}_x(fg) = \mathbb{D}_x f \, \mathbb{M}_x g + \mathbb{M}_x f \, \mathbb{D}_x g , \tag{40}$$

$$\mathbb{D}_x(f/g) = \frac{\mathbb{D}_x f \, \mathbb{M}_x g - \mathbb{D}_x g \, \mathbb{M}_x f}{\mathbb{E}_x^- f \, \mathbb{E}_x^+ f} , \tag{41}$$

$$\mathbb{M}_x(fg) = \mathbb{M}_x f \, \mathbb{M}_x g + \frac{\Delta_y^2}{4} \, \mathbb{D}_x f \, \mathbb{D}_x g, \tag{42}$$

$$\mathbb{M}_x(f/g) = \frac{\mathbb{E}_x^- f \, \mathbb{E}_x^+ g + \mathbb{E}_x^+ f \, \mathbb{E}_x^- g}{2\mathbb{E}_x^- g \, \mathbb{E}_x^+ g} . \tag{43}$$

Equation (40) has the equivalent forms:

$$\mathbb{D}_x(gf) = \mathbb{D}_x g \, \mathbb{E}_x^- f + \mathbb{D}_x f \, \mathbb{E}_x^+ g ,$$

$$\mathbb{D}_x(gf) = \mathbb{D}_x g \, \mathbb{E}_x^+ f + \mathbb{D}_x f \, \mathbb{E}_x^- g .$$

Also, one has two equivalent forms for (41):

$$\mathbb{D}_x(g/f) = \frac{\mathbb{D}_x g \, \mathbb{E}_x^- f - \mathbb{D}_x f \, \mathbb{E}_x^- g}{\mathbb{E}_x^- f \, \mathbb{E}_x^+ f} ,$$

$$\mathbb{D}_x(g/f) = \frac{\mathbb{D}_x g \mathbb{E}_x^+ f - \mathbb{D}_x f \mathbb{E}_x^+ g}{\mathbb{E}_x^- f \, \mathbb{E}_x^+ f} .$$

The operators $\mathbb{D}_x$ and $\mathbb{M}_x$ also satisfy the product rules II (see [5, Eq. (15), 4])

$$\mathbb{D}_x \mathbb{M}_x = \alpha \mathbb{M}_x \mathbb{D}_x + U_1 \mathbb{D}_x^2 , \quad \mathbb{M}_x^2 = U_1 \mathbb{M}_x \mathbb{D}_x + \alpha \frac{\Delta_y^2}{4} \mathbb{D}_x^2 + \mathbb{I}, \tag{44}$$

where $\mathbb{I}$ is the identity operator, $\mathbb{I}f(x) = f(x)$, and

$$U_1(x) = (p_1^2 - 1)x + \frac{r_1}{2}, \tag{45}$$

with p_1 and r_1 defined in (15) in the quadratic case, or in (16)–(18) in the q-quadratic case.

5.1. *The explicit parameterizations revisited*

Let us recall the conic (4), $ay^2 + 2bxy + cx^2 + 2dy + 2ex + f = 0$, $a \neq 0$, as well as its two y-roots, satisfying (5) and (6). Assuming $c \neq 0$ in (4), one defines the inverse functions of y_- and y_+, denoted by y_-^{-1} and y_+^{-1}, respectively, such that

$$y_-^{-1}(y_-(x)) = x, \quad y_+^{-1}(y_+(x)) = x,$$

together with the corresponding operators

$$\left(\mathbb{E}_x^-\right)^{-1} f(x) = f\left(y_-^{-1}(x)\right), \quad \left(\mathbb{E}_x^+\right)^{-1} f(x) = f\left(y_+^{-1}(x)\right). \tag{46}$$

Let us also define the operators $\mathbb{E} = (\mathbb{E}_x^-)^{-1}\mathbb{E}_x^+$, $\mathbb{E}^{-1} = (\mathbb{E}_x^+)^{-1}\mathbb{E}_x^-$ by (see [10])

$$\mathbb{E}f(x) = f\left(y_+(y_-^{-1}(x))\right), \quad \mathbb{E}^{-1}f(x) = f\left(y_-(y_+^{-1}(x))\right). \tag{47}$$

In order to deduce the parameterizations of the quadratic and q-quadratic cases, we first present the following lemma. The results are gathered in [10], but here we detail its proof.

Lemma 1. *Recalling the conic (4) and the operators previously defined, the following equalities hold:*

$$\mathbb{E}x + x = \frac{-2(by_-^{-1}(x) + d)}{a}, \tag{48}$$

$$\mathbb{E}^{-1}x + x = \frac{-2(by_+^{-1}(x) + d)}{a}, \tag{49}$$

$$y_-^{-1}(x) + y_+^{-1}(x) = \frac{-2(bx + e)}{c}, \tag{50}$$

$$\mathbb{E}x + \mathbb{E}^{-1}x = 2\left(\frac{2b^2}{ac} - 1\right)x + 4\left(\frac{be - cd}{ac}\right). \tag{51}$$

Proof.   Equations (48) and (49) follow by taking $x = y_-^{-1}(X)$ and $x = y_+^{-1}(X)$, respectively, in (5), $y_-(x) + y_+(x) = -2(bx + d)/a$.

To deduce (50), we start by evaluating (6) at $y_-^{-1}(x)$ as well as at $y_+^{-1}(x)$, thus getting

$$x\, y_+(y_-^{-1}(x)) = \frac{c(y_-^{-1}(x))^2 + 2ey_-^{-1}(x) + f}{a}, \tag{52}$$

$$x\, y_-(y_+^{-1}(x)) = \frac{c(y_+^{-1}(x))^2 + 2ey_+^{-1}(x) + f}{a}. \tag{53}$$

Subtracting (53) to (52) yields

$$x\left(y_+(y_-^{-1}(x)) - y_-(y_+^{-1}(x))\right)$$
$$= \frac{c\left((y_-^{-1}(x))^2 - (y_+^{-1}(x))^2\right) + 2e\left(y_-^{-1}(x) - y_+^{-1}(x)\right)}{a}.$$

Thus, we have

$$\mathbb{E}x + x - (\mathbb{E}^{-1}x + x) = \frac{(y_-^{-1}(x) - y_+^{-1}(x))}{xa}(c(y_-^{-1}(x) + y_+^{-1}(x)) + 2e). \tag{54}$$

Using (48) and (49) in (54) gives us, after simplifications, equation (50).

Equation (51) follows from the sum of (48) with (49), and using (50). $\qquad\square$

Applying $\mathbb{E}^n$ to (51), we obtain the difference equation

$$\mathbb{E}^{n+1}x + \mathbb{E}^{n-1}x = 2\left(\frac{2b^2}{ac} - 1\right)\mathbb{E}^n x + 4\left(\frac{be - cd}{ac}\right). \tag{55}$$

The solution of the equation (55) leads us to the form of the parameterizations already discussed in Section 2.1 (see [10, p. 264]; [13]). Here, the detailed proof is given in what follows.

Theorem 1. *Let q satisfy*

$$q + q^{-1} = 2\left(\frac{2b^2}{ac} - 1\right). \tag{56}$$

The solution of the difference equation (55) is given by

$$\mathbb{E}^n x = \alpha q^n + \beta q^{-n} + \frac{cd - be}{b^2 - ac}, \quad if\, q \neq 1 \tag{57}$$

or

$$\mathbb{E}^n x = \alpha + \beta n + \frac{2(be - cd)}{ac} n^2, \quad \textit{if } q = 1. \tag{58}$$

Proof. Recall that the solution of a difference equation such as (55), say,

$$X_{n+1} - \xi X_n + X_{n-1} = 4 \left(\frac{be - cd}{ac} \right), \quad \xi = 2 \left(\frac{2b^2}{ac} - 1 \right), \tag{59}$$

can be written as $X_n = X_{h,n} + X_p$, with $X_{h,n}$ the solution of the homogeneous equation

$$X_{n+1} - \xi X_n + X_{n-1} = 0 \tag{60}$$

and X_p a particular solution of the complete equation (59). Also, denoting by ξ_1, ξ_2 the two roots of the so-called associated characteristic equation of (60),

$$x^2 - \xi x + 1 = 0, \tag{61}$$

the solution of (60) is given by (see [12])

$$X_{h,n} = \begin{cases} \alpha \xi_1^n + \beta \xi_2^n & \text{if } \xi_1 \neq \xi_2, \\ \alpha \xi_1^n + \beta n \xi_1^n & \text{if } \xi_1 = \xi_2. \end{cases}$$

Note that the roots of $x^2 - \xi x + 1 = 0$ are $q_\pm := \frac{\xi \pm \sqrt{\xi^2 - 4}}{2}$. Hence, when $\xi^2 - 4 \neq 0$, we have two different roots of the quadratic equation, which satisfy indeed $q_- = (q_+)^{-1}$, and $q_- + q_+ = \xi$. Thus, we have the parameter q, say $q = q_+$, defined as in (56). If $\xi^2 - 4 = 0$, then $\xi = 2$, which implies the double root of the quadratic equation being $q := q_- = q_+ = 1$, thus, also defined as in (56).

Finally, we get (57) in the account that $\lambda := \frac{cd - be}{b^2 - ac}$ is a particular solution of the complete equation (59) in the case of two different roots of (61), and we get (58) in the account that $\lambda := \frac{2(be - cd)}{ac} n^2$ is a particular solution of the complete equation (59) in the case of a double root of (61). $\qquad \square$

5.2. *The divided-difference operators as exact lowering operators*

We now give the analogues of the well-known formulae for the continuous case $\frac{d}{dx}x^n = nx^{n-1}$, as proposed by [16]. Further details are given in the more recent approach [18].

Let $\{l_n(x;a)\}_{n=0}^{+\infty}$ be a polynomial basis of $L^2(w(x)\mathbb{D}x, G)$, where l_n is a polynomial of exact degree n and the support is $G = \{\mathbb{E}^{+k}x : k \in 2\mathbb{Z}\}$ or, if finite, $G = \{x_0, \ldots, x_{n_0}\}$, and a denotes the set of parameters characterizing the lattice. The general requirements for the polynomial basis are:

(i) $l_n(x)$ is of precise degree n in x,

(ii) $\mathbb{D}_x$ is an exact lowering operator in this basis, that is, $\mathbb{D}_x l_n(x) = c_n l_{n-1}(x)$, $n \geq 1$, where $c_n = c_n(\check{a})$ is a constant with respect to x, depending on a set of parameters $\check{a} := \{a_1, a_2, \ldots, a_{m_0}\}$, characterizing the lattice.

A general solution of the above requirements is the polynomial defined by (see [18, Sec. 2])

$$l_n(x;\check{a}) = g_n(\check{a}) \prod_{j=0}^{n-1} \left(x - \left(\mathbb{E}_x^+\right)^{2j} x(\check{a}) \right),$$

where $x(\check{a})$ denotes the so-called basal point, parameterized by $\check{a}$, and $g_n(\check{a}) \neq 0$.

We have the following:

(1) In the q-quadratic lattice $x(s) = \kappa_1 q^s + \kappa_2 q^{-s} + \kappa_3$, with $q \neq 1$ and $\kappa_1 > 0$, $\kappa_2 > 0$, the basis is

$$l_n(x(s)) = g_n \left(\frac{q^{-\frac{n}{2}+s+\frac{1}{4}}\sqrt{\kappa_1}}{\sqrt{\kappa_2}}; q \right)_n \left(\frac{q^{-\frac{n}{2}-s+\frac{1}{4}}\sqrt{\kappa_2}}{\sqrt{\kappa_1}}; q \right)_n, \quad n \geq 1, \tag{62}$$

with

$$g_n = g_n(\kappa_1, \kappa_2, q) = \left(-\frac{\kappa_1^{3/2} q^{1/4}}{\sqrt{\kappa_2}} \right)^n.$$

The divided-difference operator satisfies $\mathbb{D}_x l_n(x(s)) = c_n l_{n-1}(x(s))$, $n \geq 1$, that is,

$$\mathbb{D}_x l_n(x(s)) = \frac{l_n(x(s+1/2)) - l_n(x(s-1/2))}{x(s+1/2) - x(s-1/2)} = c_n l_{n-1}(x(s)),$$

with

$$c_n = c_n(\kappa_1, \kappa_2, q) = \frac{\kappa_1 q^{\frac{1-n}{2}} [n]_q}{\kappa_2}.$$

Here, the Pochhammer symbol is used, given by

$$(a;q)_0 = 1, \quad (a;q)_n = \prod_{j=0}^{n-1}(1 - aq^j), \quad n = 1, 2, \ldots,$$

and the number $[z]_q$ is defined by

$$[z]_q = \frac{q^z - 1}{q - 1}.$$

(2) In the quadratic lattice $x(s) = \tilde{\kappa}_2 s^2 + \tilde{\kappa}_1 s + \tilde{\kappa}_0$, with $\tilde{\kappa}_2 \neq 0$, the basis is

$$l_n(x(s)) = 4^{-n}(-\tilde{\kappa}_2)^n \left(-\frac{\tilde{\kappa}_1}{\tilde{\kappa}_2} - 2s + \frac{1}{2}\right)_n \left(\frac{\tilde{\kappa}_1}{\tilde{\kappa}_2} + 2s + \frac{1}{2}\right)_n.$$
$$n \geq 1. \tag{63}$$

The divided-difference operator satisfies $\mathbb{D}_x l_n(x(s)) = c_n l_{n-1}(x(s))$, $n \geq 1$, that is,

$$\mathbb{D}_x l_n(x(s)) = \frac{l_n(x(s+1/2)) - l_n(x(s-1/2))}{x(s+1/2) - x(s-1/2)} = c_n l_{n-1}(x(s)),$$

with

$$c_n = n.$$

Here, the Pochhammer symbol $(A)_n = A(A+1)\cdots(A+n-1)$ is used.

(3) In the q-linear lattice, the basis is

$$l_n(x) = (\breve{a}x; q)_n = \prod_{j=0}^{n-1}(1 - \breve{a}q^j x), \quad n \geq 1. \tag{64}$$

The divided-difference operator, taken in its canonical form as the $\mathcal{D}_q$ operator given in (27), satisfies $\mathcal{D}_q l_n(x) = c_n l_{n-1}(x)$, $n \geq 1$, that is,

$$\mathcal{D}_q l_n(x) = \frac{l_n(qx) - l_n(x)}{(q-1)x} = c_n l_{n-1}(x),$$

with

$$c_n = -\frac{1 - \breve{a}q^n}{q - 1}.$$

(4) In the linear lattice, the basis is

$$l_n(x) = \prod_{j=0}^{n-1}(x - j) = \frac{\Gamma(x+1)}{\Gamma(x-n+1)}, \quad n \geq 1, \tag{65}$$

where $\Gamma(\cdot)$ denotes the Gamma function. The divided-difference operator, taken in its canonical form as the forward difference operator $\Delta f(x) = f(x+1) - f(x)$, satisfies

$$\Delta l_n(x) = l_n(x+1) - l_n(x) = c_n l_{n-1}(x),$$

with

$$c_n = n.$$

5.3. *Integrals*

Let the lattice points be denoted by $G[x] = \{x(s) : s \in \mathbb{Z}\}$, with the point $x(0)$ as the basal point, and let us denote the dual lattice by $\tilde{G}[x] = \{x(s+1/2) : s \in \mathbb{Z}\}$. The $\mathbb{D}$-integral of a function defined on the x-lattice, $f : G[x] \to \mathbb{C}$ with basal point $x_0 = x(0)$, is defined by the Riemmann sum over the lattice points (see [18, Sec. 2])

$$I[f](x_0) = \int_G f(x(s))\mathbb{D}x(s) := \sum_{s \in \mathbb{Z}^*} f(x(s))(y_+(x(s)) - y_-(x(s))). \tag{66}$$

Recalling that, in the quadratic case, $y_+(x(s)) = x(s + 1/2)$, $y_-(x(s)) = x(s - 1/2)$, and also recalling the notation $x_s := x(s))$, we can write

$$I[f](x_0) = \sum_{s\in\mathbb{Z}^*} f(x(s))((x(s+1/2))-(x(s-1/2))) = \sum_{s\in\mathbb{Z}^*} f(x_s)\Delta_y(x_s).$$

Here, $\mathbb{Z}^*$ is a finite subset of $\mathbb{Z}$, namely $\{0,1,\ldots,n_0\}$, or $\mathbb{Z}_{\geq 0}$, or $\mathbb{Z}$.

Recalling that $\mathbb{E}_x^{\pm} f(x(s)) = f(x(s\pm 1/2))$, for $x(s) \in G[x]$, the following properties follow from (66) (see [18]):

(1) an analog of the fundamental theorem of calculus:

$$\int_{x_0 \leq x_s \leq x_{n_0}} \mathbb{D}_x f(x(s))\mathbb{D}x(s) = f(\mathbb{E}_x^+ x_{n_0}) - f(\mathbb{E}_x^- x_0). \qquad (67)$$

(2) an analog of integration by parts for two functions $f(x), g(x)$:

$$\int_{x_0 \leq x_s \leq x_{n_0}} f(x(s))\mathbb{D}_x g(x(s))\mathbb{D}x(s) = f(\mathbb{E}_x^{+\,2} x_{n_0})g(\mathbb{E}_x^+ x_{n_0})$$

$$- f(x_0)g(\mathbb{E}_x^- x_0)$$

$$- \int_{x_0 \leq x_s \leq x_{n_0}} \mathbb{D}_x f(\mathbb{E}_x^+ x(s))g(\mathbb{E}_x^+ x(s))\mathbb{D}\left(\mathbb{E}_x^+ x(s)\right). \qquad (68)$$

Remark 6. The definition (66) reduces to the usual definition of the difference integral and the Thomae–Jackson q-integrals in the canonical forms of the linear and q-linear lattices, respectively [8,17].

Acknowledgments

This work was partially supported by the Centre for Mathematics of the University of Coimbra (UIDB/00324/2020, funded by the Portuguese Government through FCT/MCTES).

References

1. G. E. Andrews and R. Askey, Classical orthogonal polynomials, pp. 36–62 in: "Polynômes Orthogonaux et Applications, Proceedings, Bar-le-Duc 1984", Lecture Notes Math. 1171 (C. Brezinski *et al.* Editors), Springer, Berlin 1985.

2. R. Askey and J. Wilson, Some basic hypergeometric orthogonal polynomials that generalize Jacobi polynomials. *Memoirs AMS*, **54**(319) (1985).

3. N. M. Atakishiev, M. Rahman, and S. K. Suslov, On classical orthogonal polynomials. *Construct. Approx.*, **11**(1995), 181–226.

4. M. Foupouagnigni, On difference equations for orthogonal polynomials on nonuniform lattices. *J. Diff. Eqn. Appl.*, **14**(2008), 127–174.

5. M. Foupouagnigni, M. Kenfack Nangho, and S. Mboutngam, Characterization theorem for classical orthogonal polynomials on nonuniform lattices: The functional approach. *Integral Trans. Spec. Funct.*, **22**(2011), 739–758.

6. W. Hahn, Über Orthogonalpolynome, die q-Differenzengleichungen genügen. *Math. Nachr.*, **2**(1949), 4–34.

7. M. E. H. Ismail, *Classical and Quantum Orthogonal Polynomials in One Variable*, Vol. 98 of Encyclopedia of Mathematics and its Applications. Cambridge University Press, Cambridge, 2005.

8. F. Jackson, On q-definite integrals. *Q. J. Pure Appl. Math.*, **41**(1910), 193–203.

9. R. Koekoek, P. A. Lesky, and R. F. Swarttouw, *Hypergeometric Orthogonal Polynomials and their q-Analogues*. (With a Foreword by Tom H. Koornwinder)., Springer Monographs in Mathematics. Springer-Verlag, Berlin, 2010.

10. A. P. Magnus, *Associated Askey–Wilson Polynomials as Laguerre-Hahn Orthogonal Polynomials*, Springer Lect. Notes in Math. 1329. Springer, Berlin, 1988, pp. 261–278.

11. A. P. Magnus, Special nonuniform lattice (snul) orthogonal polynomials on discrete dense sets of points. *J. Comput. Appl. Math.*, **65**(1995), 253–265.

12. P. Montel, *Leçons sur les récurrences et leurs applications*. Gauthier-Villars. Paris, 1957.

13. A. F. Nikiforov and S. K. Suslov, Classical orthogonal polynomials of a discrete variable on non uniform lattices. *Lett. Math. Phys.*, **11**(1986), 27–34.

14. A. F. Nikiforov, S. K. Suslov and V. B. Uvarov, *Classical Orthogonal Polynomials of a Discrete Variable*. Springer, Berlin, 1991.

15. A. F. Nikiforov and V. B. Uvarov, *Special Functions of Mathematical Physics: A Unified Introduction with Applications*. Birkhäuser, Basel, Boston, 1988.

16. S. K. Suslov, On the theory of difference analogues of special functions of hypergeometric type. *Usp. Mat. Nauk*, **44**(1989), 185–226.

M. d. N. Rebocho

17. J. Thomae, Beitrage zur Theorie der durch die Heinesche Reihe. *J. Reine Angew. Math.*, **70**(1869), 258–281.
18. N. S. Witte, Semi-classical orthogonal polynomial systems on nonuniform lattices, deformations of the Askey table, and analogues of isomonodromy. *Nagoya Math. J.*, **219**(2015), 127–234.

© 2022 World Scientific Publishing Europe Ltd.
https://doi.org/10.1142/9781800611368_0016

Chapter 16

Formal P-Gevrey Series Solutions of First-Order Holomorphic PDEs

Sergio A. Carrillo[*,‡] and Carlos A. Hurtado[†,§]

*Programa de matemáticas, Universidad Sergio Arboleda
Calle 74 # 14-14, Bogotá, Colombia
†Universidad Privada del Norte, Campus Breña
Avenida Tingo Maria 1122, Lima, Perú
‡sergio.carrillo@usa.edu.co
§carlos.hurtado@upn.edu.pe

We provide a complete and self-contained proof of the Gevrey character, in an analytic function P, of formal power series solutions of some families of first-order holomorphic PDEs. Our approach is based on a majorant series technique by applying Nagumo norms joint with a division algorithm.

1. Introduction

In the study of singular ordinary and partial differential equations or in the case of singular perturbation problems, a technique to obtain holomorphic solutions from formal ones is by applying certain summability methods such as Borel-summability and multisummability. These solutions represent asymptotically the formal power series solution as the variables approach the singular locus in adequate domains. In general, the first step to follow this method is to determine the existence, uniqueness and divergence rate (Gevrey order) of these series. The study of their summability is determined by the nature of the equation and it is a much harder problem.

325

We refer to [1–15] for some examples of ODEs and PDEs, which are susceptible to this type of analysis.

The goal of this chapter is to provide a self-contained proof on the Gevrey type of formal power series solutions $\widehat{y}$ of holomorphic partial differential equations of first order at a singular locus S. We will show that under a suitable geometric condition, the germ of analytic function P that generates S is the generic source of divergence: $\widehat{y}$ is a P-1-Gevrey series. Roughly speaking, this means that we can write $\widehat{y} = \sum_{n=0}^{\infty} y_n P^n$ as a power series in P, where the coefficients y_n are holomorphic in a common polydisc D at the origin and $\sup_{x \in D} |y_n(x)| \leq CA^n n!$, for some constants $C, A > 0$. More specifically, if $x = (x_1, \ldots, x_d) \in (\mathbb{C}^d, \mathbf{0})$ and $y = (y_1, \ldots, y_N) \in \mathbb{C}^N$, we consider a germ P of a non-zero holomorphic function on $(\mathbb{C}^d, \mathbf{0})$ such that $P(\mathbf{0}) = 0$, and the system of partial differential equations

$$P(x)L(y)(x) = F(x, y), \quad \text{where } L := a_1(x)\partial_{x_1} + \cdots + a_d(x)\partial_{x_d}$$

$$(1)$$

is a first-order differential operator with holomorphic coefficients a_j near the origin — not all identically zero — and F is a $\mathbb{C}^N$-valued holomorphic map defined near $(\mathbf{0}, \mathbf{0}) \in \mathbb{C}^d \times \mathbb{C}^N$. The *singular locus* of (1) is the germ at the origin of the analytic set

$$S := \{x \in (\mathbb{C}^d, \mathbf{0}) : P(x)a_j(x) = 0, \ j = 1, \ldots, d\},$$

where the nature of equation (1) changes from differential to implicit. Note that S contains the zero set of P, and both coincide if $a_j(\mathbf{0}) \neq 0$, for at least some index j. Furthermore, if $\frac{\partial F}{\partial y}(\mathbf{0}, \mathbf{0})$ is an invertible matrix, P cannot be canceled from (1), so its zero set is a non-removable singular part of (1). Under these conditions, our main result can be stated as follows.

Theorem 1. *Consider the partial differential equation* (1) *where* $F(\mathbf{0}, \mathbf{0}) = \mathbf{0}$, *and* $\mu := \frac{\partial F}{\partial y}(\mathbf{0}, \mathbf{0})$ *is an invertible matrix. If* P *divides* $L(P)$, *equation* (1) *has a unique formal power series solution* $\widehat{y} \in \mathbb{C}[[x]]^N$. *Moreover,* $\widehat{y}$ *is a* P-1-*Gevrey series.*

The notion of P-Gevrey series was introduced by J. Mozo-Fernández and R. Schäfke in Ref. [16] in the framework of asymptotic expansions and summability with respect to a germ of an analytic function, and it generalizes the notion of Gevrey series in one variable. Our aim is prove Theorem 1, resorting only on the ideas of this recent theory instead of using previous results on divergent solutions of systems of holomorphic PDEs, see e.g. [17].

Equation (1) falls into the category of singular first-order PDEs

$$L_1(\boldsymbol{y})(\boldsymbol{x}) = F(\boldsymbol{x}, \boldsymbol{y}), \tag{2}$$

where $L_1 = \sum_{j=1}^d X_j(\boldsymbol{x})\partial_{x_j}$ is a germ of a holomorphic vector field, singular at $\boldsymbol{0} \in \mathbb{C}^d$, i.e. $X_j(\boldsymbol{0}) = 0$, for all $j = 1, \ldots, d$. The convergence vs. rate of divergence of formal power series solutions of (2) has been studied extensively by several authors, see, e.g. [18–23]. These growth properties depend on conditions on $S = \{\boldsymbol{x} \in \mathbb{C}^d : X_j(\boldsymbol{x}) = 0, j = 1, \ldots, d\}$ or on its associated ideal $(X_1, \ldots, X_d) \subseteq \mathbb{C}\{\boldsymbol{x}\}$, and on non-resonance conditions on μ and the Jacobian matrix $\Lambda := (\partial_{x_i} X_j(\boldsymbol{0}))_{i,j}$ that we will explain in what follows. Then, if S is an analytic submanifold, c.f. [18,21], by choosing a suitable analytic coordinate system $\boldsymbol{\xi} = (\xi_1, \ldots, \xi_d)$ of $(\mathbb{C}^d, \boldsymbol{0})$ where S is the zero set of some of these coordinates, and Λ is in canonical Jordan form, the convergence or a Gevrey type of solutions can be obtained. For a recent account on these results, see [24] and the references therein.

Set $\mathrm{Spec}(\mu) = \{\mu_1, \ldots, \mu_d\}$ and $\mathrm{Spec}(\Lambda) = \{\lambda_1, \ldots, \lambda_m, 0, \ldots, 0\}$, where $\mu_k \neq 0$, $\lambda_j \neq 0$, and all eigenvalues are repeated according to multiplicity. If $m \geq 1$, the classical non-resonance *Poincaré condition* requests that

$$|\lambda_1\beta_1 + \cdots + \lambda_m\beta_m - \mu_k| \geq \nu|\boldsymbol{\beta}|, \quad \text{for all } \boldsymbol{\beta} \in \mathbb{N}^m, k = 1, \ldots, d, \tag{3}$$

for some constant $\nu > 0$, see, e.g. [25, p. 71; 18, p. 166]. Then, if (3) is valid and $m = d$, i.e. Λ is invertible, the solution of (2) is convergent, see [19,22,23]. Otherwise, the solution is generically divergent, but of some Gevrey order in the variable $\boldsymbol{\xi}$, depending on the sizes of the blocks of the canonical Jordan form of Λ associated to the zero eigenvalue, see [22,23]. It is worth remarking that (3) is better known in the theory of normal forms, see, e.g. [20,26], or in the problem of existence of analytic invariant manifolds, see [27], both for holomorphic vector fields defined near a singular point. In particular, in the problem of their local analytic linearization where much more complicated non-resonance conditions ([25, Theorem 2.12] or [26, p. 606]) appear.

Returning to our problem, the linear part of $L_1 = P \cdot L$ in equation (1) can be highly degenerated and Λ is generically the zero matrix. In fact, $\Lambda = (a_i(\mathbf{0})p_j)_{i,j}$, $p_j = \partial_{x_j} P(\mathbf{0})$, having the diagonal matrix $\mathrm{diag}(\mathrm{tr}(\Lambda), 0, \ldots, 0)$ as canonical Jordan form, where

$$\mathrm{tr}(\Lambda) = a_1(\mathbf{0})p_1 + \cdots + a_d(\mathbf{0})p_d = L(P)(\mathbf{0}).$$

Thus, the only case for which $m \geq 1$, in fact, $m = 1$, is when $L(P)(\mathbf{0}) \neq 0$. Furthermore, Poincaré condition (3) is satisfied if and only if

$$\mu_k - nL(P)(\mathbf{0}) \neq 0, \quad \text{for all } k = 1, \ldots, d, \ n \in \mathbb{N}.$$

In the aforementioned papers, our situation ($m = 0$ or 1) is covered in [22, Theorem 1.1]; [23, Theorem 1.2], claiming the solution is $(1, \ldots, 1)$-Gevrey while working in the variable $\boldsymbol{\xi}$, see Section 4 for definitions. For the case $m = 0$, Theorem 1 improves the divergence rate of the formal solution by showing it is $(1/k, \ldots, 1/k)$-Gevrey, where $k = o(P)$ is the order of P, see Proposition 3. But more importantly, it identifies a possible variable to study summability phenomena. Finally, in the case $m = 1$, $L(P)(\mathbf{0}) \neq 0$, the formal solution is convergent. In fact, by reordering the coordinates we can assume $a_1(\mathbf{0})p_1 \neq 0$, thus, $\xi_1 = P(\boldsymbol{x}), \xi_2 = x_2, \ldots, \xi_d = x_d$ is a local change of variables in which our differential operator takes the form

$$P \cdot L = \xi_1 \cdot (U(\boldsymbol{\xi})\partial_{\xi_1} + \overline{a}_2(\boldsymbol{\xi})\partial_{\xi_2} + \cdots + \overline{a}_d(\boldsymbol{\xi})\partial_{\xi_d}),$$

where $\overline{a}_j(\boldsymbol{\xi}) = a_j(\boldsymbol{x})$, and $U(\boldsymbol{\xi}) = \overline{a}_1\partial_{x_1}P + \cdots + \overline{a}_d\partial_{x_d}P$ is a unit since $U(\mathbf{0}) = L(P)(\mathbf{0})$. Then, a standard majorant argument by working in the variable ξ_1 proves the convergence of the solution. We can also prove this by a slight modification in the proof of Theorem 1. In this way, we find:

Theorem 2. *Assume the hypotheses of Theorem 1, but now suppose $L(P)(\mathbf{0}) \neq 0$. If $\mu - nL(P)(\mathbf{0})I_N$ is invertible, for all $n \in \mathbb{N}$, then equation (1) has a unique analytic solution at the origin $\widehat{\boldsymbol{y}} \in \mathbb{C}\{\boldsymbol{x}\}^N$.*

The technique to prove Theorems 1 and 2 is based on modified Nagumo norms for several variables, as introduced in [28], joint with a generalized Weierstrass division theorem that allows to write a power series as a series in the germ P, although the decomposition depends on the monomial order employed. Due to the compatibility of these tools, we can use a majorant series argument to establish the results.

The structure of the chapter is as follows: Sections 2 and 3 contain the technical parts of the work where we explain the properties we need on modified Nagumo norms, the Weierstrass division theorem and their compatibility. In Section 4, we recall the notions of $(s, \ldots, s)$– and P-s-Gevrey series, $s \geq 0$, and we develop some properties relating them. Sections 5 and 6 contain the proof of Theorems 1 and 2 and a simple extension to higher order systems (Corollary 2), respectively. Finally, Section 7 encloses several examples, including one showing that the hypotheses of Theorem 1 are necessary to conclude the desired Gevrey type.

2. Nagumo Norms

Let us start by fixing some notation: $\mathbb{N}$ is the set of natural numbers including 0, $\mathbb{N}^+ = \mathbb{N} \setminus \{0\}$, and $\mathbb{R}^+$ is the set of positive real numbers. For a coordinate t, we will write $\frac{\partial}{\partial t} = \partial_t$ for the corresponding derivative. If $\boldsymbol{\beta} = (\beta_1, \ldots, \beta_d) \in \mathbb{N}^d$, we use the multi-index notation $|\boldsymbol{\beta}| = \beta_1 + \cdots + \beta_d$, $\boldsymbol{\beta}! = \beta_1! \cdots \beta_d!$, $\boldsymbol{x}^{\boldsymbol{\beta}} = x_1^{\beta_1} \cdots x_d^{\beta_d}$ and
$$\frac{\partial^{\boldsymbol{\beta}}}{\partial \boldsymbol{x}^{\boldsymbol{\beta}}} = \frac{\partial^{|\boldsymbol{\beta}|}}{\partial x_1^{\beta_1} \cdots \partial x_d^{\beta_d}}.$$

Let $d \geq 1$ be an integer. We will work with $(\mathbb{C}^d, \boldsymbol{0})$ and local coordinates $\boldsymbol{x} = (x_1, \ldots, x_d)$. We also write $\boldsymbol{x}' = (x_2, \ldots, x_d)$ when removing the first variable. $\widehat{\mathcal{O}} = \mathbb{C}[[\boldsymbol{x}]]$ and $\mathcal{O} = \mathbb{C}\{\boldsymbol{x}\}$ denote the rings of formal and convergent power series in $\boldsymbol{x}$ with complex coefficients, respectively. $\mathcal{O}^* = \{U \in \mathcal{O} : U(\boldsymbol{0}) \neq 0\}$ will denote the corresponding groups of units. Given $\hat{f} = \sum a_{\boldsymbol{\beta}} \boldsymbol{x}^{\boldsymbol{\beta}} \in \widehat{\mathcal{O}}$, $o(\hat{f})$ will denote its order: if $\hat{f} = \sum_{n=0}^{\infty} f_n$, $f_n = \sum_{|\boldsymbol{\beta}|=n} a_{\boldsymbol{\beta}} \boldsymbol{x}^{\boldsymbol{\beta}}$, is written as sum of its homogeneous components, $o(\hat{f})$ is the least integer k for which $f_k \neq 0$.

For $\boldsymbol{r} = (r_1, \ldots, r_d) \in (\mathbb{R}^+)^d$, $D_{\boldsymbol{r}} = \{\boldsymbol{x} \in \mathbb{C}^d : |x_j| < r_j, j = 1, \ldots, d\}$ is the polydisc centered at the origin with polyradius $\boldsymbol{r}$. If $r_j = r$, for all j, we write $D_{\boldsymbol{r}} = D_r^d$ as a Cartesian product instead. By using the norm $|\boldsymbol{x}| := \max_{1 \leq j \leq d} |x_j|$, we can write $D_r^d = \{\boldsymbol{x} \in \mathbb{C}^d : |\boldsymbol{x}| < r\}$. Also, $\mathcal{O}(D_{\boldsymbol{r}})$ and $\mathcal{O}_b(D_{\boldsymbol{r}})$ will denote the sets of holomorphic and bounded holomorphic $\mathbb{C}$-valued functions on the given polydisc. We denote by $J : \mathcal{O}(D_{\boldsymbol{r}})^N \to \mathcal{O}^N$ the Taylor map sending a vector function to its Taylor series at the origin.

Nagumo norms were introduced originally by Nagumo in [29] in his study of analytic partial differential equations. There are other alternative versions that have been used successfully in this context, see, e.g. [1,13]. We will use a variant as it appears in [28] for the case of one complex variable. Let us fix two numbers $0 < \rho < r$ and consider the function

$$d_r(x) = \begin{cases} r - |x| & \text{if } |x| \geq \rho, \\ r - \rho & \text{if } |x| < \rho, \end{cases}$$

which satisfies

$$|d_r(x) - d_r(y)| \leq |x - y|, \quad x, y \in D_r. \tag{4}$$

The number ρ can be chosen arbitrarily, but we will always take $\rho = r/2$.

Fix a polyradius $\boldsymbol{r} = (r_1, \ldots, r_d)$. If $f \in \mathcal{O}(D_{\boldsymbol{r}})$ and $m \in \mathbb{N}$, we consider the family of *Nagumo norms*

$$\|f\|_m := \sup_{\boldsymbol{x} \in D_{\boldsymbol{r}}} |f(\boldsymbol{x})| d_{r_1}(x_1)^m \cdots d_{r_d}(x_d)^m. \tag{5}$$

These norms depend on $\boldsymbol{r}$, but to simplify the notation, we omit this dependence. There is no reason for these values to be finite, for instance, if $m = 0$, this norm is reduced to the maximum norm. Note that if $\|f\|_k$ is finite and $m > k$, then

$$\|f\|_m \leq (r_1/2)^{m-k} \cdots (r_d/2)^{m-k} \|f\|_k. \tag{6}$$

In particular, if $k = 0$, i.e. if $f \in \mathcal{O}_b(D_{\boldsymbol{r}})$, all its Nagumo norms are finite.

We collect in the next proposition the main properties of these norms we will use in the proof of Theorems 1 and 2, including their behavior under the shift operators

$$S_j(f)(\boldsymbol{x}) = \begin{cases} (f(\boldsymbol{x}) - f(x_1, \ldots, x_{j-1}, 0, x_{j+1}, \ldots, x_d)) / x_j & \text{if } x_j \neq 0, \\ \dfrac{\partial f}{\partial x_j}(\boldsymbol{x}) & \text{if } x_j = 0. \end{cases}$$
$$\tag{7}$$

Proposition 1. *Fix $m, k \in \mathbb{N}$ and $1 \leq j \leq d$. If $f, g \in \mathcal{O}(D_{\boldsymbol{r}})$, then*

(i) $\|f + g\|_m \leq \|f\|_m + \|g\|_m$ *and* $\|fg\|_{m+k} \leq \|f\|_m \|g\|_k$.

(ii) $\left\| \dfrac{\partial f}{\partial x_j} \right\|_{m+1} \leq e(m+1) \prod_{i \neq j}(r_i/2) \|f\|_m$.

(iii) $\|S_j(f)\|_m \leq \dfrac{4}{r_j} \|f\|_m$.

Proof. The inequalities in (i) are clear from the definition. We prove (ii) and (iii) for the variable x_1. If $\boldsymbol{x} = (x_1, \boldsymbol{x}') \in D_{\boldsymbol{r}}$, we have

$$|f(\boldsymbol{x})| d_{r_1}(x_1)^m \cdots d_{r_d}(x_d)^m \leq \|f\|_m. \tag{8}$$

To establish (ii), we use Cauchy's formula

$$\left| \frac{\partial f}{\partial x_1}(\boldsymbol{x}) \right| = \left| \frac{1}{2\pi} \int_{|\xi - x_1| = R} \frac{f(\xi, \boldsymbol{x}')}{(\xi - x_1)^2} d\xi \right| \leq \frac{1}{R} \sup_{|\xi - x_1| = R} |f(\xi, \boldsymbol{x}')|, \tag{9}$$

valid for $0 < R < r - |x_1|$. If $|\xi - x_1| = R$, then $d_{r_1}(x_1) - R \leq d_{r_1}(\xi)$ by applying inequality (4). In particular, if $0 < R < d_{r_1}(x_1)$ we find

$$|f(\xi, \boldsymbol{x}')| \leq \|f\|_m d_{r_2}(x_2)^{-m} \cdots d_{r_d}(x_d)^{-m} (d_{r_1}(x_1) - R)^{-m}.$$

For the case $m > 0$, choose $R = \dfrac{d_{r_1}(x_1)}{m+1}$ to find

$$\left| \frac{\partial f}{\partial x_1}(\boldsymbol{x}) \right| \leq \frac{(m+1)\|f\|_m}{d_{r_1}(x_1)^{m+1} d_{r_2}(x_2)^m \cdots d_{r_d}(x_d)^m} \left(1 + \frac{1}{m} \right)^m.$$

Therefore, using the inequality $(1 + 1/m)^m < e$, we conclude that

$$\left| \frac{\partial f}{\partial x_1}(\boldsymbol{x}) d_{r_1}(x_1)^{m+1} \cdots d_{r_d}(x_d)^{m+1} \right| \leq e(m+1) d_{r_2}(x_2) \cdots d_{r_d}(x_d) \|f\|_m$$

$$\leq e(m+1) \left(\frac{r_2}{2} \cdots \frac{r_d}{2} \right) \|f\|_m,$$

as we wanted to show. Now, if $m = 0$, taking $R = d_{r_1}(x_1)/e$ in (9), we find

$$\left| \frac{\partial f}{\partial x_1}(\boldsymbol{x}) d_{r_1}(x_1) \cdots d_{r_d}(x_d) \right| \leq e \left(\frac{r_2}{2} \cdots \frac{r_d}{2} \right) \|f\|_0,$$

as desired.

For (iii), by inequality (8), $|f(0, \boldsymbol{x}')|$ is less than

$$\|f\|_m((r_1/2)d_{r_2}(x_2)\cdots d_{r_d}(x_d))^{-m}$$
$$\leq \|f\|_m(d_{r_1}(x_1)d_{r_2}(x_2)\cdots d_{r_d}(x_d))^{-m},$$

for all $\boldsymbol{x} \in D_{\boldsymbol{r}}$. Hence, if $|x_1| \geq r_1/2$,

$$\left|\frac{f(x_1, \boldsymbol{x}') - f(0, \boldsymbol{x}')}{x_1}\right| \leq \frac{4}{r_1}\|f\|_m d_{r_1}(x_1)^{-m}\cdots d_{r_d}(x_d)^{-m}.$$

For $|x_1| < r_1/2$, we can use the maximum modulus principle and the estimate above to see that

$$|S_1(f)(\boldsymbol{x})| \leq \max_{|\xi|=r_1/2}|S_1(f)(\xi, \boldsymbol{x}')| \leq \frac{4}{r_1}\|f\|_m$$

$$((r_1/2)d_{r_2}(x_2)\cdots d_{r_d}(x_d))^{-m}.$$

Since $d_{r_1}(x_1) = r_1/2$ if $|x_1| < r_1/2$, we find in all cases that $|S_1(f)(\boldsymbol{x})d_{r_1}(x_1)^m \cdots d_{r_d}(x_d)^m| \leq (4/r_1)\|f\|_m$ as required. $\qquad\square$

For vector-valued $\boldsymbol{y} = (y_1, \ldots, y_N) \in \mathcal{O}(D_{\boldsymbol{r}})^N$, and matrix-valued $A = (A_{i,j}) \in \mathcal{O}(D_{\boldsymbol{r}})^{N\times N}$ maps, we extend Nagumo norms by the rules

$$\|\boldsymbol{y}\|_m := \max_{1\leq i\leq N}\|y_i\|_m, \quad \|A\|_m := \max_{1\leq i\leq N}\sum_{j=1}^{N}\|A_{i,j}\|_m. \qquad (10)$$

Then, it is immediate to check that

$$\|f \cdot \boldsymbol{y}\|_{m+k} \leq \|f\|_m\|\boldsymbol{y}\|_k, \quad \|A \cdot \boldsymbol{y}\|_{m+k} \leq \|A\|_m\|\boldsymbol{y}\|_k,$$
$$\|A \cdot B\|_{m+k} \leq \|A\|_m\|B\|_k,$$

for all $f \in \mathcal{O}(D_{\boldsymbol{r}})$, $\boldsymbol{y} \in \mathcal{O}(D_{\boldsymbol{r}})^N$, and $A, B \in \mathcal{O}(D_{\boldsymbol{r}})^{N\times N}$.

3.   The Division Algorithm

We recall here a generalized Weierstrass division theorem by following closely [16], and whose original version is due to Aroca, Hironaka, and Vicente, see [30]. For the sake of completeness, we include the proof for convergent series including the compatibility of the division algorithm with the Nagumo norms introduced in Section 2.

We will use the partial order $\leq$ on $\mathbb{N}^d$ defined by $\boldsymbol{\alpha} \leq \boldsymbol{\beta}$ if $\alpha_j \leq \beta_j$, for all $j = 1, \ldots, d$. Thus, $\boldsymbol{\alpha} \not\leq \boldsymbol{\beta}$ means there is an index j such that $\beta_j < \alpha_j$. We also use the notation

$$\Delta_{\boldsymbol{\alpha}} := \left\{ \sum g_{\boldsymbol{\beta}} \boldsymbol{x}^{\boldsymbol{\beta}} \in \widehat{\mathcal{O}} : g_{\boldsymbol{\beta}} = 0 \text{ if } \boldsymbol{\alpha} \leq \boldsymbol{\beta} \right\}.$$

Given $\boldsymbol{\alpha} = (\alpha_1, \ldots, \alpha_d) \in \mathbb{N}^d \setminus \{\boldsymbol{0}\}$, a power series $\hat{f} = \sum_{\boldsymbol{\beta} \in \mathbb{N}^d} f_{\boldsymbol{\beta}} \boldsymbol{x}^{\boldsymbol{\beta}} \in \widehat{\mathcal{O}}$ can be written uniquely as a series in the monomial $\boldsymbol{x}^{\boldsymbol{\alpha}}$ as

$$\hat{f} = \sum_{n=0}^{\infty} \hat{f}_{\boldsymbol{\alpha},n}(\boldsymbol{x}) \boldsymbol{x}^{n\boldsymbol{\alpha}}, \quad \hat{f}_{\boldsymbol{\alpha},n}(\boldsymbol{x}) = \sum_{\boldsymbol{\alpha} \not\leq \boldsymbol{\beta}} f_{n\boldsymbol{\alpha}+\boldsymbol{\beta}} \boldsymbol{x}^{\boldsymbol{\beta}} \in \Delta_{\boldsymbol{\alpha}}. \qquad (11)$$

This decomposition is obtained by a repeated use of the canonical division algorithm by $\boldsymbol{x}^{\boldsymbol{\alpha}}$: given $\hat{f} \in \widehat{\mathcal{O}}$, there are unique $q \in \widehat{\mathcal{O}}$, $r \in \Delta_{\boldsymbol{\alpha}}$ such that

$$\hat{f} = q \boldsymbol{x}^{\boldsymbol{\alpha}} + r, \quad \text{where } q = \sum_{\boldsymbol{\alpha} \leq \boldsymbol{\beta}} f_{\boldsymbol{\beta}} \boldsymbol{x}^{\boldsymbol{\beta}-\boldsymbol{\alpha}}, \, r = \sum_{\boldsymbol{\alpha} \not\leq \boldsymbol{\beta}} f_{\boldsymbol{\beta}} \boldsymbol{x}^{\boldsymbol{\beta}}.$$

Moreover, if $f \in \mathcal{O}(D_{\boldsymbol{r}})$, then $q, r \in \mathcal{O}(D_{\boldsymbol{r}})$. Actually, we can use the shift operators (7) to write

$$q = Q_{\boldsymbol{\alpha}}(f) := S_1^{\alpha_1} \circ \cdots \circ S_d^{\alpha_d}(f), \quad r = R_{\boldsymbol{\alpha}}(f) := f - Q_{\boldsymbol{\alpha}}(f) \cdot \boldsymbol{x}^{\boldsymbol{\alpha}}.$$

In particular, Proposition 1 (iii) shows that

$$\|Q_{\boldsymbol{\alpha}}(f)\|_m \leq \frac{4^{|\boldsymbol{\alpha}|}}{\boldsymbol{r}^{\boldsymbol{\alpha}}} \|f\|_m, \quad \|R_{\boldsymbol{\alpha}}(f)\|_m \leq (1 + 4^{|\boldsymbol{\alpha}|}) \|f\|_m, \qquad (12)$$

for all $m \in \mathbb{N}$. By taking $m = 0$, we conclude $Q_{\boldsymbol{\alpha}}, R_{\boldsymbol{\alpha}} : \mathcal{O}_b(D_{\boldsymbol{r}}) \to \mathcal{O}_b(D_{\boldsymbol{r}})$ are linear continuous maps.

The generalized Weierstrass division allows to extend the previous considerations by dividing by an element of $\widehat{\mathcal{O}} \setminus \{0\}$ with zero constant term, but not in a canonical way. We will focus on division by an analytic germ $P \in \mathcal{O} \setminus \{0\}$, $P(\boldsymbol{0}) = 0$. The division is determined by P and an injective linear form $\ell : \mathbb{N}^d \to \mathbb{R}^+$, $\ell(\boldsymbol{\alpha}) = \ell_1 \alpha_1 + \cdots + \ell_d \alpha_d$ used to order the monomials by the rule

$$\boldsymbol{x}^{\boldsymbol{\alpha}} <_\ell \boldsymbol{x}^{\boldsymbol{\beta}} \quad \text{if } \ell(\boldsymbol{\alpha}) < \ell(\boldsymbol{\beta}).$$

Then, any $\hat{f} = \sum f_{\boldsymbol{\beta}} \boldsymbol{x}^{\boldsymbol{\beta}} \in \widehat{\mathcal{O}} \setminus \{0\}$ has a *minimal exponent* $\nu_\ell(\hat{f})$ with respect to ℓ, i.e. $\nu_\ell(\hat{f}) = \boldsymbol{\alpha}$ where $\boldsymbol{x}^{\boldsymbol{\alpha}} = \min_\ell \{\boldsymbol{x}^{\boldsymbol{\beta}} : f_{\boldsymbol{\beta}} \neq 0\}$, and the minimum is taken according to $<_\ell$. The division process, for formal and convergent series, can be stated as follows, c.f., [16, Lemmas 2.4, 2.6].

Proposition 2 (Generalized Weierstrass division). *Let P and ℓ he as above. For every $\hat{g} \in \widehat{\mathcal{O}}$, there are unique $\hat{q} \in \widehat{\mathcal{O}}$, $\hat{r} \in \Delta_{\nu_\ell(P)}$ such that*

$$\hat{g} = \hat{q}P + \hat{r}.$$

Moreover, if $\rho > 0$ is sufficiently small, then for every $g \in \mathcal{O}_b(D_{\rho(\ell)})$, $\rho(\ell) = (\rho^{\ell_1}, \ldots, \rho^{\ell_d})$, there are unique $q \in \mathcal{O}_b(D_{\rho(\ell)})$, $r \in \mathcal{O}_b(D_{\rho(\ell)})$ with $J(r) \in \Delta_{\nu_\ell(P)}$ such that

$$g = qP + r, \ Q_{P,\ell}(g) := q, \ R_{P,\ell}(g) := r.$$

The corresponding operators $Q_{P,\ell}, R_{P,\ell} : \mathcal{O}_b(D_{\rho(\ell)}) \to \mathcal{O}_b(D_{\rho(\ell)})$ are linear and continuous. In fact, if ρ is sufficiently small, then

$$\|Q_{P,\ell}(g)\|_m \leq \frac{2 \cdot 4^{|\nu_\ell(P)|}}{\rho^{\ell(\nu_\ell(P))}} \|g\|_m, \ \|R_{P,\ell}(g)\|_m \leq 2(1 + 4^{|\nu_\ell(P)|})\|g\|_m,$$

for all $m \in \mathbb{N}$.

Proof. By the choice of $\rho(\ell)$, we have $|x^\beta| \leq \rho^{\ell(\beta)}$ if $x \in D_{\rho(\ell)}$. Let us write $\alpha = \nu_\ell(P)$. Without loss of generality, we can assume $P = x^\alpha + \widetilde{P}$, where $\widetilde{P} \in \mathcal{O} \setminus \{0\}$ and $\nu_\ell(\widetilde{P}) >_\ell x^\alpha$. Then, solving $g = qP + r$ or $qx^\alpha + r = g - q\widetilde{P}$ for q and r being equivalent to find a fixed point for

$$q = Q_\alpha(g - q\widetilde{P}), \tag{13}$$

joint with $r = R_\alpha(g - q\widetilde{P})$. Note that if ρ is sufficiently small, we can choose a constant $K > 0$ such that $\|\widetilde{P}\|_0 \leq K\rho^{\ell(\nu_\ell(\widetilde{P}))}$. Consider the map $\phi_g : \mathcal{O}_b(D_{\rho(\ell)}) \to \mathcal{O}_b(D_{\rho(\ell)})$ given by $\phi_g(h) = Q_\alpha(g - h\widetilde{P})$. By using the first inequality in (12) for $m = 0$, we see that

$$\|\phi_g(h_1) - \phi_g(h_2)\|_0 = \left\|Q_\alpha((h_1 - h_2)\widetilde{P})\right\|_0$$

$$\leq 4^{|\alpha|} K \rho^{\ell(\nu(\widetilde{P})) - \ell(\alpha)} \|h_1 - h_2\|_0.$$

Thus, ϕ_g defines a contraction if ρ is small enough, i.e. if $4^{|\alpha|} K \rho^{\ell(\nu_\ell(\widetilde{P})) - \ell(\alpha)} < 1$, and it has a unique fixed point q. This determines the existence and uniqueness of q and r. Finally, from equation (13) we find that

$$\|Q_{P,\ell}(g)\|_m \leq \frac{4^{|\alpha|}/\rho^{\ell(\alpha)}}{1 - 4^{|\alpha|}K\rho^{\ell(\nu_\ell(\widetilde{P}))-\ell(\alpha)}}\|g\|_m,$$

and from $r = R_\alpha(g - q\widetilde{P})$ that

$$\|R_{P,\ell}(g)\|_m \leq (1 + 4^{|\alpha|})\left(1 + \frac{\|\widetilde{P}\|_0 4^{|\alpha|}/\rho^{\ell(\alpha)}}{1 - 4^{|\alpha|}K\rho^{\ell(\nu_\ell(\widetilde{P}))-\ell(\alpha)}}\right)\|g\|_m$$

$$= \frac{(1 + 4^{|\alpha|})\|g\|_m}{1 - 4^{|\alpha|}K\rho^{\ell(\nu_\ell(\widetilde{P}))-\ell(\alpha)}}.$$

Therefore, the result follows by taking additionally $\rho > 0$ such that $4^{|\alpha|}K\rho^{\ell(\nu_\ell(\widetilde{P}))-\ell(\alpha)} < 1/2$. $\qquad\square$

By a repeated application of the previous proposition, see [16, Corollary 2.5], any $\hat{f} \in \widehat{\mathcal{O}}$ can be written uniquely as

$$\hat{f} = \sum_{n=0}^{\infty} \hat{f}_{P,\ell,n}P^n, \quad \hat{f}_{P,\ell,n} \in \Delta_{\nu_\ell(P)}. \tag{14}$$

For the convergent case, we have a similar result, see [16, Corollary 2.7].

Corollary 1. *If $s > 0$ is such that the operators $Q_{P,\ell}$ and $R_{P,\ell}$ are defined over $\mathcal{O}_b(D_{s(\ell)})$, there is $r = r(s) > 0$, depending only on s, such that for any $f \in \mathcal{O}_b(D_{s(\ell)})$ we can find a unique sequence $(f_n)_{n \in \mathbb{N}} \subset \mathcal{O}_b(D_r^d)$ with $J(f_n) \in \Delta_{\nu_\ell(P)}$, such that*

$$f = \sum_{n=0}^{\infty} f_n P^n, \quad f_n = R_{P,\ell} \circ Q_{P,\ell}^n(f), \ \text{both convergent for} |x| < r.$$

$$\tag{15}$$

Proof. By applying Proposition 2, we obtain

$$f = \sum_{n=0}^{N-1} R_{P,\ell}(Q_{P,\ell}^n(f))P^n + Q^N(f)P^N, \quad \text{for all } N \in \mathbb{N}.$$

If we choose $0 < r \leq s$ with $M = \sup_{|\boldsymbol{x}| < r} |P(\boldsymbol{x})| < s^{\ell(\nu_\ell(P))}/(2 \cdot 4^{|\nu_\ell(P)|}) = 1/b$, then we can estimate

$$\sup_{|\boldsymbol{x}| < r} \left| f(\boldsymbol{x}) - \sum_{n=0}^{N-1} R_{P,\ell}(Q_{P,\ell}^n(f))(\boldsymbol{x})P(\boldsymbol{x})^n \right| \leq (bM)^N \sup_{\boldsymbol{y} \in D_{s(\ell)}} |f(\boldsymbol{y})|.$$

The result follows by taking $N \to +\infty$. $\qquad\qquad\square$

To finish this section, we remark that if $\hat{f}_l = \sum_{n=0}^{\infty} f_{l,n}P^n$, $l = 1, \ldots, m$, where $f_{l,n} \in \mathcal{O}_b(D_r^d)$ for a common r, decomposition (14) for their product is given by

$$\hat{f}_1 \cdots \hat{f}_m = \sum_{j_1,\ldots,j_m \geq 0} f_{1,j_1} \cdots f_{m,j_m} P^{j_1 + \cdots + j_m}$$

$$= \sum_{k,j_1,\ldots,j_m \geq 0} R_{P,\ell}(Q_{P,\ell}^k(f_{1,j_1} \cdots f_{m,j_m})) P^{k+j_1+\cdots+j_m}$$

$$= \sum_{n=0}^{\infty} \left[\sum_{\substack{k+j_1+\cdots+j_m=n \\ k,j_1,\ldots,j_m \geq 0}} R_{P,\ell}(Q_{P,\ell}^k(f_{1,j_1} \cdots f_{m,j_m})) \right] P^n. \quad (16)$$

In particular, note that for $n = 0$ we find that

$$R_{P,\ell}(\hat{f}_1 \cdots \hat{f}_m) = R_{P,\ell}(f_{1,0} \cdots f_{m,0}). \quad (17)$$

Also, note that for two factors the previous sum takes the form

$$\hat{f}_1 \cdot \hat{f}_2 = \sum_{n=0}^{\infty} \left(\sum_{k=0}^{n} \sum_{j=0}^{k} R_{P,\ell}(Q_{P,\ell}^{n-k}(f_{1,j}f_{2,k-j})) \right) P^n. \quad (18)$$

4. Gevrey Series

Given $\boldsymbol{s} = (s_1, \ldots, s_d) \in \mathbb{R}_{\geq 0}^d$ and $\hat{f} = \sum_{\beta \in \mathbb{N}^d} a_\beta \boldsymbol{x}^\beta \in \hat{\mathcal{O}}$, the series $\hat{f}$ is said to be a $\boldsymbol{s}$-*Gevrey* if we can find $C, A > 0$ such that $|a_\beta| \leq CA^{|\beta|}\beta!^{\boldsymbol{s}}$, for all $\beta \in \mathbb{N}^d$. Note that $\boldsymbol{s} = \boldsymbol{0}$ means convergence. We will be interested in the case $s_1 = \cdots = s_d = s \geq 0$. Thanks to the inequalities

$$\beta! \leq |\beta|! \leq d^{|\beta|}\beta!,$$

a series $\hat{f}$ is $(s,\ldots,s)$-Gevrey if and only if there are $C, A > 0$ such that

$$|a_{\boldsymbol{\beta}}| \leq CA^{|\boldsymbol{\beta}|}|\boldsymbol{\beta}|!^s, \quad \boldsymbol{\beta} \in \mathbb{N}^d.$$

We denote by $\widehat{\mathcal{O}}_s$ the set of $(s,\ldots,s)$-Gevrey series.

Remark 1. For any $s \geq 0$, $\widehat{\mathcal{O}}_s$ is closed under sums, products, partial derivatives, composition, and it contains $\mathcal{O}$. These properties can be seen as a particular case in the setting of ultradifferentiable functions. In that framework, the Gevrey sequence $(n!^s)_{n\in\mathbb{N}}$ is generalized by a sequence of positive numbers $(M_n)_{n\in\mathbb{N}}$ satisfying log-convexity $(M_n^2 \leq M_{n-1}M_{n+1})$, stability under derivatives $(M_{n+1} \leq K^n M_n$ for some $K > 0)$ and the condition $M_n^{1/n} \to +\infty$ as $n \to \infty$, see e.g. [31,32] including other stability properties in a more general context.

According to the previous remark, if $\hat{f} \in \widehat{\mathcal{O}}_s$, the same is true for $\hat{f}(A\boldsymbol{x})$, for all matrices $A \in \mathbb{C}^{d\times d}$, cf., [22, Lemma 2.1]. In particular, we highlight the following simple statement we will need later.

Lemma 1. *Let $s \geq 0$. Then, $\hat{f} \in \widehat{\mathcal{O}}_s$ if and only if $\hat{f}(A\boldsymbol{x}) \in \widehat{\mathcal{O}}_s$, for all $A \in GL_n(\mathbb{C})$.*

Consider a germ $P \in \mathcal{O} \setminus \{0\}$ such that $P(\boldsymbol{0}) = 0$, and $s \geq 0$. There are equivalent definitions for Gevrey series with respect to the germ P, see [16, Definition/Proposition 7.5]. For simplicity, we will use the characterization given in [33, Lemma 4.1].

Definition 1. A series $\hat{f} \in \widehat{\mathcal{O}}$ is *P-s-Gevrey series* if there is a polyradius $\boldsymbol{r}$, constants $C, A > 0$ and a sequence $\{f_n\}_{n\in\mathbb{N}} \in \mathcal{O}_b(D_{\boldsymbol{r}})$ such that

$$\hat{f} = \sum_{n=0}^{\infty} f_n P^n, \quad \text{where} \ \sup_{\boldsymbol{x}\in D_{\boldsymbol{r}}} |f_n(\boldsymbol{x})| \leq CA^n n!^s. \tag{19}$$

We will use the notation $\widehat{\mathcal{O}}^{P,s}$ for the set of P-s-Gevrey series.

This generalizes the notion of s-Gevrey series in x_j, uniformly in the other variables (x_j-s-Gevrey series in our notation). In fact, setting $j = 1$ to fix ideas, the classical notion requires that when we write $\hat{f} = \sum_{n=0}^{\infty} f_n x_1^n$ as a power series in x_1, there is a polyradius $r' \in \mathbb{R}^{d-1}$ such that $f_n \in \mathcal{O}_b(D_{r'})$ and $\sup_{x' \in D_{r'}} |f_n(x')| \leq C A^n n!^s$, for adequate constants C, A.

By using the generalized Weierstrass division, we can show the notion of P-s-Gevrey series is well-defined, in the sense that it is independent of the decomposition (19). Note it is enough to check the definition for the decomposition (14) induced by a given injective linear form $\ell : \mathbb{N}^d \to \mathbb{R}^+$. In fact, if (19) holds, and since all f_n are defined in a common polydisc, we can use decomposition (15) to find $\rho > 0$ and sequences $\{f_{n,j}\}_{n \in \mathbb{N}} \subset \mathcal{O}_b(D_\rho^d)$ with $J(f_{n,j}) \in \Delta_\ell(P)$, such that $f_n = \sum_{j=0}^{\infty} f_{n,j} P^j$, valid for $|x| < \rho$, where $f_{n,j} = R_{P,\ell} \circ Q_{P,\ell}^j(f_n)$. Therefore, the decomposition (14) of $\hat{f}$ is given by

$$\hat{f} = \sum_{n=0}^{\infty} g_n P^n, \quad g_n = \sum_{j=0}^{n} f_{j,n-j} \in \mathcal{O}_b(D_\rho^d), \quad J(g_n) \in \Delta_\ell(P),$$

and the sequence $(g_n)_{n \in \mathbb{N}}$ exhibits s-Gevrey bounds since

$$|g_n(x)| \leq \sum_{j=0}^{n} \|R_{P,\ell}\| \|Q_{P,\ell}\|^{n-j} \sup_{|y| \leq \rho} |f_j(y)|$$

$$\leq \sum_{j=0}^{n} \|R_{P,\ell}\| \|Q_{P,\ell}\|^{n-j} C A^j j!^s,$$

for $|x| < \rho$, as we wanted to show.

From the previous definition, it is easy to deduce many properties on this type of series. We recall the following, valid for $s \geq 0$ and $P, Q \in \mathcal{O} \setminus \{0\}$ such that $P(\mathbf{0}) = Q(\mathbf{0}) = 0$, c.f., [33, Corollary 4.2, Lemma 4.3]:

(i) $\widehat{\mathcal{O}}^{P,s}$ is stable under sums, products and partial derivatives.

(ii) $\mathcal{O} \subset \widehat{\mathcal{O}}^{P,s}$.

(iii) For any $k \in \mathbb{N}^+$, $\widehat{\mathcal{O}}^{P^k,ks} = \widehat{\mathcal{O}}^{P,s}$.

(iv) If Q divides P, then $\widehat{\mathcal{O}}^{P,s} \subseteq \widehat{\mathcal{O}}^{Q,s}$. In particular, if $Q = U \cdot P$, $U \in \mathcal{O}^*$, then $\widehat{\mathcal{O}}^{P,s} = \widehat{\mathcal{O}}^{Q,s}$.

(v) Let $\phi : (\mathbb{C}^d, \mathbf{0}) \to (\mathbb{C}^d, \mathbf{0})$ be analytic, $\phi(\mathbf{0}) = \mathbf{0}$, and assume $P \circ \phi$ is not identically zero. If $\hat{f} \in \widehat{\mathcal{O}}^{P,s}$, then $\hat{f} \circ \phi \in \widehat{\mathcal{O}}^{P \circ \phi, s}$.

(vi) If $P(\boldsymbol{x}) = \boldsymbol{x}^{\boldsymbol{\alpha}}$, $\boldsymbol{\alpha} \in \mathbb{N}^d \setminus \{\mathbf{0}\}$, then $\hat{f} = \sum f_{\boldsymbol{\beta}} \boldsymbol{x}^{\boldsymbol{\beta}} \in \widehat{\mathcal{O}}^{\boldsymbol{x}^{\boldsymbol{\alpha}}, s}$ if and only if there are constants $C, A > 0$ satisfying

$$|f_{\boldsymbol{\beta}}| \leq C A^{|\boldsymbol{\beta}|} \min\{\beta_j!^{s/\alpha_j} : j = 1, \ldots, d, \alpha_j \neq 0\}, \quad \boldsymbol{\beta} \in \mathbb{N}^d. \quad (20)$$

It follows from (20) that if $\hat{f} \in \widehat{\mathcal{O}}^{\boldsymbol{x}^{\boldsymbol{\alpha}}, s}$, then $\hat{f} \in \widehat{\mathcal{O}}_{s/|\boldsymbol{\alpha}|}$. Indeed, this is a consequence of the inequality $\min\{a_1, \ldots, a_d\} \leq a_1^{\tau_1} \cdots a_d^{\tau_d}$, valid for all $a_j > 0$ and $\tau_j \geq 0$ such that $\tau_1 + \cdots + \tau_d = 1$, by applying it to $\tau_j = \alpha_j/|\boldsymbol{\alpha}|$. This property can be generalized to an arbitrary germ and we have the following new inclusion of rings of Gevrey series.

Proposition 3. *Consider $P \in \mathcal{O}$ with $o(P) = k \geq 1$. Then, a P-s-Gevrey series is a $(s/k, \ldots, s/k)$-Gevrey series. In symbols,*

$$\widehat{\mathcal{O}}^{P,s} \subseteq \widehat{\mathcal{O}}_{s/k}.$$

Proof. Write $P = \sum_{j=k}^{\infty} P_j$ as sum of homogeneous polynomials. Since $P_k \neq 0$, we can find $\boldsymbol{a} \neq \mathbf{0}$ such that $P_k(\boldsymbol{a}) \neq 0$. Choose $A \in \mathrm{GL}_n(\mathbb{C})$ having $\boldsymbol{a}$ as first column. If we set $Q(\boldsymbol{x}) = P(A\boldsymbol{x})$ and we write it as sum of its homogeneous components $Q = \sum Q_j$, then $Q_j(\boldsymbol{x}) = P_j(A\boldsymbol{x})$, and $Q_k(\boldsymbol{x}) = P_k(\boldsymbol{a})x_1^k + \cdots$, i.e. $o(Q) = k$ and $Q_k(1, 0, \ldots, 0) \neq 0$.

Consider a P-s-Gevrey series $\hat{f}$. Then $\hat{f}_0(\boldsymbol{x}) = \hat{f}(A\boldsymbol{x}) = \sum a_{\boldsymbol{\beta}} \boldsymbol{x}^{\boldsymbol{\beta}}$ is a Q-s-Gevrey series, thanks to (v) above. We consider the change of variables

$$x_1 = z_1, \quad x_2 = z_1 z_2, \ldots, x_d = z_1 z_d, \quad (21)$$

that geometrically corresponds to a local expression for the blow-up of the origin in $\mathbb{C}^d$, see, e.g. [34]. If $R(\boldsymbol{z}) = Q(\boldsymbol{x})$ and $\hat{f}_1(\boldsymbol{z}) = \hat{f}_0(\boldsymbol{x})$, we see $\hat{f}_1$ is a R-s-Gevrey series, again by (v). On the other hand,

$$R(\boldsymbol{z}) = Q(z_1, z_1 z_2, \ldots, z_1 z_d) = \sum_{j=k}^{\infty} z_1^j Q_j(1, z_2, \ldots, z_d) = z_1^k U(\boldsymbol{z}),$$

where U is a unit, since $U(\mathbf{0}) = Q_k(1, 0, \ldots, 0) \neq 0$. Thus, we conclude $\hat{f}_1$ is z_1^k-s-Gevrey, or equivalently, a z_1-s/k-Gevrey series.

Now, since

$$\hat{f}_1(z) = \sum_{\beta \in \mathbb{N}^d} a_\beta z_1^{|\beta|} z_2^{\beta_2} \cdots z_d^{\beta_d}$$

$$= \sum_{\substack{(n,\gamma) \in \mathbb{N} \times \mathbb{N}^{d-1} \\ n \geq |\gamma|}} a_{n-|\gamma|,\gamma} z_1^n z'^\gamma,$$

we can find constants $C, A > 0$ such that $|a_{n-|\gamma|,\gamma}| \leq CA^{n+|\gamma|} n!^{s/k}$. Therefore, in the index $\beta = (n, \gamma)$, we find the bound

$$|a_\beta| \leq CA^{\beta_1 + 2\beta_2 + \cdots + 2\beta_d} |\beta|!^{s/k}, \quad \text{for all } \beta \in \mathbb{N}^d.$$

This means $\hat{f}_0$ and $\hat{f}$ are $(s/k, \ldots, s/k)$-Gevrey series, due to Lemma 1. $\qquad\square$

Remark 2. Proposition 3 and Lemma 1 show that if $\hat{f} \in \hat{\mathcal{O}}^{x^\alpha, s}$, then $\hat{f}(Ax) \in \hat{\mathcal{O}}_{s/|\alpha|}$, for all $A \in \mathbb{C}^{d \times d}$. However, x^α-s-Gevrey is not stable under linear changes of variable, and we illustrate this by a simple example: the series $\hat{f}(x_1, x_2) = \sum_{n=0}^\infty n!(x_1 x_2)^n$ is $x_1 x_2$-1-Gevrey, but

$$\hat{f}_0(\xi_1, \xi_2) = \hat{f}(\xi_1 + \xi_2, \xi_1 - \xi_2) = \sum_{j,k \geq 0} \binom{j+k}{j}(j+k)!(-1)^k \xi_1^{2j} \xi_2^{2k},$$

is $(1/2, 1/2)$-Gevrey in ξ_1, ξ_2, but not $\xi_1 \xi_2$-1-Gevrey.

5.   Proof of the Main Results

The idea of both proofs is to find a formal solution of the form

$$\widehat{y} = \sum_{n=0}^\infty y_n P^n, \tag{22}$$

according to decomposition (14) associated to an injective linear form $\ell : \mathbb{N}^d \to \mathbb{R}^+$, $\ell(\alpha) = \ell_1 \alpha_1 + \cdots + \ell_d \alpha_d$ that we fix from now on. Then we find recursively the coefficients y_n such that $J(y_n) \in \Delta_{\nu_\ell(P)}^N$, and use majorant series employing the Nagumo norms to establish the derides bounds.

It is important to remark that under the conditions of Theorems 1 and 2, equation (1) admits a unique formal power series solution $\widehat{y} \in \mathbb{C}[[x]]^N$ such that $\widehat{y}(0) = 0$. This can be seen directly by writing $\widehat{y}$ as the sum of its homogeneous components in x and then finding these terms recursively after plugging $\widehat{y}$ in (1). Hence, if we are able to find a formal solution of (1) of the form (22), it coincides with the unique formal power series solution of the problem.

Proof of Theorem 1. We divide it in to several steps. Before starting we note that if $F(x, 0) \equiv 0$, the unique formal power series solution is the zero series. Thus we assume $c(x) := F(x, 0) \not\equiv 0$.

Step 0 (Preliminaries): Since F is analytic and $\mu = \frac{\partial F}{\partial y}(0, 0)$, we can write it as a convergent power series in y, say

$$F(x, y) = c(x) + (\mu + A(x))y + \sum_{|I| \geq 2} A_I(x) y^I,$$

where $c, A_I \in \mathcal{O}_b(D_{r'}^d)^N$, $A \in \mathcal{O}_b(D_{r'}^d)^{N \times N}$, $A(0) = 0$, and the summation is taken over all $I = (i_1, \ldots, i_N) \in \mathbb{N}^N$ such that $|I| = i_1 + \cdots + i_N \geq 2$. Furthermore, we can find $K, \delta > 0$ such that

$$\|A_I\|_0 \leq K\delta^{|I|}, \quad \text{for all } I \in \mathbb{N}^N. \tag{23}$$

Note that all the previous coefficients are defined in the common polydisc of polyradius $(r', \ldots, r')$. By reducing r' if necessary, we can also assume that the coefficients of $L = a_1 \partial_{x_1} + \cdots + a_d \partial_{x_d}$ in (1) — which are not all identically zero — belong to $\mathcal{O}_b(D_{r'}^d)$.

Given the injective linear form $\ell : \mathbb{N}^d \to \mathbb{R}^+$, take $\rho' > 0$ such that $Q_{P,\ell}, R_{P,\ell} : \mathcal{O}_b(D_{\rho'(\ell)}) \to \mathcal{O}_b(D_{\rho'(\ell)})$ are defined (Proposition 2). We consider the linear operators $Q, R : \mathcal{O}_b(D_{\rho'(\ell)})^N \to \mathcal{O}_b(D_{\rho'(\ell)})^N$ given by $Q(f_1, \ldots, f_N) = (Q_{P,\ell}(f_1), \ldots, Q_{P,\ell}(f_N))$, and $R(f_1, \ldots, f_N) = (R_{P,\ell}(f_1), \ldots, R_{P,\ell}(f_N))$. Then, by using the norms (10) we see that

$$\|Q(f)\|_m \leq \|Q\| \cdot \|f\|_m, \quad \|R(f)\|_m \leq \|R\| \cdot \|f\|_m, \tag{24}$$

for all $m \in \mathbb{N}$ and $f \in \mathcal{O}_b(D_{\rho'(\ell)})^N$, where to simplify notation we write $\|Q\| = 2 \cdot 4^{|\nu_\ell(P)|}/\rho'^{\ell(\nu_\ell(P))}$ and $\|R\| = 2(1 + 4^{|\nu_\ell(P)|})$, according to the values found in Proposition 2. The same considerations and inequalities are valid for matrix-valued maps. It will be important for later to note that $\|R\|$ is independent of the radius, since we will shrink ρ' during this proof.

Now, choose $0 < s < \rho'$ with $s^{l_j} \leq r'$, for all j, in order to apply Corollary 1 to the previous functions. Then, we conclude there is $r > 0$ such that the maps can be written as

$$a_j = \sum_{n=0}^{\infty} a_{j,n} P^n, \quad c = \sum_{n=0}^{\infty} c_n P^n, \quad A = \sum_{n=0}^{\infty} A_n P^n, \quad A_I = \sum_{n=0}^{\infty} A_{I,n} P^n,$$

$$\tag{25}$$

where $A_n \in (\mathcal{O}_b(D_r^d) \cap \Delta_{\nu_\ell(P)})^{N \times N}$, $c_n, A_{I,n} \in (\mathcal{O}_b(D_r^d) \cap \Delta_{\nu_\ell(P)})^N$, $a_{j,n} \in \mathcal{O}_b(D_r^d) \cap \Delta_{\nu_\ell(P)}$, for all $j = 1, \ldots, d$, $I \in \mathbb{N}^N$ and $n \in \mathbb{N}$.

Step 1 (The coefficient y_0): First we determine the term y_0 in (22). Note that $\widehat{\boldsymbol{y}}(\boldsymbol{0}) = y_0(\boldsymbol{0}) = \boldsymbol{0}$ since $F(\boldsymbol{0}, \boldsymbol{0}) = \boldsymbol{0}$ and $P(\boldsymbol{0}) = 0$. When we plug $\widehat{\boldsymbol{y}}$ into equation (1) and equate in common powers of P, we find y_0 must be an analytic solution of

$$0 = R \left(c + (\mu + A)y + \sum_{|I| \geq 2} A_I y^I \right)$$

$$= c_0 + \mu y_0 + R(A_0 y_0) + \sum_{|I| \geq 2} R(A_{I,0} y_0^I), \tag{26}$$

satisfying $J(y_0) \in \Delta_{\nu_\ell(P)}^N$. Note that in the last equality we have used (17) joint with $R(c) = c_0$, $R(A) = A_0$ and $R(A_I) = A_{I,0}$.

We will prove that (26) has a unique solution in $\mathcal{O}_b(D_r^d)$ if $r > 0$ is taken small enough. In order to proceed, we write (26) as the fixed point equation

$$y = G(y), \quad G(y) := -\mu^{-1} R \left(c_0 + A_0 y + \sum_{|I| \geq 2} A_{I,0} y^I \right).$$

By reducing $r > 0$, we show there is $\epsilon > 0$ such that $G : \overline{B}_\epsilon \to \overline{B}_\epsilon$ is well-defined and a contraction, where $\overline{B}_\epsilon := \{y \in \mathcal{O}_b(D_r^d)^N : \|y\|_0 \leq \epsilon, y(\boldsymbol{0}) = \boldsymbol{0}\}$ which is closed. Then, by Banach's fixed point theorem, G has a unique fixed point.

Let us check first that G maps $\overline{B}_{1/2\delta}$ to $\mathcal{O}_b(D_r^d)^N$, where δ is as in (23): by using (24) for R and (23) we see that

$$\|G(y)\|_0 \leq \|\mu^{-1}\|_0\|R\| \left(\|c_0\|_0 + \|A_0\|_0\|y\|_0 + \sum_{|I|\geq 2} \|A_{I,0}\|_0\|y\|_0^{|I|} \right)$$

$$\leq \|\mu^{-1}\|_0\|R\|^2 \left(\|c\|_0 + \|A\|_0\|y\|_0 + \sum_{|I|\geq 2} K\delta^{|I|}\|y\|_0^{|I|} \right).$$

But the identity $\sum_{|I|\geq 1} \alpha^{|I|} = (1-\alpha)^{-N} - 1$, $|\alpha| < 1$, shows that

$$\|G(y)\|_0 \leq \|\mu^{-1}\|_0\|R\|^2\left(\|c\|_0 + \|A\|_0\|y\|_0 + K\delta\left(2^N - 1\right)\|y\|_0 \right),$$

thus, $\|G(y)\|_0$ is finite as desired. Now, if $y + h, y \in \overline{B}_{1/2\delta}$, we also have

$$\|G(y+h) - G(y)\|_0 \leq \|\mu^{-1}\|_0\|R\|^2$$

$$\times \left(\|A\|_0\|h\|_0 + \sum_{|I|\geq 2} K\delta^{|I|}\|(y+h)^I - y^I\|_0 \right).$$

Taking into account the inequality

$$\|(y+h)^I - y^I\|_0 \leq |I|(\|y\|_0 + \|h\|_0)^{|I|-1}\|h\|_0,$$

that follows readily by induction on $|I|$, we obtain $\|G(y+h)-G(y)\|_0$ is bounded by

$$\|\mu^{-1}\|_0\|R\|^2\left(\|A\|_0 + K\sum_{|I|\geq 2} |I|\delta^{|I|}(\|y\|_0 + \|h\|_0)^{|I|-1} \right)\|h\|_0$$

$$= \|\mu^{-1}\|_0\|R\|^2\left(\|A\|_0 + K\delta g\big(\delta(\|y\|_0 + \|h\|_0)\big) \right)\|h\|_0,$$

where $g(\alpha) = \sum_{|I|\geq 2} |I|\alpha^{|I|-1} = \frac{d}{d\alpha}\left(\sum_{|I|\geq 2} \alpha^{|I|} \right) = N((1-\alpha)^{-N-1} - 1)$, for $|\alpha| < 1$. Since g is continuous and $g(0) = 0$, we can choose $0 < \epsilon < \min\{1, 1/2\delta\}$ such that $\|\mu^{-1}\|_0\|R\|^2K\delta \cdot g(\epsilon) < 1/4$. Also, since $A(\mathbf{0}) = \mathbf{0}$, we can reduce r to assure that $\|\mu^{-1}\|_0\|R\|^2\|A\|_0 < 1/4$. Therefore, if $\|y\|_0 + \|h\|_0 \leq \epsilon$, we conclude

$$\|G(y+h) - G(y)\|_0 \leq \frac{1}{2}\|h\|_0.$$

But $c(\mathbf{0}) = F(\mathbf{0},\mathbf{0}) = \mathbf{0}$, and since ϵ has been fixed, we can reduce r again to have $\|\mu^{-1}\|_0\|R\|^2\|c\|_0 < \epsilon/2$. By applying the previous inequality to $y = 0$ we find $\|G(h)\|_0 \leq \|G(h) - G(0)\|_0 + \|G(0)\|_0 \leq \frac{1}{2}\|h\|_0 + \epsilon/2$. Thus, $G : \overline{B}_\epsilon \to \overline{B}_\epsilon$ has the desired properties.

Several remarks are at hand. First, if y_0 is the solution of equation (26), $J(y_0)$ will be the unique formal solution of (26) and it is convergent. But $R(y_0)$ is another analytic solution of (26), thus $J(y_0) = J(R(y_0)) \in \Delta^N_{\nu_\ell(P)}$. Second, there is a direct way to find a solution of (26) as follows: we find first a solution $Y_0(\boldsymbol{x})$ of $F(\boldsymbol{x}, \boldsymbol{y}(\boldsymbol{x})) = \mathbf{0}$, with the aid of the holomorphic implicit function theorem — it can be applied since $F(\mathbf{0},\mathbf{0}) = \mathbf{0}$ and $\frac{\partial F}{\partial \boldsymbol{y}}(\mathbf{0},\mathbf{0}) = \mu$ is invertible. Then, it follows by applying R to the previous equation that $y_0 = R(Y_0)$ is the solution of (26), since we already know it is unique.

Step 2 (Recurrence equations for y_n): We can now assume $y_0 = 0$ by making the change of variables $y \mapsto y - y_0$ in the initial equation (1). In fact, after doing so, we obtain a similar PDE such that P divides c and we search for a formal solution $\widehat{\boldsymbol{y}} = \sum_{n=1}^{\infty} y_n P^n$ which is divisible by P.

To find the recurrence equations satisfied by the y_n, we start with the right-hand side of (1). By using the identity (18) we find

$$A \cdot \widehat{\boldsymbol{y}} = \sum_{n=1}^{\infty} \left(\sum_{k=1}^{n} \sum_{j=1}^{k} RQ^{n-k}(A_{k-j}y_j) \right) P^n.$$

For the nonlinear term, analogously to (16), we have the decomposition

$$\sum_{|I|\geq 2} A_I \widehat{\boldsymbol{y}}^I = \sum_{k=2}^{\infty} \sum_{*_k} A_{I,m} \prod_{\substack{1\leq l\leq N \\ 1\leq j\leq i_l}} y_{l,n_{l,j}} P^k$$

$$= \sum_{k=2}^{\infty} \sum_{p=0}^{\infty} \sum_{*_k} RQ^p \left(A_{I,m} \prod_{\substack{1 \le l \le N \\ 1 \le j \le i_l}} y_{l,n_{l,j}} \right) P^{k+p}$$

$$= \sum_{n=2}^{\infty} \left[\sum_{k=2}^{n} \sum_{*_k} RQ^{n-k} \left(A_{I,m} \prod_{\substack{1 \le l \le N \\ 1 \le j \le i_l}} y_{l,n_{l,j}} \right) \right] P^n,$$

where the sum $\sum_{*_k}$ is taken over all $I \in \mathbb{N}^N$ such that $2 \le |I| \le k$, m satisfying $0 \le m \le k - |I|$, and $n_{l,j} \ge 1$ such that $k = m + n_{1,1} + \cdots + n_{1,i_1} + \cdots + n_{N,1} + \cdots + n_{N,i_N}$. Note in particular that $n_{l,j} < k \le n$ and thus no component of $y_n = (y_{n,1}, \ldots, y_{n,d})$ appears in the coefficient corresponding to P^n.

For the left-hand side of (1), using the hypothesis $L(P) = P \cdot h$, for some $h \in \mathcal{O}_b(D_r^d)$, we can write

$$P \cdot L(\widehat{y}) = \sum_{n=1}^{\infty} \left(L(y_{n-1}) + n y_n L(P) \right) P^n = \sum_{n=2}^{\infty} (L(y_{n-1})$$

$$+ (n-1) h y_{n-1}) P^n. \tag{27}$$

Now, equating both sides of (1) in common power series of P, we obtain the recurrence

$$\sum_{k=2}^{n} RQ^{n-k}(L(y_{k-1})) + \sum_{k=2}^{n} (k-1) RQ^{n-k}(h y_{k-1}) = c_n$$

$$+ \mu y_n + \sum_{k=1}^{n} \sum_{j=1}^{k} RQ^{n-k}(A_{k-j} y_j)$$

$$+ \sum_{k=2}^{n} \sum_{*_k} RQ^{n-k} \left(A_{I,m} \prod_{\substack{1 \le l \le N \\ 1 \le j \le i_l}} y_{l,n_{l,j}} \right).$$

Equivalently, we have

$$\mu y_n + R(A_0 y_n) = b_n := e_n + \sum_{k=2}^{n} (k-1) RQ^{n-k}(h y_{k-1}), \tag{28}$$

for all $n \geq 1$, where

$$e_n = - c_n + \sum_{k=2}^{n} RQ^{n-k}(L(y_{k-1})) - \sum_{k=1}^{n-1}\sum_{j=1}^{k} RQ^{n-k}(A_{k-j}y_j)$$

$$- \sum_{j=1}^{n-1} R(A_{n-j}y_j) - \sum_{k=2}^{n}\sum_{*_k} RQ^{n-k}\left(A_{I,m} \prod_{\substack{1\leq l\leq N \\ 1\leq j\leq i_l}} y_{l,n_{l,j}} \right).$$

Note in particular that $b_1 = -c_1$.

Equation (28) can be solved as follows: consider $Y_n = (\mu + A_0)^{-1}(b_n)$, where we have, if necessary, reduced r to ensure $\mu + A_0(x)$ is invertible for all $|x| \leq r$. Then $R(Y_n)$ solves (28), as we see by applying R to $(\mu + A_0)Y_n = b_n$ and recalling that $R(b_n) = b_n$. To check uniqueness, note that if y_n and w_n are solutions, then $R((\mu + A_0)(y_n - w_n)) = 0$, so $(\mu + A_0)(y_n - w_n) = h_1 P$, for some $h_1 \in \mathcal{O}$. Thus, $y_n - w_n = R(y_n - w_n) = R((\mu + A_0)^{-1}h_1 P) = 0$. In conclusion, we can find recursively the coefficients y_n by means of the formulas

$$y_n = R\left((\mu + A_0)^{-1}(b_n)\right), \tag{29}$$

and equation (1) has a unique formal power series solution.

Step 3 (Majorant series): We use the majorant series technique to show that $\widehat{y}$ is P-1-Gevrey by proving that $\sum_{n=1}^{\infty} \|y_n\|_n \tau^n$ is 1-Gevrey in τ.

We have chosen $r > 0$ satisfying all previous requirements in order to find y_n recursively. Now, we take $0 < \rho < \min\{r, 1\}$ satisfying $\rho^{l_j} < r$, $j = 1, \ldots, d$, in order to apply the bounds (24) for functions in $\mathcal{O}_b(D_{\rho(\ell)})^N$.

Let $M = \|(\mu + A_0)^{-1}\|_0 > 0$. By applying the Nagumo norm $\|\cdot\|_n$ to equation (29) and taking into account the properties developed in Propositions 1 and 2, we find that

$$\frac{\|y_n\|_n}{M\|R\|} \leq \|c_n\|_n + \sum_{k=2}^{n} \|R\|\|Q\|^{n-k}(\|L(y_{k-1})\|_n + (k-1)\|hy_{k-1}\|_n)$$

$$+ \sum_{k=1}^{n-1} \|R\|\|Q\|^{n-k} \sum_{j=1}^{k} \|A_{k-j}\|_{k-j}\|y_j\|_j$$

$$+ \sum_{j=1}^{n-1} \|R\| \|A_{n-j}\|_{n-j} \|y_j\|_j$$

$$+ \sum_{k=2}^{n} \|R\| \|Q\|^{n-k} \sum_{*_k} \|A_{I,m}\|_m \prod_{\substack{1 \le l \le N \\ 1 \le j \le i_l}} \|y_{l,n_{l,j}}\|_{n_{l,j}}.$$

To bound the term $\|L(y_{k-1})\|_n$ use Proposition 1 (ii) and ensure $\rho < 1$ to get

$$\left\| a_j \frac{\partial y_{k-1}}{\partial x_j} \right\|_n \le \|a_j\|_{n-k} \left\| \frac{\partial y_{k-1}}{\partial x_j} \right\|_k \le ek \prod_{i \ne j} (\rho^{\ell_i}/2) \|a_j\|_{n-k} \|y_{k-1}\|_{k-1}$$

$$\le ek \|a_j\|_{n-k} \|y_{k-1}\|_{k-1}.$$

But inequality (6) implies that $\|a_j\|_{n-k} \le \|a_j\|_0$. If $a = \|a_1\|_0 + \cdots + \|a_d\|_0$, by hypothesis $a > 0$, and

$$\|L(y_{k-1})\|_n \le eak \|y_{k-1}\|_{k-1}.$$

On the other hand, $\|hy_{k-1}\|_n \le \|h\|_{n-k+1} \|y_{k-1}\|_{k-1} \le \|h\|_0 \|y_{k-1}\|_{k-1}$. Thus, we find that

$$\frac{\|y_n\|_n}{M\|R\|} \le \|c_n\|_n + \|R\|(ea + \|h\|_0) \sum_{k=2}^{n} k \|Q\|^{n-k} \|y_{k-1}\|_{k-1}$$

$$+ \|R\| \sum_{k=1}^{n-1} \|Q\|^{n-k} \sum_{j=1}^{k} \|A_{k-j}\|_{k-j} \|y_j\|_j$$

$$+ \|R\| \sum_{j=1}^{n-1} \|A_{n-j}\|_{n-j} \|y_j\|_j$$

$$+ \|R\| \sum_{k=2}^{n} \|Q\|^{n-k} \sum_{*_k} \|A_{I,m}\|_m \prod_{\substack{1 \le l \le N \\ 1 \le j \le i_l}} \|y_{n_{l,j}}\|_{n_{l,j}}. \tag{30}$$

348   *S. A. Carrillo & C. A. Hurtado*

If we divide by $n!$ and using that $m!k! \leq (m+k)!$, we conclude that

$$\frac{\|y_n\|_n}{M\|R\|n!} \leq \frac{\|c_n\|_n}{n!} + \|R\|(ea + \|h\|_0) \sum_{k=2}^{n} \frac{\|Q\|^{n-k}}{(n-k)!} \frac{\|y_{k-1}\|_{k-1}}{(k-1)!}$$

$$+ \|R\| \sum_{k=1}^{n-1} \frac{\|Q\|^{n-k}}{(n-k)!} \sum_{j=1}^{k} \frac{\|A_{k-j}\|_{k-j}}{(k-j)!} \frac{\|y_j\|_j}{j!}$$

$$+ \|R\| \sum_{j=1}^{n-1} \frac{\|A_{n-j}\|_{n-j}}{(n-j)!} \frac{\|y_j\|_j}{j!}$$

$$+ \|R\| \sum_{k=2}^{n} \frac{\|Q\|^{n-k}}{(n-k)!} \sum_{*_k} \frac{\|A_{I,m}\|_m}{m!} \prod_{\substack{1 \leq l \leq N \\ 1 \leq j \leq i_l}} \frac{\|y_{n_{l,j}}\|_{n_{l,j}}}{n_{l,j}!}.$$

Let us define the sequence z_n recursively by

$$\frac{z_n}{M\|R\|} = \frac{\|c_n\|_n}{n!} + \|R\|(ea + \|h\|_0) \sum_{k=2}^{n} \frac{\|Q\|^{n-k}}{(n-k)!} z_{k-1}$$

$$+ \|R\| \sum_{k=1}^{n-1} \frac{\|Q\|^{n-k}}{(n-k)!} \sum_{j=1}^{k} \frac{\|A_{k-j}\|_{k-j}}{(k-j)!} z_j$$

$$+ \|R\| \sum_{j=1}^{n-1} \frac{\|A_{n-j}\|_{n-j}}{(n-j)!} z_j$$

$$+ \|R\| \sum_{k=2}^{n} \frac{\|Q\|^{n-k}}{(n-k)!} \sum_{*_k} \frac{\|A_{I,m}\|_m}{m!} \prod_{\substack{1 \leq l \leq N \\ 1 \leq j \leq i_l}} z_{n_{l,j}}, \tag{31}$$

where $z_1 = M\|R\|\|c_1\|_1$. Since the terms of the previous equation are all nonnegative real numbers, we find inductively that

$$\frac{\|y_n\|_n}{n!} \leq z_n. \tag{32}$$

On the other hand, we consider the generating power series

$$\bar{c} = \sum_{n=1}^{\infty} \frac{\|c_n\|_n}{n!} \tau^n, \quad \overline{A} = \sum_{n=0}^{\infty} \frac{\|A_n\|_n}{n!} \tau^n, \quad \overline{F} = \sum_{m \geq 0, |I| \geq 2} \frac{\|A_{I,m}\|_m}{m!} \tau^m Z^{|I|},$$

$$\tag{33}$$

which are convergent. In fact, recalling the expansions in (25), Corollary 1 and the inequalities in (24) and (6) — recall $\rho < 1$ — we have

$$\|c_n\|_n = \|R \circ Q^n(c)\|_n \leq \|R\|\|Q\|^n\|c\|_n \leq \|R\|\|Q\|^n\|c\|_0, \qquad (34)$$

thus $\bar{c}$ is actually entire. The same argument applies for $\overline{A}$. Now, for $\overline{F}$, we use inequality (23) to write

$$\|A_{I,m}\|_m = \|R \circ Q^m(A_I)\|_m \leq \|R\|\|Q\|^m\|A_I\|_0 \leq \|R\|\|Q\|^m K\delta^{|I|},$$

and thus

$$\sum_{|I|=j} \|A_{I,m}\|_m \leq K\|R\|\|Q\|^m \sum_{|I|=j} \delta^{|I|}$$

$$= K\|R\|\|Q\|^m \binom{j+N-1}{N-1}\delta^j \leq (K\|R\|2^{N-1})$$

$$\times \|Q\|^m(2\delta)^j, \qquad (35)$$

since the number of solutions $I \in \mathbb{N}^N$ of $|I| = j$ is $\binom{j+N-1}{N-1}$, which is less than 2^{j+N-1}. Therefore, the coefficient in $\tau^m Z^j$ of $\overline{F}$ is bounded by $(K\|R\|2^{N-1})(2\delta)^j\|Q\|^m/m!$ proving the convergence of $\overline{F}$.

Using these series and equation (31), we find $Z(\tau) = \sum_{n=1}^{\infty} z_n \tau^n$ is a formal solution of the analytic equation

$$E(\tau, Z(\tau)) = 0,$$

where

$$E(\tau, Z) := -\frac{Z}{M\|R\|} + \bar{c}(\tau) + \|R\|(ea + \|h\|_0)e^{\|Q\|\tau}\tau Z$$

$$+ \|R\|(e^{\|Q\|\tau}\overline{A}(\tau) - \|A_0\|_0)Z + \|R\|e^{\|Q\|\tau}\overline{F}(\tau, Z).$$

But (33) shows that $\bar{c}(0) = \overline{F}(0,0) = \frac{\partial \overline{F}}{\partial Z}(0,0) = 0$ and $\overline{A}(0) = \|A_0\|_0$. Therefore, $E(0,0) = \bar{c}(0) + \|R\|\overline{F}(0,0) = 0$ and $\frac{\partial E}{\partial Z}(0,0) = -1/M\|R\| + \|R\|(\overline{A}(0) - \|A_0\|_0) + \|R\|\frac{\partial \overline{F}}{\partial Z}(0,0) = -1/M\|R\| \neq 0$. Then the holomorphic implicit function theorem implies this equation has a unique convergent power series solution at the origin, thus it must be $Z(\tau)$, so it is convergent. By (32), $\sum_{n=1}^{\infty} \|y_n\|_n \tau^n$ is 1–Gevrey in τ as desired. $\qquad \square$

Proof of Theorem 2. Regarding the previous proof, only some minor changes are required so establish the result. While Step 0 and Step 1 remain the same, in Step 2 the recurrence for y_n takes the form

$$\mu y_n - nR(L(P)y_n) + R(A_0 y_n) = d_n := e_n + \sum_{k=1}^{n-1} kRQ^{n-k}(L(P)y_k),$$
(36)

for all $n \geq 1$, where e_n is as before. Then d_n and b_n differ only in the previous sum, that we bound by shifting one index, as follows

$$\left\| \sum_{k=2}^{n} (k-1)RQ^{n-k+1}(L(P)y_{k-1}) \right\|_n \leq C \sum_{k=2}^{n} k\|Q\|^{n-k}\|y_{k-1}\|_{k-1},$$
(37)

where $C = \|R\|\|Q\|\|L(P)\|_0$. In the current case, the solution of (36) is

$$y_n = R\left((\mu - nL(P)I_N + A_0)^{-1}(d_n)\right),$$

where I_N is the identity matrix of size N. To make this formula meaningful, it is enough to prove $\mu - nL(P)(\boldsymbol{x})I_N + A_0(\boldsymbol{x})$ is invertible for all $n \geq 1$ and $|\boldsymbol{x}| \leq r$, if r is sufficiently small. To proceed, let us recall that if $B \in \mathbb{C}^{N \times N}$ is such that $|B| < 1$ for a matrix norm $|\cdot|$, then $I_N - B$ is invertible, $(I_N - B)^{-1} = \sum_{n=0}^{\infty} B^n$, and $|(I_N - B)^{-1}| \leq (1 - |B|)^{-1}$. Here, as before, we use $|B| = \max_{1 \leq i \leq N} \sum_{j=1}^{N} |B_{i,j}|$. Now, since $L(P)(\boldsymbol{0}) \neq 0$, we can choose a small $r > 0$ such that $\alpha = \inf_{|\boldsymbol{x}| \leq r} |L(P)(\boldsymbol{x})| > 0$. Thus, if $n > \|\mu + A_0\|_0/\alpha$, we see that $|\mu + A_0(\boldsymbol{x})|/|nL(P)(\boldsymbol{x})| \leq \|\mu + A_0\|_0/n\alpha < 1$, for all $|\boldsymbol{x}| \leq r$, so $\mu - nL(P) + A_0$ is invertible and

$$|(\mu - nL(P)(\boldsymbol{x}) + A_0(\boldsymbol{x}))^{-1}|$$

$$= \frac{1}{n|L(P)(\boldsymbol{x})|} \left| \left(I_N - \frac{1}{nL(P)(\boldsymbol{x})}(\mu + A_0(\boldsymbol{x})) \right)^{-1} \right|$$

$$\leq \frac{1/\alpha n}{1 - \frac{\|\mu + A_0\|_0}{\alpha n}} = \frac{1}{\alpha n - \|\mu + A_0\|_0}.$$

For $n \leq \|\mu + A_0\|_0/\alpha$, by hypothesis the remaining finite number of matrices $\mu - nL(P)(\boldsymbol{x}) + A_0(\boldsymbol{x})$ are invertible at the origin. Thus, we can shrink r and assume they are invertible for all $|\boldsymbol{x}| \leq r$. In conclusion, all these matrices are invertible, and we can find $M > 0$ such that

$$\|(\mu - nL(P) + A_0)^{-1}\|_0 \leq M/n, \quad \text{for all } n \geq 1. \tag{38}$$

At this stage, we can proceed with Step 3 by using the Nagumo norms and taking into account (37). However, the factor M/n in (38) improves our bounds and shows that

$$\frac{\|y_n\|_n}{M\|R\|} \leq \|c_n\|_n + \|R\|(ea + \|Q\|\|L(P)\|_0)$$

$$\times \sum_{k=2}^{n} \|Q\|^{n-k} \|y_{k-1}\|_{k-1} + \cdots,$$

where the dots indicate the remaining terms are the same as in (30). In this case, it is not necessary to divide by $n!$, just by defining z_n accordingly we find $\|y_n\|_n \leq z_n$, for all $n \geq 1$, and $Z(\tau)$ satisfies the analytic equation

$$\widetilde{E}(\tau, Z(\tau)) = 0,$$

where

$$\widetilde{E}(z, Z) := - \frac{Z}{M\|R\|} + \widetilde{c}(\tau) + \|R\|(ea + \|Q\|\|L(P)\|_0)\frac{\tau Z}{1 - \|Q\|\tau}$$

$$+ \|R\| \left(\frac{\widetilde{A}(\tau)}{1 - \|Q\|\tau} - \|A_0\|_0 \right) Z + \|R\|\frac{\widetilde{F}(\tau, Z)}{1 - \|Q\|\tau},$$

with coefficients $\widetilde{c}(\tau) = \sum_{n=1}^{\infty} \|c_n\|_n \tau^n$, $\widetilde{A}(\tau) = \sum_{n=0}^{\infty} \|A_n\|_n \tau^n$, and $\widetilde{F}(\tau, Y) = \sum_{m \geq 0, |I| \geq 2} \|A_{I,m}\|_m \tau^m Y^{|I|}$, which are again convergent as justified by the inequalities (34) and (35). Since $\widetilde{E}(0, 0) = \widetilde{c}(0) + \|R\|\widetilde{F}(0, 0) = 0$ and $\frac{\partial \widetilde{E}}{\partial Z}(0, 0) = -1/M\|R\| + \|R\|(\widetilde{A}(0) - \|A_0\|_0) + \|R\|\frac{\partial \widetilde{F}}{\partial Z}(0, 0) = -1/M\|R\| \neq 0$, the holomorphic implicit function theorem proves that $Z(\tau)$ and therefore $\sum_{n=1}^{\infty} \|y_n\|_n \tau^n$ are convergent. This proves the convergence of $\widehat{\boldsymbol{y}} = \sum_{n=1}^{\infty} y_n P^n$ as required. $\qquad\square$

6. A Simple Extension to Higher-Order Systems

There is a straightforward way to extend our theorems for systems of PDEs of higher order by augmenting the size of the given equation.

Corollary 2. *Let P, L and F be as in Theorem 1, fix $u_1, \ldots, u_{k-1} \in \mathcal{O}$, and consider the system of PDEs*

$$(P \cdot L)^k(\boldsymbol{y})(\boldsymbol{x}) + u_{k-1}(\boldsymbol{x})(P \cdot L)^{k-1}(\boldsymbol{y})(\boldsymbol{x}) + \cdots + u_1(\boldsymbol{x})$$
$$\times (P \cdot L)(\boldsymbol{y})(\boldsymbol{x}) = F(\boldsymbol{x}, \boldsymbol{y}). \tag{39}$$

Then, the following statements hold:

(i) *If P divides $L(P)$, (39) has a unique formal power series solution which is P-1-Gevrey.*

(ii) *If $L(P)(\boldsymbol{0}) \neq 0$, and $\sigma - nL(P)(\boldsymbol{0}) \neq 0$, for all $n \in \mathbb{N}$ and all solutions σ of the polynomial equation*

$$p_\mu \left(\sigma^k + u_{k-1}(\boldsymbol{0})\sigma^{k-1} + \cdots + u_2(\boldsymbol{0})\sigma^2 + u_1(\boldsymbol{0})\sigma \right) = 0, \tag{40}$$

where p_μ is the characteristic polynomial of μ, then (39) has a unique convergent power series solution.

Proof.   In the variable $\boldsymbol{w} = (\boldsymbol{w}_0, \boldsymbol{w}_1, \ldots, \boldsymbol{w}_{k-1}) \in \mathbb{C}^{Nk}$, where $\boldsymbol{w}_0 = \boldsymbol{y}$, $\boldsymbol{w}_1 = (P \cdot L)(\boldsymbol{w}_0), \boldsymbol{w}_2 = (P \cdot L)(\boldsymbol{w}_1), \ldots, \boldsymbol{w}_{k-1} = (P \cdot L)(\boldsymbol{w}_{k-2})$, (39) can be written as

$$P \cdot L(\boldsymbol{w}) = G(\boldsymbol{x}, \boldsymbol{w})$$
$$= (\boldsymbol{w}_1, \boldsymbol{w}_2 \ldots, \boldsymbol{w}_{k-1}, F(\boldsymbol{x}, \boldsymbol{w}_0)$$
$$- u_1(\boldsymbol{x})\boldsymbol{w}_1 - \cdots - u_{k-1}(\boldsymbol{x})\boldsymbol{w}_{k-1}),$$

which has the form of equation (1). Then, the results follow from Theorem 1 and 2 by noting that

$$\frac{\partial G}{\partial \boldsymbol{w}}(\boldsymbol{0}, \boldsymbol{0}) = \begin{pmatrix} 0 & I_N & 0 & \cdots & 0 \\ 0 & 0 & I_N & \cdots & 0 \\ \vdots & \vdots & \vdots & \ddots & \vdots \\ 0 & 0 & 0 & \cdots & I_N \\ \mu & -u_1(\boldsymbol{0})I_N & -u_2(\boldsymbol{0})I_N & \cdots & -u_{k-1}(\boldsymbol{0})I_N \end{pmatrix} \in \mathbb{C}^{Nk} \times \mathbb{C}^{Nk}$$

is invertible with eigenvalues given by the solutions of (40), see, e.g. [35, p. 293; 36].

7. Examples

Theorem 1 has a general nature and recovers many examples of Gevrey formal power series solutions of ODEs and PDEs that have been treated in the literature. We conclude this chapter explaining some of them.

Example 1. The solution of (1) is generically divergent, but there are cases where it can be convergent. This is evidenced already in the case of one variable: while Euler's equation $x^2 y' + y = x$ has the x-1-Gevrey solution $\widehat{y}(x) = \sum_{n=0}^{\infty} (-1)^n n! x^{n+1}$, the equation $x^2 y' + y = x + x^2$ has $\widehat{y}(x) = x$ as analytic solution. More examples can be obtained by taking $f \in \mathbb{C}\{z\}$ and $P \in \mathcal{O}$, $P(\mathbf{0}) = 0$. If $L(P) = 0$, then the solution of

$$P(\boldsymbol{x})L(y) = y - f(P(\boldsymbol{x})), \quad \text{is } \widehat{y}(\boldsymbol{x}) = f(P(\boldsymbol{x})),$$

which is convergent.

Example 2. We consider the equation

$$x_1 x_2 \frac{\partial y}{\partial x_1} = \mu y - \frac{x_1}{1 - x_1},$$

where $\mu \neq 0$ is constant. A way to find its unique formal power series solution is to plug $\widehat{y} = \sum_{n=0}^{\infty} y_n(x_2) x_1^n$ into the equation and then equate common powers of x_1. Thus, we find $y_0(x_2) = 0$, $y_n(x_2) = (\mu - nx_2)^{-1}$, $n \geq 1$, and the formal solution is equal to

$$\widehat{y}(x_1, x_2) = \sum_{n \geq 1, m \geq 0} \frac{n^m}{\mu^{m+1}} x_1^n x_2^m.$$

We see $\widehat{y}$ is x_2-1-Gevrey by direct inspection or by applying Theorem 1 to $P = x_2$ and $L = x_1 \frac{\partial}{\partial x_1}$ since $L(P) = 0$. However, $\widehat{y}$ is not x_1-1-Gevrey, i.e. $P = x_1$, $L = x_2 \frac{\partial}{\partial x_1}$ is not a valid choice: as a power series in x_1, $y_n(x_2)$ is analytic on the disc $\{x_2 \in \mathbb{C} : |x_2| < |\mu|/n\}$, so there is no common neighborhood of the origin where all the $y_n(x_2)$ are defined.

Example 3. Equation (1) includes the case of singularly perturbed and doubly singular ODEs

$$Q(\varepsilon)x^{k+1}\frac{\partial \boldsymbol{y}}{\partial x}(x,\boldsymbol{\varepsilon}) = F(x,\boldsymbol{\varepsilon},\boldsymbol{y}), \tag{41}$$

where $x \in (\mathbb{C},0)$, $\boldsymbol{\varepsilon} = (\varepsilon_1,\ldots,\varepsilon_m) \in (\mathbb{C}^m,\boldsymbol{0})$, Q is analytic at the origin and $k \geq -1$ is an integer. In the regular case $k = -1$ or 0 and $Q(\boldsymbol{0}) \neq 0$, if there exists a formal solution, it is convergent. In the irregular case, if $Q(\boldsymbol{0}) \neq 0$, we can interpret $\boldsymbol{\varepsilon}$ as regular parameters and the classical theory establishes that the formal solution of (41) is $1/k$-Gevrey in x, uniformly in $\boldsymbol{\varepsilon}$, see [37], i.e. it is a x^k-1-Gevrey series. Equation (41) was studied by Balser and Kostov in [38] for $m = 1$, $k = 0$, and by Balser and Mozo-Fernández in [2] for $m = 1$, $k \geq 1$, both when $Q(\varepsilon) = \varepsilon$ and in the linear case $F(x,\varepsilon,y) = A(x,\varepsilon)y - f(x,\varepsilon)$, proving the summability of the formal solution in the perturbation parameter ε, in adequate domains of x. On the other hand, Canalis-Durand *et al.* in [28] studied this equation when $m = 1$, $k = -1$ and $Q(\varepsilon) = \varepsilon^\sigma$, $\sigma \geq 1$ a positive integer. In particular, they showed that the solution is $1/\sigma$-Gevrey in ε, uniformly in x. Later on, Canalis-Durand *et al.* in [3] considered the case $m = 1$, $Q(\varepsilon) = \varepsilon^q$, and $k, q \geq 1$, and they proved the $\varepsilon^q x^k$-1-summability of the formal power series solution and the singular directions are determined by the solutions of $\det\left(k\eta^q\xi^k I_N - \mu\right) = 0$, in the two-dimensional (ξ,η)-Borel space. We can recover all these divergence rates using Theorem 1:

(i) If $k = -1$ and $Q(\boldsymbol{0}) = 0$, by choosing $P(x,\boldsymbol{\varepsilon}) = Q(\boldsymbol{\varepsilon})$ and $L = \partial_x$, we have $L(P) = 0$. Thus, the solution is $Q(\boldsymbol{\varepsilon})$-1-Gevrey.

(ii) If $k \geq 0$, we take $P(x,\boldsymbol{\varepsilon}) = x^k Q(\boldsymbol{\varepsilon})$ and $L = x\partial_x$, since $L(P) = kx^k Q(\boldsymbol{\varepsilon}) = kP$. Thus, the solution is $x^k Q(\boldsymbol{\varepsilon})$-1-Gevrey. If $Q(\boldsymbol{0}) \neq 0$, this means the solution is a x^k-1-Gevrey series.

Example 4. Let $\boldsymbol{\varepsilon}$ and Q be as in the previous example and assume $P(\boldsymbol{0}) = 0$. If $L = a_1(\boldsymbol{x},\boldsymbol{\varepsilon})\partial_{x_1} + \cdots + a_d(\boldsymbol{x},\boldsymbol{\varepsilon})\partial_{x_d}$, the system of PDEs

$$Q(\boldsymbol{\varepsilon})P(\boldsymbol{x})L(\boldsymbol{y})(\boldsymbol{x},\boldsymbol{\varepsilon}) = F(\boldsymbol{x},\boldsymbol{\varepsilon},\boldsymbol{y}), \tag{42}$$

can be seen as a singularly perturbed problem where the perturbation is given by Q when $Q(\boldsymbol{0}) = 0$. If P divides $L(P)$, then $L(QP) = QL(P)$ is divisible by QP and we can apply Theorem 1 to conclude the system has a unique formal power series solution which is $Q(\varepsilon)P(\boldsymbol{x})$-1–Gevrey. Let us consider several instances of this situation. First, assume $P \in \mathbb{C}[\boldsymbol{x}]$ is a quasi-homogeneous polynomial, i.e. there are rational numbers $\lambda, \lambda_1, \ldots, \lambda_d > 0$ such that $P(t^{\lambda_1}x_1, \ldots, t^{\lambda_d}x_d) = t^{\lambda}P(\boldsymbol{x})$. Then

$$L_{\boldsymbol{\lambda}} := \lambda_1 x_1 \partial_{x_1} + \cdots + \lambda_d x_d \partial_{x_d},$$

satisfies $L_{\boldsymbol{\lambda}}(P) = \lambda P$, and the solution of (42) is a $Q(\varepsilon)P(\boldsymbol{x})$-1-Gevrey. Second, consider the choice

$$P(\boldsymbol{x}) = \boldsymbol{x}^{\alpha} \quad \text{and} \quad L = \sum_{j=1}^{d} b_j(\boldsymbol{x})x_j \partial_{x_j},$$

with b_j holomorphic near the origin. Then $L(\boldsymbol{x}^{\alpha}) = \boldsymbol{x}^{\alpha} \sum_{j=1}^{d} \alpha_j b_j(\boldsymbol{x})$, thus the solution of (42) is $Q(\varepsilon)\boldsymbol{x}^{\alpha}$-1-Gevrey. Third, families of PDEs with normal crossings given by

$$\varepsilon^{\alpha'} \boldsymbol{x}^{\alpha} L_{\boldsymbol{\lambda}}(\boldsymbol{y})(\boldsymbol{x}, \varepsilon) = F(\boldsymbol{x}, \varepsilon, \boldsymbol{y}), \tag{43}$$

$L_{\boldsymbol{\lambda}}$ and $\boldsymbol{\alpha}$ as before, $\boldsymbol{\alpha}' \in \mathbb{N}^m$, and $\boldsymbol{\lambda} = (\lambda_1, \ldots, \lambda_d) \in (\mathbb{C}^*)^d$. Since

$$L_{\boldsymbol{\lambda}}(\varepsilon^{\alpha'} \boldsymbol{x}^{\alpha}) = \langle \boldsymbol{\lambda}, \boldsymbol{\alpha} \rangle \, \varepsilon^{\alpha'} \boldsymbol{x}^{\alpha},$$

where $\langle \boldsymbol{\lambda}, \boldsymbol{\alpha} \rangle := \lambda_1 \alpha_1 + \cdots + \lambda_d \alpha_d$, we obtain a $\varepsilon^{\alpha'} \boldsymbol{x}^{\alpha}$-1-Gevrey solution. We remark these equations have been studied by Yamazawa and Yoshino in [15] in the case $m = 1$, $\boldsymbol{\alpha} = \boldsymbol{0}$, $\mu = \mathrm{diag}(\mu_1, \ldots, \mu_d)$ a diagonal matrix and $\lambda_j, \mathrm{Re}(\mu_k) > 0$, for all $j, k = 1, \ldots, d$. In fact, the authors proved the 1-summability in $\varepsilon = \eta$ of the formal solution, uniformly in $\boldsymbol{x}$. In this trend, and assuming that $\boldsymbol{\lambda}$ has, up to a non-zero constant, positive entries, Mozo-Fernández and the first author in [39] studied these equations for the case $d = 2$ and $m = 0$ proving the solution is actually $x_1^{\alpha_1} x_2^{\alpha_2}$-1-summable. Later on, this was generalized by the first author in [4] for any $d \geq 2$ and m by using an adapted Borel–Laplace method: the formal solution is $\varepsilon^{\alpha'} \boldsymbol{x}^{\alpha}$-1-summable and the singular directions are determined by the solutions of $\det(\langle \boldsymbol{\lambda}, \boldsymbol{\alpha} \rangle \, \boldsymbol{\xi}^{\alpha} \boldsymbol{\eta}^{\alpha'} I_N - \mu) = 0$, in the $(d+m)$-dimensional $(\boldsymbol{\xi}, \boldsymbol{\eta})$-space.

Finally, another instance of equation (42) is the family of scalar singular first-order linear PDEs of nilpotent type

$$(\alpha(x) + \beta(x, y))y\partial_x u + (a + b(x, y))y^2 \partial_y u + (1 + \mathrm{a}(x, y)y)u = f(x, y),$$

where $\alpha(0) \neq 0$ and $\beta(x,0) \equiv b(x,0) \equiv 0$. We obtain a unique y-1-Gevrey series solution by taking $P(x,y) = y$ and $L = (\alpha + \beta)\partial_x + (a + b)y\partial_y$. These equations were studied by Hibino in [7,8] proving the 1-summability in y, uniformly in x, under conditions on α, and on the analytic continuation and exponential growth of β, b, a and f.

Example 5. We examine a particular case of equation (43) without the singular parameter ε but this time decomposing the series involved as a power series with coefficients series in a monomial. Taking $\boldsymbol{\alpha} \in \mathbb{N}^d \setminus \{\mathbf{0}\}$, $\mu \in \mathbb{C}^*$, $\boldsymbol{\lambda} = (\lambda_1, \ldots, \lambda_d) \in (\mathbb{C}^*)^d$, and $c(\boldsymbol{x}) = \sum_{\boldsymbol{\beta} \in \mathbb{N}^d} a_{\boldsymbol{\beta}} \boldsymbol{x}^{\boldsymbol{\beta}} \in \mathbb{C}\{\boldsymbol{x}\}$, consider

$$\boldsymbol{x}^{\boldsymbol{\alpha}} L_{\boldsymbol{\lambda}}(\boldsymbol{y}) = \boldsymbol{x}^{\boldsymbol{\alpha}} \left(\lambda_1 x_1 \partial_{x_1} \boldsymbol{y} + \cdots + \lambda_d x_d \partial_{x_d} \boldsymbol{y} \right) = \mu \boldsymbol{y} - c(\boldsymbol{x}).$$

This equation has generically a $\boldsymbol{x}^{\boldsymbol{\alpha}}$-1-Gevrey formal solution $\widehat{y}$. To find it we can reduce the problem to solve a family of ODEs as follows: write

$$\widehat{y}(\boldsymbol{x}) = \sum_{\boldsymbol{\alpha} \nleq \boldsymbol{\beta}} \boldsymbol{x}^{\boldsymbol{\beta}} \widehat{y}_{\boldsymbol{\beta}}(\boldsymbol{x}^{\boldsymbol{\alpha}}), \ c(\boldsymbol{x}) = \sum_{\boldsymbol{\alpha} \nleq \boldsymbol{\beta}} \boldsymbol{x}^{\boldsymbol{\beta}} c_{\boldsymbol{\beta}}(\boldsymbol{x}^{\boldsymbol{\alpha}}), \tag{44}$$

as power series with coefficients series in $\boldsymbol{x}^{\boldsymbol{\alpha}}$, according to the decomposition $c_{\boldsymbol{\beta}}(t) = \sum_{n=0}^{\infty} a_{n\boldsymbol{\alpha}+\boldsymbol{\beta}} t^n$. Then, plug $\widehat{y}$ into the equation and equate the common terms in $\boldsymbol{x}^{\boldsymbol{\beta}}$. It follows the initial problem is equivalent to solve the family of independent ODEs

$$\langle \boldsymbol{\lambda}, \boldsymbol{\alpha} \rangle t^2 \widehat{y}'_{\boldsymbol{\beta}}(t) = (\mu - \langle \boldsymbol{\lambda}, \boldsymbol{\beta} \rangle t) \widehat{y}_{\boldsymbol{\beta}}(t) - c_{\boldsymbol{\beta}}(t), \quad \boldsymbol{\alpha} \nleq \boldsymbol{\beta}.$$

Thus, each $\widehat{y}_{\boldsymbol{\beta}}$ is uniquely determined and generically t-1-Gevrey. For instance, if $c(\boldsymbol{x}) = \boldsymbol{x}^{\boldsymbol{\beta}}$, we have two cases: if $\langle \boldsymbol{\lambda}, \boldsymbol{\alpha} \rangle = 0$, the solution is $\widehat{y}(\boldsymbol{x}) = \frac{\boldsymbol{x}^{\boldsymbol{\beta}}}{1 - \langle \boldsymbol{\lambda}, \boldsymbol{\beta} \rangle \boldsymbol{x}^{\boldsymbol{\alpha}}}$, which is convergent. Otherwise, after some calculations we find the $\boldsymbol{x}^{\boldsymbol{\alpha}}$-1–Gevrey solution

$$\widehat{y}(\boldsymbol{x}) = \sum_{n=0}^{\infty} \binom{-\langle \boldsymbol{\lambda}, \boldsymbol{\beta} \rangle / \langle \boldsymbol{\lambda}, \boldsymbol{\alpha} \rangle}{n} (-1)^n n! \frac{\langle \boldsymbol{\lambda}, \boldsymbol{\alpha} \rangle^n}{\mu^{n+1}} \boldsymbol{x}^{n\boldsymbol{\alpha}+\boldsymbol{\beta}}.$$

Note $\widehat{y}$ reduces to a polynomial if $\langle \boldsymbol{\lambda}, m\boldsymbol{\alpha} + \boldsymbol{\beta} \rangle = 0$, for some $m \geq 0$.

As an explicit example in two dimensions, consider

$$x_1 x_2 \left(x_1 \partial_{x_1} y - x_2 \partial_{x_2} y \right) = \mu y - (1 - x_1)^{-1}(1 - x_2)^{-1}.$$

Then, decomposition (44) takes the form

$$\widehat{y}(x_1, x_2) = \sum_{n=0}^{\infty} y_{(n,0)}(x_1 x_2) x_1^n + \sum_{n=1}^{\infty} y_{(0,n)}(x_1 x_2) x_2^n,$$

$$\frac{1}{(1 - x_1)(1 - x_2)} = \sum_{n,m \geq 0} x_1^n x_2^m = \frac{1}{1 - x_1 x_2} + \sum_{n=1}^{\infty} \frac{x_1^n + x_2^n}{1 - x_1 x_2},$$

and we find the coefficients are equal to

$$y_{(n,0)}(t) = \frac{1}{(1 - t)(\mu - nt)},$$

$$y_{(0,n)}(t) = \frac{1}{(1 - t)(\mu + nt)}, \quad \text{valid for } |t| < \frac{|\mu|}{n}.$$

By using the Taylor series at the origin of the previous functions, we can determine $\widehat{y}$. The relation between $\widehat{y}$ and the solution

$$y_0(x_1, x_2) = \frac{1}{1 - x_1 x_2} \sum_{n=0}^{\infty} \frac{x_1^n}{\mu - n x_1 x_2} + \frac{1}{1 - x_1 x_2} \sum_{n=1}^{\infty} \frac{x_2^n}{\mu + n x_1 x_2},$$

which is analytic on $\{(x_1, x_2) \in \mathbb{C}^2 : |x_1|, |x_2| < 1, x_1 x_2 \neq \mu/n, n \geq 1\}$, is that y_0 is the $x_1 x_2$-1-sum of $\widehat{y}$, see [40, Example 2.1] for more details on similar calculations with these series and their monomial summability.

Theorem 1 can also be applied in other situations after ramifications and punctual blow-ups. We illustrate this fact with two final examples.

Example 6. Consider the system of PDEs given by

$$x_1^{p_1 + 1} c_1(\boldsymbol{x}) \partial_{x_1} \boldsymbol{y} + \cdots + x_d^{p_d + 1} c_d(\boldsymbol{x}) \partial_{x_d} \boldsymbol{y} = \boldsymbol{F}(\boldsymbol{x}, \boldsymbol{y}), \tag{45}$$

where $\mu = \frac{\partial \boldsymbol{F}}{\partial \boldsymbol{y}}(\boldsymbol{0}, \boldsymbol{0})$ is invertible, and we assume $1 \leq p_1 \leq p_j$, for all j. Then the system has a unique formal power series solution $\widehat{\boldsymbol{y}}$ which is $(1/p_1, \ldots, 1/p_1)$-Gevrey. For instance, an explicit example is the multidimensional Euler's equation

$$x_1^2 \partial_{x_1} y + \cdots + x_d^2 \partial_{x_d} y + y = \boldsymbol{x^1}, \quad \widehat{\boldsymbol{y}}(\boldsymbol{x}) = \sum_{\boldsymbol{\beta} \in \mathbb{N}^d} (-1)^{|\boldsymbol{\beta}|} |\boldsymbol{\beta}|! \boldsymbol{x}^{\boldsymbol{\beta} + \boldsymbol{1}},$$

where $\boldsymbol{1} = (1, \ldots, 1)$.

To prove the claim, we use the punctual blow-up (21) to find that

$$z_1\partial_{z_1} = x_1\partial_{x_1} + \cdots + x_d\partial_{x_d}, \quad z_j\partial_{z_j} = x_j\partial_{x_j}, \quad j = 2,\ldots,d,$$

and thus (45) takes the form

$$z_1^{p_1}L'(\boldsymbol{u}) = z_1^{p_1+1}c_1'\partial_{z_1}\boldsymbol{u} + z_1^{p_1}\sum_{j=2}^{d}(z_1^{p_j-p_1}z_j^{p_j}c_j' - c_1')z_j\partial_{z_j}\boldsymbol{u} = \boldsymbol{F}'(\boldsymbol{z},\boldsymbol{u}),$$

where $c_j'(\boldsymbol{z}) = c_j(\boldsymbol{x})$, $\boldsymbol{F}'(\boldsymbol{z},\boldsymbol{u}) = \boldsymbol{F}(\boldsymbol{x},\boldsymbol{y})$, and $\boldsymbol{u}(\boldsymbol{z}) = \boldsymbol{y}(\boldsymbol{x})$. Since $L'(z_1^{p_1}) = p_1 z_1^{p_1}c_1'(\boldsymbol{z})$ and $\frac{\partial\boldsymbol{F}'}{\partial\boldsymbol{z}}(\boldsymbol{0},\boldsymbol{0}) = \mu$ is invertible, Theorem 1 implies that $\widehat{\boldsymbol{u}}(\boldsymbol{z}) = \widehat{\boldsymbol{y}}(\boldsymbol{x})$ is a $z_1^{p_1}$-1-Gevrey series. Then the proof of Proposition 3 shows $\widehat{\boldsymbol{y}}$ is a $(1/p_1,\ldots,1/p_1)$-Gevrey series as desired.

On the other hand, it is worth remarking (45) has been recently studied by Luo *et al.* in [11] for the case

$$x_1^2 c_1(\boldsymbol{x})\partial_{x_1}u + x_2^2 c_2(\boldsymbol{x})\partial_{x_2}u = b(\boldsymbol{x})u - a(\boldsymbol{x}),$$

where a, b, c_1, c_2 are analytic near $\boldsymbol{0} \in \mathbb{C}^2$ and $b(\boldsymbol{0})c_1(\boldsymbol{0})c_2(\boldsymbol{0}) \neq 0$. This scalar equation has a unique formal power series solution $\hat{u}$ which is Borel-summable in the variables (x_1, x_2). In particular, $\hat{u}$ is $(1,1)$-Gevrey as we have shown.

Example 7. Consider the family of ODEs unfolding $k+1$ singularities

$$(x^{k+1} - \varepsilon)\frac{d\boldsymbol{y}}{dx} = \mu\boldsymbol{y} - f(x,\varepsilon,\boldsymbol{y}), \tag{46}$$

where k is a positive integer, μ is an invertible matrix, f is analytic at the origin in $\mathbb{C} \times \mathbb{C} \times \mathbb{C}^d$, $\frac{\partial f}{\partial\boldsymbol{y}}(0,0,\boldsymbol{0}) = \boldsymbol{0}$, and $\varepsilon \in (\mathbb{C},0)$ is a small parameter. These systems have been studied by Klimeš in [9] for the case $k = 1$ by using an adapted (unfolded) Borel–Laplace method to obtain parametric solutions bounded on certain ramified domains attached to both singularities $x = \pm\sqrt{\varepsilon}$, at which they possess a limit in a spiraling manner.

As it is remarked in [9, Section 2.4], the system above has a unique formal power series solution $\widehat{\boldsymbol{y}}$ which is $\left(\frac{1}{k}, \frac{k+1}{k}\right)$-Gevrey in (x,ε). We can prove this readily as follows: consider the ramification $\varepsilon = \eta^{k+1}$, and afterwards the punctual blow-up $x = z$, $\eta = z\zeta$. In these coordinates, equation (46) takes the form

$$z^k \left(z\frac{\partial u}{\partial z} - \zeta\frac{\partial u}{\partial \zeta} \right) = (1 - \eta^{k+1})^{-1} \left(\mu u - f(z, z^{k+1}\zeta^{k+1}, u) \right),$$

where $u(z, \zeta) = y(z, z^{k+1}\zeta^{k+1}) = y(x, \varepsilon)$. By applying Theorem 1 to $P = z^k$ and $L = z\partial_z - \zeta\partial_\zeta$ we find $\widehat{u}(z, \zeta) = \widehat{y}(x, \varepsilon)$ is z^k-1–Gevrey, since $L(P) = kz^k$. Thus, $\widehat{y}(x, \eta^{k+1})$ is $\left(\frac{1}{k}, \frac{1}{k}\right)$–Gevrey in (x, η), and therefore $\widehat{y}(x, \varepsilon)$ is $\left(\frac{1}{k}, \frac{k+1}{k}\right)$–Gevrey in (x, ε) as claimed.

Acknowledgments

The authors wish to thank the organizers of FASnet20 for their efforts to host this online workshop and for giving us the opportunity to present there the results discussed in this chapter. We also thank the referee for the careful reading and the many suggestions that improved the chapter.

The first author is supported by Ministerio de Economía y Competitividad from Spain, under the Project "Métodos asintóticos, algebraicos y geométricos en foliaciones singulares y sistemas dinámicos" (Ref.: PID2019-105621GB-I00) and Universidad Sergio Arboleda project IN.BG.086.20.002.

References

1. W. Balser and M. Loday-Richaud, Summability of solutions of the heat equation with inhomogeneous thermal conductivity in two variables. *Adv. Dyn. Syst. Appl.*, **4**(2) (2009), 159–177, ISSN 0973-5321.
2. W. Balser and J. Mozo-Fernández, Multisummability of formal solutions of singular perturbation problems. *J. Diff. Eqn.*, **183**(2) (2002), 526–545, ISSN 0022-0396.
3. M. Canalis-Durand, J. Mozo-Fernández, and R. Schäfke, Monomial summability and doubly singular differential equations. *J. Diff. Eqn.*, **233**(2) (2007), 485–511, ISSN 0022-0396.
4. S. A. Carrillo, Summability in a monomial for some classes of singularly perturbed partial differential equations. *Publ. Mat.*, **65**(1) (2021), 83–127, ISSN 0210-2978.

5. O. Costin and S. Tanveer, Nonlinear evolution PDEs in $\mathbb{R}^+ \times \mathbb{C}^d$: Existence and uniqueness of solutions, asymptotic and Borel summability properties. *Ann. I. H. Poincaré (C).*, **24**(5) (2007), 795–823, ISSN 0294-1449.

6. O. Costin and S. Tanveer, Short time existence and Borel summability in the Navier-Stokes equation in $\mathbb{R}^3$. *Commun. Part. Diff. Eqns.*, **34**(7–9) (2009), 785–817, ISSN 3605302.

7. M. Hibino, Borel summability of divergent solutions for singular first–order partial differential equations with variable coefficients. Part I., *J. Diff. Eqn.*, **227**(2) (2006), 499–533. ISSN 0022-0396.

8. M. Hibino, Borel summability of divergent solutions for singular first–order partial differential equations with variable coefficients. Part II, *J. Diff. Eqns.*, **227**(2), 534–563 (2006). ISSN 0022-0396.

9. M. Klimeš, Confluence of singularities of nonlinear differential equations via Borel-Laplace transformations. *J. Dyn. Control Syst.*, **22**(2016), 499–533, ISSN 1079-2724.

10. A. Lastra and S. Malek, On parametric Gevrey asymptotics for some nonlinear initial value Cauchy problems. *J. Diff. Eqns.*, **259**(10) (2015), 5220–5270, ISSN 0022-0396.

11. Z. Luo, H. Chen, and C. Zhang, On the summability of divergent power series satisfying singular PDEs. *C. R. Math. Acad. Sci. Paris*, **357**(3) (2019), 258–262, ISSN 1631-073X.

12. S. Ōuchi, Multisummability of formal solutions of some linear partial differential equations. *J. Diff. Eqn.*, **185**(2) (2002), 513–549, ISSN 0022-0396.

13. P. Remy, Gevrey properties and summability of formal power series solutions of some inhomogeneous linear Cauchy-Goursat problems. *J. Dyn. Control Syst.*, **26**(1) (2020), 69–108, ISSN 1573-8698.

14. H. Tahara and H. Yamazawa, Multisummability of formal solutions to the Cauchy Problem for some linear partial differential equations. *J. Diff. Eqns.*, **255**(10) (2013), 3592–3637, ISSN 0022-0396.

15. H. Yamazawa and M. Yoshino, Parametric Borel summability for some semilinear system of partial differential equations. *Opuscula Math.*, **35**(5) (2015), 825–845, ISSN 232-9274.

16. J. Mozo-Fernández and R. Schäfke, Asymptotic expansions and summability with respect to an analytic germ. *Publ. Mat.*, **63**(1) (2019), 3–79, ISSN 0210-2978.

17. R. Gérard and H. Tahara, *Singular Nonlinear Partial Differential Equations*, Aspects of Mathematics. Friedr. Vieweg & Sohn, Braunschweig, 1996, ISSN 0179-2156.

18. S. Kaplan, Formal and convergent power series solutions of singular partial differential equations. *Trans. Amer. Math. Soc.*, **256**(1979), 163–183, ISSN 0002-9947.

19. T. Oshima, On the theorem of Cauchy-Kowalevsky for first order linear differential equations with degenerate principal symbols. *Proc. Japan Acad. Ser. A Math. Sci.*, **49**(1973), 83–87, ISSN 0021-4280.

20. S. Ōuchi, Borel summability of formal solutions of some first order singular partial differential equations and normal forms of vector fields, *J. Math. Soc. Japan*, **57**(2), 415–460 (2005). ISSN 0025-5645.

21. H. Yamazawa, Formal Gevrey Class of formal power series solution for singular first order linear partial differential operators. *Tokyo J. Math.*, **23**(2) (2000), 537–561, ISSN 0387-3870.

22. M. Hibino, Divergence property of formal solutions for singular first order linear partial differential equations. *Publ. Res. Inst. Math. Sci.*, **35**(6) (1999), 893–919, ISSN 0034-5318.

23. M. Hibino, Formal Gevrey theory for singular first order semi-linear partial differential equations. *Osaka J. Math.*, **41**(1), 159–191 (2004). ISSN 0030-6126.

24. A. Lastra and H. Tahara, Maillet type theorem for nonlinear totally characteristic partial differential equations. *Math. Ann.*, **377**(2020), 1603–1641, ISSN 0025-5831.

25. S. N. Chow, C. Li, and D. Wang, *Normal Forms and Bifurcation of Planar Vector Fields*. Cambridge University Press, New York, 1994.

26. B. L. J. Braaksma and L. Stolovitch, Small divisors and large multipliers. *Ann. Inst. Fourier*, **57**(2) (2007), 603–628, ISSN 0373-0956.

27. S. A. Carrillo and F. Sanz, Briot-Bouquet's theorem in high dimension. *Publ. Mat.*, **58**(Suppl.) (2014), 135–152, ISSN 0210-2978.

28. M. Canalis-Durand, J. P. Ramis, R. Schäfke, and Y. Sibuya, Gevrey solutions of singularly perturbed differential equations. *J. Reine Angew. Math.*, **518**(2000), 95–129, ISSN 0075-4102.

29. M. Nagumo, Über das anfangswertproblem partieller differentialgleichunge. *Jap. J. Math.*, **18**(1942), 41–47, ISSN 0289-2316.

30. J. M. Aroca, H. Hironaka, and J. L. Vicente, The theory of the maximal contact, *Memorias de Matemática del Instituto "Jorge Juan"*, Vol. 29. Instituto "Jorge Juan" de Matemáticas, Consejo Superior de Investigaciones Cientícas, Madrid, 1975.

31. A. Rainer and G. Schindl, Equivalence of stability properties for ultradifferentiable function classes. *Rev. R. Acad. Cienc. Exactas Fis. Nat. Ser. A Math. RACSAM.*, **110**(1) (2016), 17–32, ISSN 1578-7303.

32. V. Thilliez, On quasianalytic local rings. *Expo. Math.*, **26**(1) (2008), 1–23, ISSN 0723-0869.

33. S. A. Carrillo, J. Mozo–Fernández, and R. Schäfke, Tauberian theorems for summability in analytic functions. *J. Math. Anal. Appl.*, **489**(2) (2020), 124174, ISSN 0022-247X.

34. J.-F. Mattei and R. Moussu, Holonomie et intégrales premières. *Ann. Sci. Éc. Norm. Supér.*, **4**(13) (1980), 469–523, ISSN 0012-9593.

35. S. Barnett, Congenial matrices. *Linear Algebra Appl.*, **41**(1981), 277–298, ISSN 0024-3795.

36. S. A. Carrillo, The composition of polynomials is a determinant. *Am. Math. Month*, Accepted.

37. Y. Sibuya, Convergence of formal solutions of meromorphic differential equations containing parameters. *Funkcial. Ekvac.*, **37**(1994), 395–400, ISSN 0532-8721.

38. W. Balser and V. Kostov, Singular perturbation of linear systems with a regular singularity. *J. Dyn. Control Syst.*, **8**(3) (2002), 313–322, ISSN 1079-2724.

39. S. A. Carrillo and J. Mozo-Fernández, An extension of Borel-Laplace methods and monomial summability, *J. Math. Anal. Appl.*, **457**(1) (2018), 461–477, ISSN 0022-247X.

40. S. A. Carrillo and J. Mozo-Fernández, Tauberian properties for monomial summability with applications to Pfaffian systems. *J. Diff. Eqns.*, **261**(12) (2016), 7237–7255, ISSN 0022-0396.

© 2022 World Scientific Publishing Europe Ltd.
https://doi.org/10.1142/9781800611368_0017

Chapter 17

Some Notes on the Parametric Gevrey Asymptotics in Two Complex Time Variables through Truncated Laplace Transforms

Guoting Chen*,§, Alberto Lastra†,¶ and Stéphane Malek‡,‖

*School of Science, Harbin Institute of Technology
518055 Shenzhen, China
†University of Alcalá, Departamento de Física y Matemáticas
Ap. de Correos 20, E-28871 Alcalá de Henares (Madrid), Spain
‡University of Lille, Laboratoire Paul Painlevé
59655 Villeneuve d'Ascq cedex, France
§chenguoting@hit.edu.cn
¶alberto.lastra@uah.es
‖stephane.malek@univ-lille.fr

This chapter is a slightly modified, abridged version of a previous work "Parametric Gevrey asymptotics in two complex time variables through truncated Laplace transforms" [1] motivated by our contribution in the conference "Formal and Analytic Solutions of Diff. (differential, partial differential, difference, q-difference, q-difference-differential, ...) Equations on the Internet" (FASnet20). It aims to clarify and give further details on some crucial points concerning the asymptotic behavior of the solutions of the problems studied in that work.

1. Introduction

The main aim of this revision is to clarify some points and give answers to some questions on a recent work on truncated Laplace transform [1] which was presented at the conference "Formal and Analytic Solutions of Diff. (differential, partial differential, difference, q-difference, q-difference-differential, ...) Equations on the Internet" (FASnet20), held virtually during the last week of June, 2020. Moreover, we motivate future problems in this direction.

All technical difficulties have been simplified, whose proof can be found in detail in our work [1], in order to focus on the mentioned details. In this revision, we highlight two aspects that have not been detailed in that paper. Namely, we utterly explain the necessity to use a truncated Laplace transform instead of a complete one in order to extract asymptotic information from our constructed solutions. Furthermore, we provide a more geometric description (with the help of enlightening drawings) of a technical part needed for the study of the difference of consecutive solutions in the framework of the Ramis-Sibuya approach.

The motivation on the use of truncated Laplace transform leans on previous recent works in which truncated Laplace transform has been applied in different settings: in the study of sharp lower estimates [2], or stability estimates [3] of truncated Laplace transform. Also, some recent advances have been made on its numerical properties [4].

Truncated Laplace transform appears in the classical theory of asymptotic expansions of complex functions at the time of constructing a function with prescribed Gevrey asymptotic behavior at the origin [5,6] and also in the study of singularities of canard solutions to singularly perturbed equations [7]. Further applications can be found in many other recent publications [1].

In Section 2, we recall the main problem under study, and the main steps to provide analytic solutions to the problem, as well as formal solutions which are related by means of certain asymptotic expansions. Further remarks on some steps are also shown. In Section 3, we give further details on the geometric aspects on the difference of two consecutive solutions, and describe further properties and future research.

2. Review of the Main Results

This section is devoted to review the main results in the work "Parametric Gevrey asymptotics in two complex time variables through truncated Laplace transforms" [1], stated without proof. We have also decided not to enter into many details and refer to the precise expression or result in the original text.

We focus our attention on the next initial value problem which involves two complex time variables t_1, t_2 and a small complex parameter ϵ,

$$
\begin{aligned}
Q(\partial_z) & u(t_1, t_2, z, \epsilon) \\
&= \epsilon^{\Delta_{D_1 D_2}} (t_1^{k_1+1} \partial_{t_1})^{\delta_{D_1}} (t_2^{k_2+1} \partial_{t_2})^{\tilde{\delta}_{D_2}} R_{D_1 D_2}(\partial_z) u(t_1, t_2, z, \epsilon) \\
&\quad + \sum_{\substack{1 \le \ell_1 \le D_1 - 1 \\ 1 \le \ell_2 \le D_2 - 1}} \epsilon^{\Delta_{\ell_1 \ell_2}} (t_1^{k_1+1} \partial_{t_1})^{\delta_{\ell_1}} t_2^{d_{\ell_2}} \partial_{t_2}^{\tilde{\delta}_{\ell_2}} c_{\ell_1 \ell_2}(z, \epsilon) \\
&\quad \times R_{\ell_1 \ell_2}(\partial_z) u(t_1, t_2, z, \epsilon) + f(t_1, t_2, z, \epsilon),
\end{aligned}
\tag{1}
$$

under given initial conditions $u(0, t_2, z, \epsilon) \equiv u(t_1, 0, z, \epsilon) \equiv 0$.

Here, we assume that $Q, R_{D_1 D_2}, R_{\ell_1 \ell_2}$ are polynomials, $k_1, k_2 \ge 1$ are integers, the coefficients $c_{\ell_1 \ell_2}(z, \epsilon)$ are bounded holomorphic functions on some horizontal strip

$$
H_\beta = \{ z \in \mathbb{C} : |\mathrm{Im}(z)| < \beta \},
$$

for some $\beta > 0$, with respect to z, and holomorphic with respect to ϵ on a disc $D(0, \epsilon_0)$, $\epsilon_0 > 0$.

The forcing term $f(t_1, t_2, z, \epsilon)$ is a holomorphic function in t_1, t_2 on $\mathbb{C}^\star \times D(0, h')$, for some radius $h' > 0$, bounded holomorphic with respect to z on H_β and on any given open sector $\mathcal{E}$ centered at 0, $\mathcal{E} \subseteq D(0, \epsilon_0)$ for some $\epsilon_0 > 0$, with respect to the perturbation parameter ϵ.

Our goal is the construction of holomorphic solutions $u(t_1, t_2, z, \epsilon)$ of (1) where t_1, t_2, ϵ are located on sectors in $\mathbb{C}$, together with the analysis of their asymptotic expansions with respect to ϵ,

$$
u(t_1, t_2, z, \epsilon) \sim_{\epsilon \to 0} \hat{u}(t_1, t_2, z, \epsilon) = \sum_{n \ge 0} u_n(t_1, t_2, z) \epsilon^n.
$$

We search for solutions as a double Laplace and Fourier transform

$$u(t_1, t_2, z, \epsilon) = \frac{k_1 k_2}{(2\pi)^{1/2}} \int_{-\infty}^{\infty} \int_{L_{d_1}} \int_{L_{d_2}} \omega(u_1, u_2, m, \epsilon)$$

$$\times \exp\left(-\left(\frac{u_1}{\epsilon t_1}\right)^{k_1} - \left(\frac{u_2}{\epsilon t_2}\right)^{k_2}\right) e^{izm} \frac{du_2}{u_2} \frac{du_1}{u_1} dm,$$

$$(2)$$

along half-lines $L_{d_j} = [0, \infty)e^{\sqrt{-1}d_j}$ for suitable directions $d_j \in \mathbb{R}$, $j = 1, 2$.

This approach has been successfully applied in two previous works [8,9] for other singularly perturbed families of PDEs with two complex time variables that can be expressed (in the linear setting) in the form

$$Q(\partial_z)\partial_{t_1}\partial_{t_2} y(t_1, t_2, z, \epsilon) = \mathcal{P}_1(\epsilon, t_1, t_2, \partial_{t_1}, \partial_{t_2}, \partial_z) y(t_1, t_2, z, \epsilon)$$

$$+ f(t_1, t_2, z, \epsilon),$$

where the differential operators with polynomial coefficients

$$\mathcal{P}_2(\epsilon, t_1, t_2, \partial_{t_1}, \partial_{t_2}, \partial_z) := Q(\partial_z)\partial_{t_1}\partial_{t_2} - L_1(\epsilon, t_1, t_2, \partial_{t_1}, \partial_{t_2}, \partial_z)$$

(for L_1 being an operator which comprises the leading terms of $\mathcal{P}_1$):

- can be factorized in a special manner as

$$\mathcal{P}_2 = \mathcal{P}_{2.1}(\epsilon, t_1, \partial_{t_1}, \partial_z) \cdot \mathcal{P}_{2.2}(\epsilon, t_2, \partial_{t_2}, \partial_z)$$

  with factors that only depend on one time variable. In this case, the "Borel map" $\omega(u_1, u_2, m, \epsilon)$ is defined w.r.t. its first two variables on domains of the form $(S_{d_1} \cup D(0, \rho_1)) \times (S_{d_2} \cup D(0, \rho_2))$, where S_{d_j} stands for an infinite sector with vertex at the origin and bisecting direction d_j for $j = 1, 2$, and $\rho_j > 0$ is small enough [8].
- cannot be factorized in the previous manner, and are of a special shape. The related domains for the "Borel map" $(u_1, u_2) \mapsto \omega(u_1, u_2, m, \epsilon)$ are of the form $S_{d_1} \times (S_{d_2} \cup D(0, \rho_2))$ or $(S_{d_1} \cup D(0, \rho_1)) \times S_{d_2}$, together with a polydisc at the origin [9].

In the study of (1), none of the solutions provided for the two previous problems hold and $\omega(u_1, u_2, m, \epsilon)$ in (2) can only be defined on products of unbounded sectors $S_{d_1} \times S_{d_2}$ with respect to (u_1, u_2). Therefore, the actual solution $u(t_1, t_2, z, \epsilon)$ can be built up, whereas no asymptotic features with respect to ϵ can be obtained.

Indeed, $u(t_1, t_2, z, \epsilon)$ solves (1) provided that $\omega(u_1, u_2, m, \epsilon)$ solves a convolution equation of the form

$$P_m(u_1, u_2)\omega(u_1, u_2, m, \epsilon) = \text{convolution terms in } \omega(u_1, u_2, m, \epsilon)$$

$$+ \text{ entire forcing term,} \tag{3}$$

where

$$P_m(u_1, u_2) = Q(im) - k_1^{\delta_{D_1}} k_2^{\tilde{\delta}_{D_2}} u_1^{k_1 \delta_{D_1}} u_2^{k_2 \tilde{\delta}_{D_2}} R_{D_1 D_2}(im), \tag{4}$$

whose precise shape is detailed in our work [1].

Under the assumption that the quotient $Q(im)/R_{D_1 D_2}(im)$ remains inside certain sectorial annulus, then, for every fixed $\rho_0 > 0$, the map $u_1 \mapsto P_m(u_1, u_2)$ has $k_1 \delta_{D_1}$ complex roots in $D(0, \rho_0)$, provided that $u_2 \in S_{d_2}$ with large enough $|u_2|$.

The observation above on ω follows from the fact that in solving (3), one needs to invert $P_m(u_1, u_2)$.

As a result, we need to follow another approach in order to analyze the asymptotic expansions with respect to ϵ. Our idea consists of the next construction: instead of a solution expressed as a double Laplace transform, we search for a genuine solution in the form of a Fourier, truncated Laplace and Laplace transform, namely

$$u(t_1, t_2, z, \epsilon) = \frac{1}{(2\pi)^{1/2}} \int_{-\infty}^{\infty} \int_{L_{1,\epsilon}} \int_{L_{d_2}} \omega(u_1, u_2, m, \epsilon)$$

$$\times \exp\left(-\left(\frac{u_1}{\epsilon t_1}\right)^{k_1} - \left(\frac{u_2}{\epsilon t_2}\right)^{k_2}\right) e^{izm} \frac{du_2}{u_2} \frac{du_1}{u_1} dm,$$

$$\tag{5}$$

where $L_{d_2} = [0, \infty)e^{\sqrt{-1}d_2}$ is such that $d_2 \in \mathbb{R}$ is a suitable direction, and $L_{1,\epsilon}$ stands for a segment of the form $\left[0, \left(C_1/\epsilon^{\lambda k_2 \tilde{\delta}_{D_2}}\right) e^{\sqrt{-1}\theta_1}\right]$, for some $C_1, \lambda > 0$ with $\lambda < (k_1 \delta_{D_1})^{-1}$, and an appropriate angle $\theta_1 \in \mathbb{R}$.

The first important feature is that the solution (5) remains close, as $\epsilon \to 0$, to a double Laplace transform in both time variables as mentioned earlier, since $C_1/\epsilon^{\lambda k_2 \tilde{\delta}_{D_2}} \to \infty$ as $\epsilon \to 0$. The second important property is that the asymptotic expansions relatively to ϵ can be reached out for this solution. Accordingly, we impose that the forcing term shares the same shape as the solution, namely

$$f(t_1, t_2, z, \epsilon) = \frac{1}{(2\pi)^{1/2}} \int_{-\infty}^{\infty} \int_{L_{1,\epsilon}} \int_{L_{d_2}} \psi(u_1, u_2, m, \epsilon)$$

$$\times \exp\left(-\left(\frac{u_1}{\epsilon t_1}\right)^{k_1} - \left(\frac{u_2}{\epsilon t_2}\right)^{k_2}\right) e^{izm} \frac{du_2}{u_2} \frac{du_1}{u_1} dm,$$

$$(6)$$

where ψ is a polynomial in u_1, entire in u_2 with at most exponential growth of order k_2, continuous with respect to m with exponential decay on $\mathbb{R}$ and holomorphic with respect to ϵ on $D(0, \epsilon_0)$. The resulting function $f(t_1, t_2, z, \epsilon)$ is holomorphic on $\mathbb{C}^\star \times D(0, h') \times H_\beta \times \mathcal{E}$, for any sector $\mathcal{E} \subseteq D(0, \epsilon_0)$ centered at the origin. From the fact that f approaches a double Laplace transform in t_1, t_2 as $\epsilon \to 0$, observe that f remains close to a polynomial in t_1 (on some sector) as $\epsilon \to 0$, $\epsilon \in \mathcal{E}$.

It is worth noting that the approach proposed in this situation differs from that previously worked out [10] which concerns a subclass of (1) where the differential operators in t_2 belong to a less general class of operators of the form $(t_2^{k_2+1} \partial_{t_2})^{\tilde{\delta}_{\ell_2}}$, and the solutions are built up as Fourier and single Laplace transform along appropriate half-lines, and with a special kernel.

The precise set of conditions we impose to (1) are the following:

$$\Delta_{D_1 D_2} = k_1 \delta_{D_1} + k_2 \tilde{\delta}_{D_2}, \text{ and for all } 1 \le \ell_j \le D_j - 1, \quad j = 1, 2,$$

$$\Delta_{\ell_1 \ell_2} > k_1 \delta_{\ell_1} + \frac{k_2 \tilde{\delta}_{D_2} \delta_{\ell_1}}{\delta_{D_1}}, \quad d_{\ell_2} > \tilde{\delta}_{\ell_2}(k_2 + 1),$$

$$\tilde{\delta}_{D_2} \delta_{\ell_1} \ge \delta_{D_1}(\tilde{\delta}_{\ell_2} + 1/k_2). \tag{7}$$

The quotient $Q(im)/R_{D_1 D_2}(im)$ remains inside a fixed unbounded sector $S_{Q, R_{D_1 D_2}}$ with positive distance to the origin. The first main result concerns the construction of actual holomorphic solutions to (1).

Theorem 1 (First statement of Theorem 1 [1]). *There exist:*

(a) *a finite set of bounded sectors covering a punctured disc at the origin, $\{\mathcal{E}_p\}_{0 \le p \le \iota-1}$, with $\mathcal{E}_p \subseteq D(0,\epsilon_0)$,*
(b) *a set of directions $d_{2,p} \in \mathbb{R}$, with $0 \le p \le \iota - 1$,*
(c) *A pair of bounded sectors $\mathcal{T}_1, \mathcal{T}_2$,*

such that a holomorphic solution

$$u_p(t_1, t_2, z, \epsilon) = \frac{1}{(2\pi)^{1/2}} \int_{-\infty}^{\infty} \int_{L_{1,p,\epsilon}} \int_{L_{d_{2,p}}} \omega_p(u_1, u_2, m, \epsilon)$$

$$\times \exp\left(-\left(\frac{u_1}{\epsilon t_1}\right)^{k_1} - \left(\frac{u_2}{\epsilon t_2}\right)^{k_2}\right) e^{izm} \frac{du_2}{u_2} \frac{du_1}{u_1} dm$$

$$\tag{8}$$

of (1) is defined on $\mathcal{T}_1 \times \mathcal{T}_2 \times H_\beta \times \mathcal{E}_p$, for all $0 \le p \le \iota - 1$.

For all $0 \le p \le \iota - 1$, the paths of integration $L_{1,p,\epsilon}$ depend on $\epsilon \in \mathcal{E}_p$, and represent the segment $[0, (C_1/\epsilon^{\lambda k_2 \tilde{\delta}_{D_2}}) e^{\sqrt{-1}\theta_{1,p}}]$, for some $\theta_{1,p} \in \mathbb{R}$, and $L_{d_{2,p}} = [0, \infty) e^{\sqrt{-1}d_{2,p}}$. The Borel function $\omega_p(u_1, u_2, m, \epsilon)$ turns out to be continuous with respect to m on $\mathbb{R}$ with exponential decay at infinity, and holomorphic with exponential growth with respect to (u_1, u_2) on domains which depend on ϵ given by:

- The polydisc $D(0, r_1(\epsilon)) \times D(0, r_2(\epsilon))$, where $r_1(\epsilon) = C_1/|\epsilon|^{\lambda k_2 \tilde{\delta}_{D_2}}$ and $r_2(\epsilon) = \frac{1}{2}|\epsilon|^{\lambda k_1 \delta_{D_1}}$.
- a product $S_{\theta_{1,p}, r_1(\epsilon)} \times (S_{d_{2,p}} \cup D(0, r_2(\epsilon)))$, where $S_{\theta_{1,p}, r_1(\epsilon)}$ is a sector centered at 0 with small aperture, bisecting direction $\theta_{1,p} - \lambda k_2 \tilde{\delta}_{D_2} \arg(\epsilon)$, and radius $r_1(\epsilon)$; $S_{d_{2,p}}$ is an unbounded sector centered at 0, with small opening and bisecting direction $d_{2,p}$.

The previous domains appear in the resolution of the related convolution problem for ω_p since $P_m(u_1, u_2)$ is invertible on these domains with appropriate lower bounds.

Our second main result deals with the asymptotic expansions of u_p relatively to ϵ.

Theorem 2 (Theorem 3 [1]). *Let*

$$\alpha = \min\{k_2(1 - \lambda k_1 \delta_{D_1}), k_1(1 + \lambda k_2 \tilde{\delta}_{D_2})\}. \tag{9}$$

There exists a formal power series

$$\hat{u}(t_1, t_2, z, \epsilon) = \sum_{m \geq 0} H_m(t_1, t_2, z)\frac{\epsilon^m}{m!},$$

where $H_m(t_1, t_2, z)$ are bounded holomorphic functions on $\mathcal{T}_1 \times \mathcal{T}_2 \times H_\beta$ for all $m \geq 0$, which solves (1) and is the common asymptotic expansion of Gevrey order $1/\alpha$ with respect to ϵ on $\mathcal{E}_p$ of $u_p(t_1, t_2, z, \epsilon)$ for all $0 \leq p \leq \iota - 1$, i.e.

$$\sup_{(t,z)\in\mathcal{T}_1\times\mathcal{T}_2\times H_\beta} \left| u_p(t, z, \epsilon) - \sum_{m=0}^{N-1} H_m(t, z)\frac{\epsilon^m}{m!} \right| \leq CM^N\Gamma\left(1 + \frac{N}{\alpha}\right)|\epsilon|^N,$$

for all $\epsilon \in \mathcal{E}_p$, $N \geq 1$ and $0 \leq p \leq \iota - 1$, for well chosen constants $C, M > 0$, where $t := (t_1, t_2)$.

This result leans on the application of the cohomological approach given by Ramis–Sibuya theorem and appropriate bounds on the difference of two consecutive solutions of (1) stated in the second statement of Theorem 1 [1], and described in Section 3.

3.   Further Comments and Open Problems

It turns out that the location of the complex singularities of the partial Borel map $u_2 \mapsto w_p(u_1, u_2, m, \epsilon)$ merging at 0 as ϵ tends to 0 (due to our construction by means of a parameter depending on truncated Laplace transform) has a direct effect on the Gevrey order $1/\alpha$ of the asymptotic expansions of our solutions.

The first occurrence of such an interplay between the accumulating singularities in the Borel plane at the origin and the asymptotics in the physical space appears in the important work [11] by Braaksma and Stolovitch where normal forms of singular holomorphic vector fields are studied. Later on, the authors and their colleagues have observed related phenomena in the study of

- PDEs with Fuchsian and irregular singularities in [12], singularly perturbed problems related to parametric summability in [13] and parametric multisummability in [14,15],

- q-difference-differential equations with Fuchsian and irregular singularities in [16,17], singularly perturbed Cauchy problems related to parametric q-Gevrey asymptotics in [18] and initial value problems mixing both Gevrey and q-Gevrey expansions in [19].

The Gevrey asymptotic expansion appearing in Theorem 2 is explained in terms of the difference of two consecutive solutions (in the sense that the solutions are related to consecutive sectors in $(\mathcal{E}_p)_{0 \leq p \leq \iota-1}$) of (1) on the intersection of their domains. More precisely for all $0 \leq p \leq \iota - 1$, identifying the indices ι and 0, we have shown the existence of two constants $K, M > 0$ with

$$\sup_{(t_1,t_2,z)\in\mathcal{T}_1\times\mathcal{T}_2\times H_\beta} |u_{p+1}(t_1,t_2,z,\epsilon) - u_p(t_1,t_2,z,\epsilon)| \leq K \exp\left(-\frac{M}{|\epsilon|^\alpha}\right),$$

for all $\epsilon \in \mathcal{E}_p \cap \mathcal{E}_{p+1}$, and $0 \leq p \leq \iota - 1$. In order to provide these bounds, we use deformations of the integration paths involved in the solution (5). More precisely, the following three cases occur:

Case 1: One can choose $L_{1,p,\epsilon} \equiv L_{1,p+1,\epsilon}$, i.e. $\theta_{1,p} = \theta_{1,p+1}$ and $L_{d_2,p}$ differs from $L_{d_2,p+1}$. Then, the path

$$L_{1,p,\epsilon} \times L_{d_2,p+1} - L_{1,p,\epsilon} \times L_{d_2,p} = L_{1,p,\epsilon} \times (L_{d_2,p+1} - L_{d_2,p}),$$

represented in Fig. 1 is deformed into the path displayed in Fig. 2.

We notice that this deformation can be performed since the function $(u_1,u_2) \mapsto \omega_j(u_1,u_2,m,\epsilon)$, $j = p,\ p+1$ is holomorphic on $D(0,r_1(\epsilon)) \times D(0,r_2(\epsilon))$. The integration along this new configuration gives rise to bounds of exponential decay at the origin of order $k_2(1 - \lambda k_1 \delta_{D_1})$ with respect to the perturbation parameter, in the

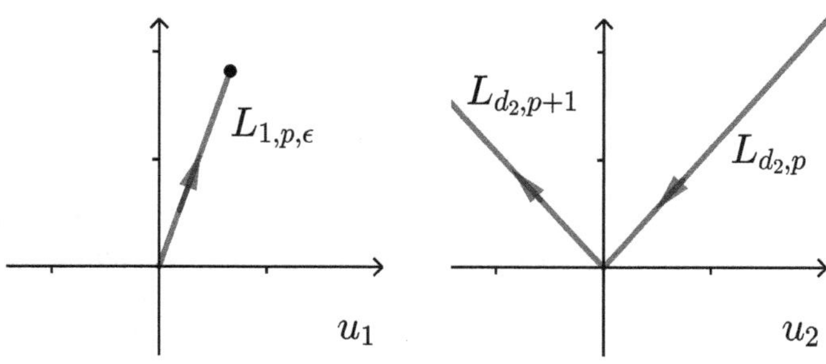

Fig. 1.   Integration path. Case 1.

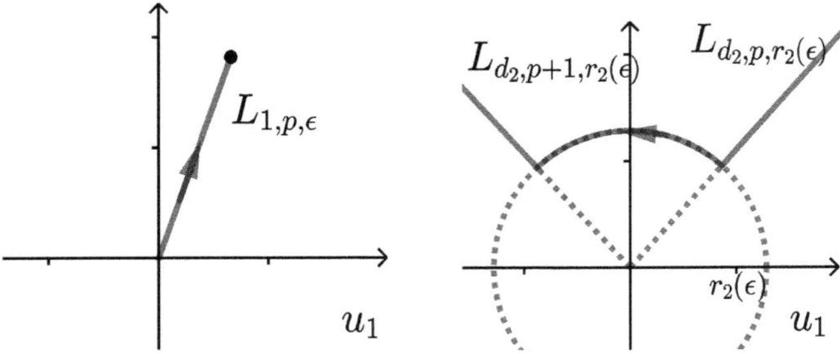

Fig. 2.   Deformation of the integration path. Case 1.

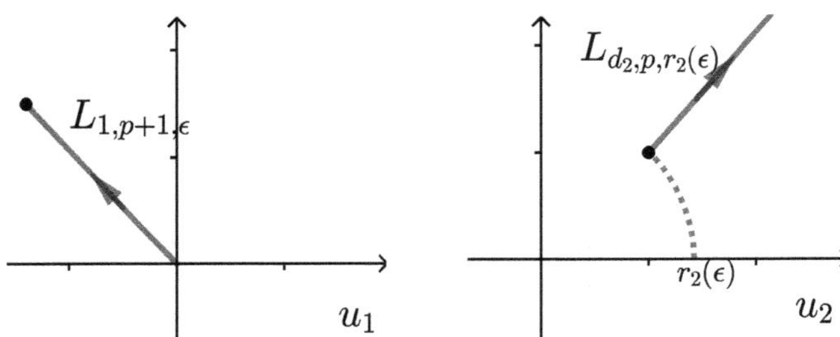

Fig. 3.   Integration path. Case 2. Piece 2.1.

intersection of the corresponding sectors of the difference of the solutions, uniformly with respect to the other variables.

Case 2: One can choose $L_{d_2,p} \equiv L_{d_2,p+1}$, but $L_{1,p,\epsilon}$ differs from $L_{1,p+1,\epsilon}$. Then, the path

$$L_{1,p+1,\epsilon} \times L_{d_2,p} - L_{1,p,\epsilon} \times L_{d_2,p}$$

is split into three pieces, and deformed to the concatenation of the following paths:

Piece 2.1: $L_{1,p+1,\epsilon} \times L_{d_2,p,r_2(\epsilon)}$ (see Fig. 3).

Piece 2.2: $-L_{1,p,\epsilon} \times L_{d_2,p,r_2(\epsilon)}$ (see Fig. 4).

Piece 2.3: $(L_{1,p+1,\epsilon} - L_{1,p,\epsilon}) \times L_{r_2(\epsilon),d_2,p}$ (see Fig. 5). Note that the deformation involved in this piece can be performed since $(u_1, u_2) \mapsto \omega_j(u_1, u_2, m, \epsilon)$, $j = p,\ p + 1$ is holomorphic on $D(0, r_1(\epsilon)) \times D(0, r_2(\epsilon))$. The deformation path is displayed in Fig. 6.

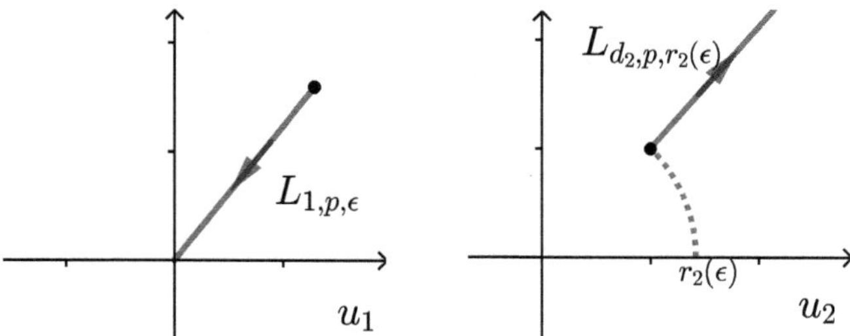

Fig. 4.   Integration path. Case 2. Piece 2.2.

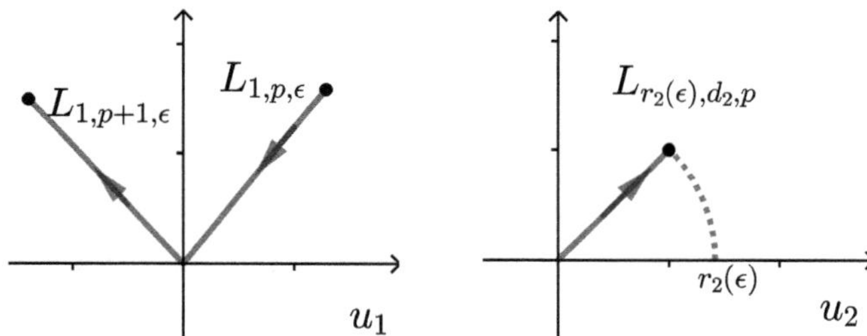

Fig. 5.   Integration path. Case 2. Piece 2.3.

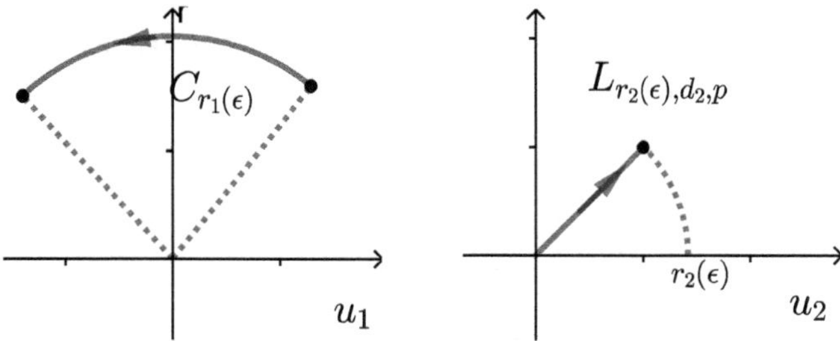

Fig. 6.   Deformation of the integration path. Case 2. Piece 2.3.

As in the first case, the integration along this new configuration in the pieces 2.1 and 2.2 gives rise to bounds of exponential decay at the origin of order $k_2(1 - \lambda k_1 \delta_{D_1})$ with respect to the perturbation parameter, in the intersection of the corresponding sectors of the difference of the solutions, uniformly with respect to the other variables. On the other hand, the integration along the arrangement in the piece 2.3 provides exponential decay at the origin of order $k_1(1 + \lambda k_2 \tilde{\delta}_{D_2})$.

Case 3: Assume that $L_{d_2,p}$ does not coincide with $L_{d_2,p+1}$, and $L_{1,p,\epsilon}$ differs from $L_{1,p+1,\epsilon}$. In this case, the path

$$L_{1,p+1,\epsilon} \times L_{d_2,p+1} - L_{1,p,\epsilon} \times L_{d_2,p}$$

is split into four pieces, and deformed as follows:

Piece 3.1: $L_{1,p+1,\epsilon} \times L_{r_2(\epsilon),d_2,p+1}$ (see Fig. 7).

Piece 3.2: $L_{1,p+1,\epsilon} \times L_{d_2,p+1,r_2(\epsilon)}$ (see Fig. 8).

Piece 3.3: $-L_{1,p,\epsilon} \times L_{r_2(\epsilon),d_2,p}$ (see Fig. 9).

Piece 3.4: $-L_{1,p,\epsilon} \times L_{d_2,p,r_2(\epsilon)}$ (see Fig. 10).

The piece 3.3 is deformed into two further blocks, namely $-L_{1,p,\epsilon} \times C_{r_2(\epsilon)}$ and $-L_{1,p,\epsilon} \times L_{r_2(\epsilon),d_2,p+1}$, say piece 3.3 (1) and piece 3.3 (2), represented in Figs. 11 and Fig. 12, respectively.

Finally, the piece 3.1 together with the piece 3.3 (2) can be deformed into $C_{r_1(\epsilon)} \times L_{r_2(\epsilon),d_2,p+1}$, shown in Fig. 13.

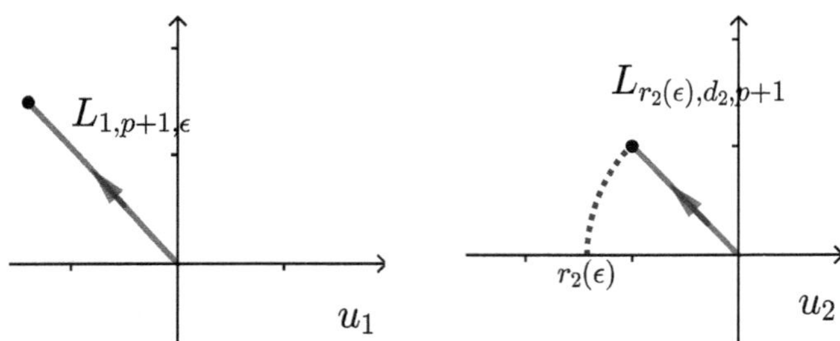

Fig. 7.   Integration path. Case 3. Piece 3.1.

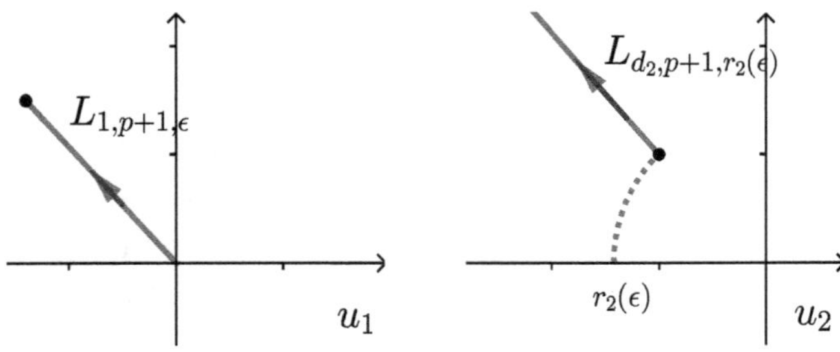

Fig. 8.   Integration path. Case 3. Piece 3.2.

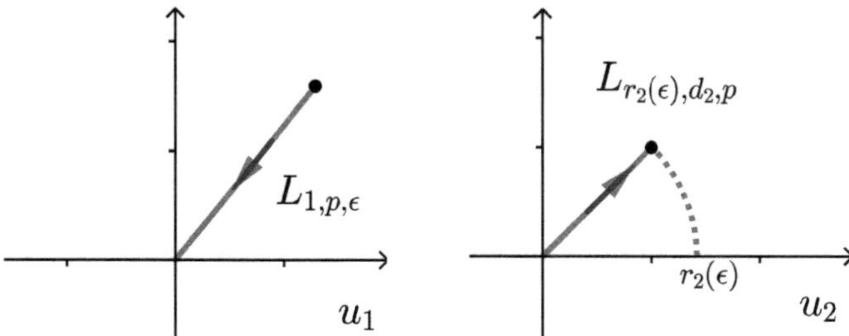

Fig. 9.   Integration path. Case 3. Piece 3.3.

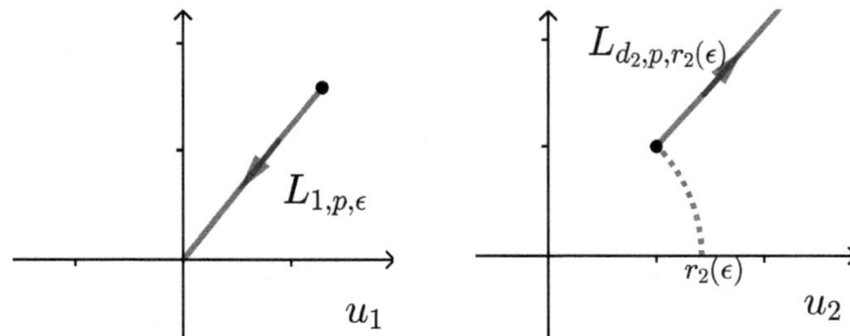

Fig. 10.   Integration path. Case 3. Piece 3.4.

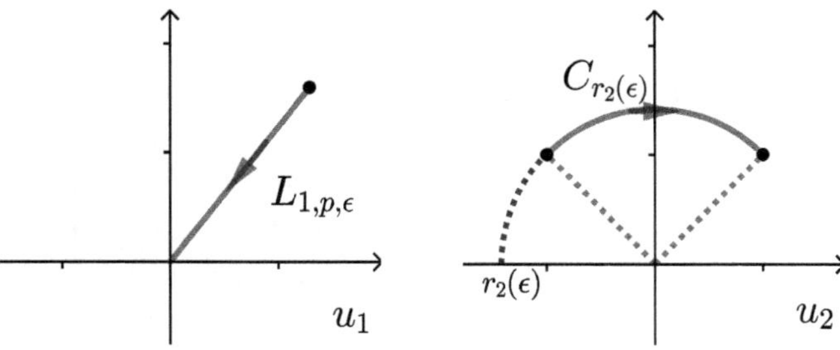

Fig. 11.   Deformation of the integration path. Case 3. Piece 3.3 (1).

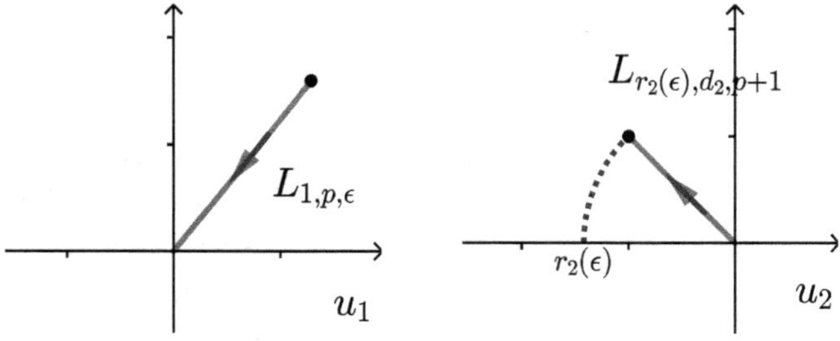

Fig. 12.   Deformation of the integration path. Case 3. Piece 3.3 (2).

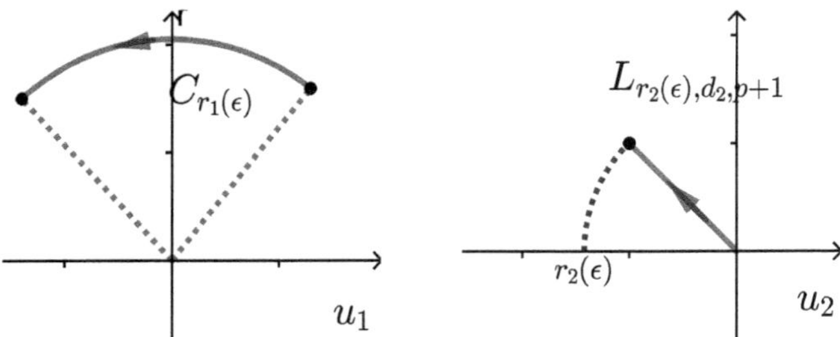

Fig. 13. Deformation of the integration path. Case 3. Piece 3.1 together with piece 3.3 (2).

Note that the deformation from the path related to the piece 3.3, into piece 3.3 (1) and piece 3.3 (2); and the piece 3.1 together with the piece 3.3 (2) can be performed since the map $(u_1, u_2) \mapsto \omega_j(u_1, u_2, m, \epsilon)$, $j = p$, $p + 1$ is holomorphic on the polydisc $D(0, r_1(\epsilon)) \times D(0, r_2(\epsilon))$. As a result, the integration along the concatenation of the pieces 3.2, 3.4 and 3.3 (1) leads to bounds of exponential decay at 0 with respect to the perturbation parameter in the intersection of the corresponding sectors, uniformly with respect to the other variables, of order $k_2(1 - \lambda k_1 \delta_{D_1})$, and the integration along the deformed piece drawn in Fig. 13 gives rise to bounds of exponential decay of order $k_1(1 + \lambda k_2 \tilde{\delta}_{D_2})$.

Since the truncated Laplace transform does not behave properly under products (unlike the complete one), one major challenging problem would be to generalize our statement to the case of nonlinear equations. Of course, the construction of solutions by means of double complete Laplace transforms remains possible in that extended situation. But the extraction of asymptotic information out of the solutions, which is the core of our study, remains an unsolved question left for future research.

Acknowledgment

The second and third authors are partially supported by the project MTM2016-77642-C2-1-P of Ministerio de Economía y Competitividad, Spain; the second author is partially supported by Dirección General de Investigación e Innovación, Consejería de Educación e Investigación of Comunidad de Madrid (Spain), and Universidad de Alcalá under grant CM/JIN/2019-010.

References

1. G. Chen, A. Lastra, and S. Malek, Parametric gevrey asymptotics in two complex time variables through truncated laplace transforms. *Adv. Differ. Eqn.*, **307**(2020). https://doi.org/10.1186/s13662-020-02773-z.

2. R. R. Lederman and S. Steinerberger, Lower bounds for truncated fourier and laplace transforms. *Integr. Equ. Oper. Theory.*, **87**(2017), 529–543.

3. R. R. Lederman and S. Steinerberger, Stability estimates for truncated fourier and laplace transforms, 2016.

4. R. R. Lederman and V. Rokhlin, On the analytical and numerical properties of the truncated laplace transform. *SIAM J. Numer. Analy.*, **53**(2015), 1214–1235.

5. W. Balser, *Formal Power Series and Linear Systems of Meromorphic Ordinary Differential Equations*. Universitext. Springer-Verlag, New York, 2000.

6. M. Loday-Richaud, *Divergent Series, Summability and Resurgence. II. Simple and Multiple Summability*, Lecture Notes in Mathematics Springer, 2016.

7. P. P. d'Escurac. The borel transform of canard values and its singularities. In: *Formal and Analytic Solutions of Diff. Equations. FASdiff 2017*. pp. 149–176, Cham, 2018. *Springer Proceedings in Mathematics and Statistics*.

8. A. Lastra and S. Malek, On parametric gevrey asymptotics for some nonlinear initial value problems in symmetric complex time variables. *Asympt. Anal.*, **118**(1–2) (2020), 49–79.

9. A. Lastra and S. Malek, On parametric gevrey asymptotics for some initial value problems in two asymmetric complex time variables. *Results Math.*, **73**(2018), 155.

10. A. Lastra and S. Malek, Boundary layer expansions for initial value problems with two complex time variables. *Adv. Differ. Eqn.*, **20**(2020). https://doi.org/10.1186/s13662-020-2496-3.

11. B. Braaksma and L. Stolovitch, Laurent small divisors and large multipliers. *Ann. Inst. Fourier (Grenoble)*, **57**(2) (2007), 603–628.

12. S. Malek, On gevrey functional solutions of partial differential equations with fuchsian and irregular singularities. *J. Dyn. Control Syst.*, **15**(2) (2009), 277–305.

13. A. Lastra, S. Malek, and J. Sanz, On gevrey solutions of threefold singular nonlinear partial differential equations. *J. Diff. Eqns.*, **255**(10) (2013), 3205–3232.

14. A. Lastra and S. Malek, Multi-level gevrey solutions of singularly perturbed linear partial differential equations. *Adv. Diff. Eqns.*, **21**(7–8) (2016), 767–800.

15. A. Lastra and S. Malek, Some notes on the multi-level gevrey solutions of singularly perturbed linear partial differential equations. In: *Analytic, Algebraic and Geometric Aspects of Differential Equations. Trends in Mathematics*, G. Filipuk, Y. Haraoka, S. Michalik (eds.), Birkhäuser, Cham, 2017. https://doi.org/10.1007/978-3-319-528 42-7_13.

16. A. Lastra, S. Malek, and J. Sanz, On q-asymptotics for linear q-difference-differential equations with fuchsian and irregular singularities. *J. Diff. Eqns.*, **252**(10) (2012), 5185–5216.

17. A. Lastra, S. Malek, and J. Sanz, On q-asymptotics for q-difference-differential equations with fuchsian and irregular singularities. *Banach Center Publ.; Polish Acad. Sci. Inst. Math.*, **97**(2012), 73–90.

18. A. Lastra and S. Malek, On parametric gevrey asymptotics for singularly perturbed partial differential equations with delays. *Abstr. Appl. Anal.*, **Art. ID 723040**, (2013), 18.

19. A. Lastra and S. Malek, On multiscale gevrey and q-gevrey asymptotics for some linear q-difference differential initial value cauchy problems, *J. Diff. Eq. Appl.*, **23**(8) (2017), 1397–1457.

© 2022 World Scientific Publishing Europe Ltd.
https://doi.org/10.1142/9781800611368_bmatter

Index

CPSIA information can be obtained
at www.ICGtesting.com
Printed in the USA
BVHW051642200322
631697BV00002B/12